Bt Resistance – Characterization and Strategies for GM Crops Producing *Bacillus thuringiensis* Toxins

_____

FSC
www.fsc.org
MIX
Paper from
responsible sources
FSC® C013604

CABI Biotechnology Series

Biotechnology, in particular the use of transgenic organisms, has a wide range of applications including agriculture, forestry, food and health. There is evidence that it could make a major impact in producing plants and animals that are able to resist stresses and diseases, thereby increasing food security. There is also potential to produce pharmaceuticals in plants through biotechnology, and provide foods that are nutritionally enhanced. Genetically modified organisms can also be used in cleaning up pollution and contamination. However, the application of biotechnology has raised concerns about biosafety, and it is vital to ensure that genetically modified organisms do not pose new risks to the environment or health. To understand the full potential of biotechnology and the issues that relate to it, scientists need access to information that not only provides an overview of and background to the field, but also keeps them up to date with the latest research findings.

This series, which extends the scope of CABI's successful 'Biotechnology in Agriculture' series, addresses all topics relating to biotechnology including transgenic organisms, molecular analysis techniques, molecular pharming, *in vitro* culture, public opinion, economics, development and biosafety. Aimed at researchers, upper-level students and policy makers, titles in the series provide international coverage of topics related to biotechnology, including both a synthesis of facts and discussions of future research perspectives and possible solutions.

Titles Available

1. Animal Nutrition with Transgenic Plants
   Edited by G. Flachowsky
2. Plant-derived Pharmaceuticals: Principles and Applications for Developing Countries
   Edited by K.L. Hefferon
3. Transgenic Insects: Techniques and Applications
   Edited by M.Q. Benedict
4. Bt Resistance – Characterization and Strategies for GM Crops producing *Bacillus thuringiensis* Toxins
   Edited by Mario Soberón, Yulin Gao and Alejandra Bravo

Bt Resistance – Characterization and Strategies for GM Crops Producing *Bacillus thuringiensis* Toxins

_____

Edited by

Mario Soberón

Universidad Nacional Autónoma de México, Cuernavaca, Mexico

Yulin Gao

Chinese Academy of Agricultural Sciences, Beijing, People's Republic of China

Alejandra Bravo

Universidad Nacional Autónoma de México, Cuernavaca, Mexico

www.cabi.org

CABI is a trading name of CAB International

CABI	CABI
Nosworthy Way	38 Chauncy Street
Wallingford	Suite 1002
Oxfordshire OX10 8DE	Boston, MA 02111
UK	USA
Tel: +44 (0)1491 832111	T: +1 800 552 3083 (toll free)
Fax: +44 (0)1491 833508	E-mail: cabi-nao@cabi.org
E-mail: info@cabi.org	
Website: www.cabi.org	

A catalogue record for this book is available from the British Library, London, UK.

Library of Congress Cataloging-in-Publication Data

Bt resistance : characterization and strategies for GM crops producing bacillus thuringiensis toxins / Mario Soberón, Yulin Gao and Alejandra Bravo (eds).
    pages cm. -- (CABI biotechnology series ; 4)
  Includes bibliographical references and index.
  ISBN 978-1-78064-437-0 (hbk : alk. paper) 1. Transgenic plants--Insect resistance.
2. Bacillus thuringiensis. I. Soberón, Mario, editor. II. Title: Bacillus thuringiensis resistance. III. Series: CABI biotechnology series ; 4.
  SB123.57.B76 2015
  632'.8--dc23
                        2014041666

ISBN-13: 978 1 78064 437 0

Commissioning editor: David Hemming
Editorial assistant: Emma McCann
Production editor: James Bishop

Typeset by Columns Design XML Limited, Reading, Berkshire
Printed and bound in the UK by CPI Group (UK) Ltd, Croydon, CR0 4YY

Contents

Contributors

Raffi V. Aroian, Program in Molecular Medicine, University of Massachusetts Medical School, Biotech Two, Suite 219, 373 Plantation Street, Worcester, MA 01605-2377, USA. E-mail: raffi.aroian@umassmed.edu

Alejandra Bravo, Departamento Microbiología Molecular, Instituto de Biotecnología, Universidad Nacional Autónoma de México, Av. Universidad 2001, CP 62210, Cuernavaca, Morelos, Mexico. E-mail: bravo@ibt.unam.mx

Pascal Campagne, Institute of Integrative Biology, University of Liverpool, Crown Street, Liverpool L69 7ZB, UK. E-mail: Pascal.Campagne@liverpool.ac.uk

Yves Carrière, Department of Entomology, University of Arizona, Forbes 410, PO Box 210036, Tucson, AZ 85721-0036, USA. E-mail: ycarrier@ag.arizona

Elena N. Elpidina, A.N. Belozersky Institute of Physico-Chemical Biology, Moscow State University, Vorobjevy Gory, Moscow 119991, Moscow, Russia. E-mail: elp@belozersky.msu.ru

Jeffrey A. Fabrick, USDA Agricultural Research Service, US Arid Land Agricultural Research Center, 21881 North Cardon Lane, Maricopa, AZ 85138, USA. E-mail: jeff.fabrick@ars.usda.gov

Juan Ferré, Departamento de Genética, Facultad de Ciencias Biológicas, Universitat de València, Dr Moliner 50, 46100 Burjassot, Valencia, Spain. E-mail: Juan.Ferre@uv.es

Michelle Franklin, Institute for Sustainable Horticulture, Kwantlen Polytechnic University, 12666 72 Avenue, Surrey, BC V3W 2M8, Canada. E-mail: michelle.franklin@kpu.ca

Yulin Gao, State Key Laboratory for Biology of Plant Disease and Insect Pests, Institute of Plant Protection, Chinese Academy of Agricultural Sciences, No. 2 Yuanmingyuan West Road, Haidian, District Beijing 100193, People's Republic of China. E-mail: ylgao@ippcaas.cn

Blanca Ines García-Gómez, Departamento Microbiología Molecular, Instituto de Biotecnología, Universidad Nacional Autónoma de México, Av. Universidad 2001, CP 62210, Cuernavaca, Morelos, Mexico. E-mail: ines@ibt.unam.mx

Meztlli Gaytán, Departamento Microbiología Molecular, Instituto de Biotecnología, Universidad Nacional Autónoma de México, Av. Universidad 2001, CP 62210, Cuernavaca, Morelos, Mexico. E-mail: ogaytan@email.ifc.unam.mx

Isabel Gómez, Departamento Microbiología Molecular, Instituto de Biotecnología, Universidad Nacional Autónoma de México, Av. Universidad 2001, CP 62210, Cuernavaca, Morelos, Mexico. E-mail: isabelg@ibt.unam.mx

David G. Heckel, Max Planck Institute for Chemical Ecology, Hans-Knoell-Str. 8, D-07745 Jena, Germany. E-mail: heckel@ice.mpg.de

Fangneng Huang, Department of Entomology, Louisiana State University Agricultural Center, A513 Life Sciences Bldg, Baton Rouge, LA 70803, USA. E-mail: FHuang@agcenter.lsu.edu

William D. Hutchison, Department of Entomology, University of Minnesota, 219 Hodson Hall, 1980 Folwell Ave., St Paul, MN 55108, USA. E-mail: hutch002@umn.edu

Siva Jakka, Department of Entomology and Plant Pathology, University of Tennessee, 370 Plant Biotechnology Building 2505, E. J. Chapman Drive, Knoxville, TN 37996, USA. Current address: Department of Entomology, Iowa State University, 13 Insectary Bldg, Ames, IA 50011, USA. E-mail: sjakka@iastate.edu

Alida Janmaat, Biology Department, University of the Fraser Valley, 33844 King Road, Abbotsford, BC V2S 7M8, Canada, E-mail: alida.janmaat@ufv.ca

Juan Luis Jurat-Fuentes, Department of Entomology and Plant Pathology, University of Tennessee, 370 Plant Biotechnology Building 2505, E. J. Chapman Drive, Knoxville, TN 37996, USA. E-mail: jurat@utk.edu

Chenxi Liu, State Key Laboratory for Biology of Plant Diseases and Insect Pests, Institute of Plant Protection, Chinese Academy of Agricultural Sciences, No. 2 Yuanmingyuan West Road, Haidian District, Beijing 100193, People's Republic of China. E-mail: liuchenxi@caas.cn

Erica S. Martins, Instituto Mato-Grossense do Algodão, Primavera do Leste, Av. Rubens de Mendonça, 157. Sala 100, Ed. Mestre Ignácio. Baú, Cuiabá-MT. 78008-000, Brazil. E-mail: ericamartins@imamt.com.br

Gretel Mendoza, Departamento Microbiología Molecular, Instituto de Biotecnología, Universidad Nacional Autónoma de México, Av. Universidad 2001, CP 62210, Cuernavaca, Morelos, Mexico. E-mail: gretelm@ibt.unam.mx

Rose Monnerat, Embrapa Recursos Genéticos e Biotecnologia (CENARGEN), Caixa Postal 02372, CEP 70849-970 Brasilia, DF, Brazil. E-mail: rose.monnerat@embrapa.br

Judith H. Myers, Department of Zoology, University of British Columbia, 2370-6270 University Boulevard, Vancouver, BC V6T 1Z4, Canada. E-mail: myers@zoology.ubc.ca

Brenda Oppert, USDA Agricultural Research Service, Center for Grain and Animal Health Research, 1515 College Ave. Manhattan, KS 66502, USA. E-mail: brenda.oppert@ars.usda.gov, bso@k-state.edu

Sabino Pacheco, Departamento Microbiología Molecular, Instituto de Biotecnología, Universidad Nacional Autónoma de México, Av. Universidad 2001, CP 62210, Cuernavaca, Morelos, Mexico. E-mail: pacheco@ibt.unam.mx

Lilian B. Praça, Embrapa Recursos Genéticos e Biotecnologia (CENARGEN), Caixa Postal 02372, CEP 70849-970 Brasilia, DF, Brazil. E-mail: lilian.praca@embrapa.br

Paulo Queiroz, Instituto Mato-Grossense do Algodão, Primavera do Leste, Av. Rubens de Mendonça, 157. Sala 100, Ed. Mestre Ignácio. Baú, Cuiabá-MT. 78008-000, Brazil. E-mail: pauloqueiroz@imamt.com.br

Manchikatla V. Rajam, Department of Genetics, University of Delhi, South Campus, Benito Juarez Road, New Delhi 110021, India. E-mail: venkat.rajam@south.du.ac.in, rajam.mv@gmail.com

Jorge Sánchez, Departamento Microbiología Molecular, Instituto de Biotecnología, Universidad Nacional Autónoma de México, Av. Universidad 2001, CP 62210, Cuernavaca, Morelos, Mexico. E-mail: jsanchez@ibt.unam.mx

Changlong Shu, State Key Laboratory for Biology of Plant Diseases and Insect Pests, Institute of Plant Protection, Chinese Academy of Agricultural Sciences, No. 2 Yuanmingyuan West Road, Haidian District, Beijing 100193, People's Republic of China. E-mail: clshu@ippcaas.cn

Anand Sitaram, Program in Molecular Medicine, University of Massachusetts Medical School, Biotech Two, Suite 219, 373 Plantation Street, Worcester, MA 01605-2377, USA. E-mail: anand.sitaram@umassmed.edu

Carlos Marcelo Soares, Instituto Mato-Grossense do Algodão, Primavera do Leste, Av. Rubens de Mendonça, 157. Sala 100, Ed. Mestre Ignácio. Baú, Cuiabá-MT. 78008-000, Brazil. E-mail: carlosmarcelo@imamt.com.br

Mario Soberón, Departamento Microbiología Molecular, Instituto de Biotecnología, Universidad Nacional Autónoma de México, Av. Universidad 2001, CP 62210, Cuernavaca, Morelos, Mexico. E-mail: mario@ibt.unam.mx

Bruce E. Tabashnik, Department of Entomology, University of Arizona, Forbes 410, PO Box 210036, Tucson, AZ 85721-0036, USA. E-mail: tabashnb@email.arizona.edu; brucet@cals.arizona.edu

Johnnie Van den Berg, Unit for Environmental Sciences and Management, North-West University, Potchefstroom Campus, Private Bag X6001, Potchefstroom 2520, South Africa. E-mail: Johnnie.VanDenBerg@nwu.ac.za

Ping Wang, Department of Entomology, Cornell University, 630 West North Street, Barton Lab, New York State Agricultural Experiment Station, Geneva, NY 14456, USA. E-mail: pingwang@cornell.edu

Zeyu Wang, State Key Laboratory for Biology of Plant Diseases and Insect Pests, Institute of Plant Protection, Chinese Academy of Agricultural Sciences, No. 2 Yuanmingyuan West Road, Haidian District, Beijing 100193, People's Republic of China. E-mail: wangzeyu198@sohu.com

Kongming Wu, State Key Laboratory for Biology of Plant Disease and Insect Pests, Institute of Plant Protection, Chinese Academy of Agricultural Sciences, No. 2 Yuanmingyuan West Road, Haidian District, Beijing 100193, People's Republic of China. E-mail: kmwu@ippcaas.cn

Yidong Wu, Department of Entomology, College of Plant Protection, Nanjing Agricultural University, Nanjing 210095, People's Republic of China. E-mail: wyd@njau.edu.cn

Sneha Yogindran, Department of Genetics, University of Delhi, South Campus, Benito Juarez Road, New Delhi 110021, India. E-mail: snehayogindran@gmail.com

Igor A. Zalunin, The State Research Institute for Genetics and Selection of Industrial Microorganisms, 1 Dorozhny pr., 113545, Moscow, Russia. E-mail: ingvarzal@mail.ru

Jie Zhang, State Key Laboratory for Biology of Plant Diseases and Insect Pests, Institute of Plant Protection, Chinese Academy of Agricultural Sciences, No. 2 Yuanmingyuan West Road, Haidian District, Beijing 100193, People's Republic of China. E-mail: jiezhang@caas.net.cn; jzhang@ippcaas.cn

Preface

Transgenic crops expressing *Bacillus thuringiensis* (Bt) Cry (or crystal) toxins were commercially released in 1996. Bt bacteria are insect pathogens that rely on insecticidal Cry pore-forming proteins to kill their insect larval hosts. Different countries have made use of this technology to control insect pests of important crops such as maize, cotton or soybean. The land area planted with Bt crops has increased every year, reaching a total of 76 million ha in 2013. Bt crops have proven to be useful in the reduction of the use of pesticides for insect control and in some cases also in promoting an increase in crop yields. However, this important technology is being endangered by the evolution of resistance in different insect pests in countries that have adopted this technology. To date, at least eight different insect species have developed resistance to Bt toxins under field conditions. Without doubt, insects possess diverse mechanisms to rapidly evolve resistance.

This book focuses on descriptions of the extent of use of Bt crops and the emerging problem of resistance, recent progress in elucidating the mechanism of action of Bt toxins and describing the different resistance mechanisms and strategies for coping with resistance in the field. There are four sections.

In the first section, 'The Extent of Use of Bt Crops and the Emerging Problem of Resistance', five chapters summarize the successes and failures of this technology both worldwide and in particular countries that have adopted this technology. Chapter 1 gives a comprehensive overview of the global monitoring of insect resistance in the six continents of the world and provides explanations for the rapid evolution of insect resistance for particular insect pests in some regions and for a more sustainable use of this technology in other regions. Chapters 2 to 4 provide detailed information on the monitoring and evolution of resistance to Bt crops in three different countries that have adopted this technology: China, Brazil and Latin America, and South Africa, respectively. Chapter 5 describes the evolution of resistance to Bt sprays in greenhouses in Canada, which has important implications for the evolution of resistance to Bt crops in tropical regions. These first five chapters place emphasis on agricultural practices and insect pest behaviours that contribute to the reduced or enhanced rate at which insects evolve resistance to Bt crops.

The second part of the book, 'Mechanism of Action of Bt Toxins and Different Resistance Mechanisms', describes different models of the mode of action of Bt Cry (or crystal) toxins and the different mechanisms of resistance that have been described to date. Chapter 6 discusses three different models describing the mode of action of Cry1A toxins in spite of the different resistant mechanisms that have already been reported. Chapters 7 to 10 describe different resistant mechanisms involving mutations affecting the expression of different Cry receptor molecules such as cadherins (Chapter 7) and aminopeptidase Ns (Chapter 8), or ABCC2 transporters (Chapter 8 and 9), while Chapter 10 describes resistance associated with altered midgut proteases that affect the processing of toxins. Chapter 11 describes important contributions on Cry resistance mechanisms that have been uncovered using the nematode *Caenorhabditis elegans* as an experimental model.

The third section, 'Strategies to Counter Resistance' describes novel methodologies to counter or delay resistance to Bt crops. Chapter 12 examines the discovery of new Cry toxins with the potential to target resistant insect populations, while Chapter 13 describes how membrane-binding competition experiments have provided the fundamental basis for selecting novel toxins with toxicity to resistant insect populations. Chapter 14 describes genetically modified CryMod toxins that circumvent specific receptor-binding interactions and how such CryMod toxins may counter resistance under field conditions. The use of other pest control strategies, such as RNAi expression in transgenic plants, is described in Chapter 15. Overall, this section provides insight into future technology that may be available to maintain the efficacy of Bt crops for pest control.

The final section, entitled 'Insect Resistance Management and Integrated Pest Management', discusses the use of insect pest management – or IPM – with emphasis on only using pest control strategies when needed, such as when pest populations exceed pre-determined economic injury levels (Chapters 16 and 17).

We believe that the information provided in this book not only provides an overview of the state-of-the-art of transgenic Bt technology, but will also enlighten and enable researchers, government officials and regulators, industry practitioners and growers to consider the problem of resistance to Bt crops and to seek a vision for the long-term sustainability of this critical technology.

Alejandra Bravo, Mario Soberón and Yulin Gao

1

Successes and Failures of Transgenic Bt Crops: Global Patterns of Field-evolved Resistance

Bruce E. Tabashnik* and Yves Carrière

Department of Entomology, University of Arizona, Tucson, Arizona, USA

Summary

Farmers planted genetically engineered crops that produce insecticidal proteins from the bacterium *Bacillus thuringiensis* (Bt) on a cumulative total of 570 million ha worldwide from 1996 to 2013. These Bt crops kill some key insect pests, yet they are not toxic to most other organisms, including people. Bt crops can suppress pests, reduce the use of insecticide sprays and increase farmer profits, but their benefits are diminished or lost when pests evolve resistance. Here we review data monitoring resistance to seven Bt proteins in 13 major pest species targeting Bt maize and Bt cotton on six continents. Of the 27 sets of monitoring data analysed, seven show severe field-evolved resistance in 2 to 8 years with practical consequences for pest control (i.e. practical resistance), eight show statistically significant but less severe field-evolved resistance and 12 show no evidence of decreased susceptibility after 2 to 15 years. The surge in cases of practical resistance since 2005 is associated with increased planting of Bt crops, increased cumulative exposure of pests to Bt crops and increased monitoring. In addition, practical resistance to Bt crops is associated with a scarcity of refuges, which consist of host plants that do not produce Bt proteins. To maximize the benefits of Bt crops, we encourage collaboration between growers and scientists in industry, academia and government to implement large refuges of non-Bt host plants, particularly when the inheritance of resistance is not recessive and alleles conferring resistance are not rare.

1.1 Introduction

The widespread bacterium *Bacillus thuringiensis* (Bt) produces proteins that kill some devastating insect pests, but are not toxic to most other organisms, including people (Mendelsohn *et al.*, 2003; Pardo-López *et al.*, 2013). These proteins, called Bt toxins, have been used for decades in sprays to control insects that attack crops, damage forests and vector human diseases (Sanahuja *et al.*, 2011). More recently, to provide another way to control insect pests, the genes encoding Bt toxins have been incorporated in the genomes of maize, cotton and soybean (Tabashnik *et al.*, 2009a; James 2013).

The area planted to transgenic crops producing Bt toxins grew from 1 million ha in 1996 to 76 million ha in 2013 (Fig. 1.1).

---

* Corresponding author. E-mail address: tabashnb@email.arizona.edu

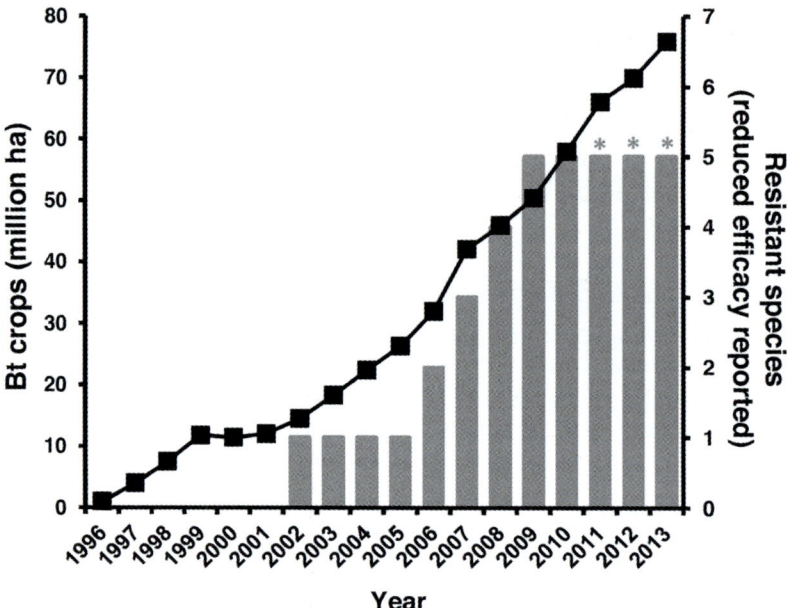

Fig. 1.1. Global area planted to Bt crops (black) and cumulative number of insect species with practical resistance to Bt crops (grey). Adapted from Tabashnik *et al.* (2013). Practical resistance, which entails one or more field populations with >50% resistant individuals and reduced efficacy reported, has been documented in five major target pests (year first detected given in parentheses): *Helicoverpa zea* (2002), *Spodoptera frugiperda* (2006), *Busseola fusca* (2007), *Pectinophora gossypiella* (2008) and *Diabrotica virgifera virgifera* (2009) (see Table 1.1). The asterisks indicate that the number of species with resistant populations may be underestimated for 2011 to 2013 because reports of field-evolved resistance typically are published 2 or more years after resistance is first detected. Although the number of pest species with practical resistance was five from 2009 to 2013, the number of cases of practical resistance increased from five to seven because of *S. frugiperda* resistance to the Cry1F Bt toxin in two countries (Brazil and USA) and *D. v. virgifera* resistance to two Bt toxins (Cry3Bb and mCry3A) (Table 1.1).

The cumulative worldwide total of Bt crops planted from 1996 to 2013 was 570 million ha (1.4 billion acres), with Bt maize and Bt cotton accounting for more than 99% of this total. In 2013, Bt maize accounted for 76% of maize in the USA, while Bt cotton accounted for 75% of cotton in the USA, 90% in China and 95% in India (James, 2013; USDA ERS, 2013). In the first large scale use of Bt soybean, farmers planted 2 million ha of this crop in Brazil during 2013 (James, 2013). In some cases, these Bt crops have suppressed pests, reduced the use of insecticide sprays, increased crop yield and increased farmers' profits (Hutchison *et al.*, 2010; National Research Council 2010,

Tabashnik *et al.*, 2010; Edgerton *et al.*, 2012; Lu *et al.*, 2012). However, these substantial environmental and economic benefits are diminished or even eliminated when pests evolve resistance to Bt toxins (Storer *et al.*, 2012; Tabashnik *et al.*, 2013; Van den Berg *et al.*, 2013).

Here we review the definition of field-evolved resistance, five categories of field-evolved resistance ranging in severity from incipient to practical resistance, the status of resistance or susceptibility of 13 pests to seven toxins used in Bt crops on six continents and factors associated with the successes and failures of Bt crops in the field.

1.2 Definition and Categories of Field-evolved Resistance

Field-evolved (or field-selected) resistance is a genetically based decrease in susceptibility of a population to a toxin that is caused by exposure of the population to the toxin in the field (Tabashnik et al., 2013). Tabashnik et al. (2014) and references therein provide detailed discussion of this and other definitions of resistance. Recognizing that field-evolved resistance is not 'all or none', Tabashnik et al. (2014) described five categories of field-evolved resistance, which each entail a genetically based decrease in susceptibility to a toxin in one or more field populations:

- incipient resistance, <1% resistant individuals;
- early warning of resistance, 1 to 6% resistant individuals;
- >6 to <50% resistant individuals (no cases reported);
- >50% resistant individuals and reduced efficacy expected, but not reported; and
- practical resistance, >50% resistant individuals and reduced efficacy reported.

Practical resistance is the only category where the practical implications of resistance for reducing the control of pests by Bt crops have been confirmed.

1.3 Global Status of Field-evolved Resistance to Bt Crops

Here we update previous reviews (e.g. Tabashnik et al., 2013) by classifying 27 cases of pest responses to Bt crops (Tables 1.1 and 1.2, Fig. 1.2). Each case represents the responses of one pest species to one Bt toxin used in transgenic crops in one country. For each case, we evaluated the percentage of resistant individuals based on survival in bioassays at a diagnostic concentration of the relevant Bt toxin. A diagnostic concentration of toxin kills all or nearly all susceptible individuals, but few or no resistant individuals (Tabashnik et al., 2014). We categorize the data based on the evidence for the most severe level of

resistance for each case. For example, if practical resistance to a toxin has been detected in one or more populations of a pest in a certain country, while some other populations in the country remain susceptible to that toxin, we classify the case as an example of practical resistance. Thus, even in cases where practical resistance to a Bt toxin occurs in some populations in a given country, the same Bt toxin may still be effective against other populations in that country.

We considered 27 cases based on resistance monitoring data from nine countries for responses to seven Bt toxins by 13 major pest species (12 lepidopterans and one coleopteran). In 12 of 27 (44%) cases, the monitoring data show no statistically significant decrease in susceptibility after 2 to 15 years (median = 8 years) of exposure to Bt crops (Table 1.2). Eleven of these cases are reviewed in Tabashnik et al. (2013). The additional case is sustained susceptibility of Ostrinia nubilalis (European corn borer) to Cry1F produced by Bt maize in the USA, which reflects a relatively low adoption of maize plants producing only Cry1F (Siegfried et al., 2014). Below we review the 15 cases of field-evolved resistance, including seven cases of practical resistance (Table 1.1).

1.3.1 Incipient resistance and early warning of resistance

All three cases of incipient resistance are from Australia, where a rigorous, proactive monitoring programme enabled the early detection of resistance to Bt toxins in Helicoverpa punctigera and H. armigera (Downes et al., 2010; Downes and Mahon, 2012a, b) (Table 1.1). Downes et al. (2010) used the term 'incipient resistance' to describe a statistically significant increase in the frequency of alleles conferring resistance to Bt toxin Cry2Ab in H. punctigera from Australia. Results from the 2008/09 field season showed that the frequency of alleles conferring resistance to Cry2Ab was eight times higher in areas where Bt cotton producing this toxin was grown compared

Table 1.1. Field-evolved resistance to Bt crops: 15 cases in nine pest species from seven countries with severity ranging from incipient to practical resistance. (Adapted from Tabashnik *et al.*, 2013, 2014.)

Pest[a]	Crop	Bt toxin	Country	Years[b]	Reference/s
Incipient resistance: <1% resistant individuals					
Helicoverpa armigera	Cotton	Cry1Ac	Australia	15	Downes and Mahon, 2012b; Tabashnik *et al.*, 2013
		Cry2Ab	Australia	8	
H. punctigera	Cotton	Cry2Ab	Australia	8	Downes *et al.*, 2010; Downes and Mahon, 2012a
Early warning of resistance: 1 to 6% resistant individuals					
Diatraea saccharalis	Maize	Cry1Ab	USA	10	Huang *et al.*, 2012
H. armigera	Cotton	Cry1Ac	China	13	Zhang *et al.*, 2011, 2012; Jin *et al.*, 2013
Ostrinia furnacalis	Maize	Cry1Ab	The Philippines	5	Alcantara *et al.*, 2011
Pectinophora gossypiella	Cotton	Cry1Ac	China	13	Wan *et al.*, 2012
>50% resistant individuals and reduced efficacy expected[c]					
H. zea[d]	Cotton	Cry2Ab	USA	2	Ali and Luttrell, 2007; Tabashnik *et al.*, 2009a, 2013
Practical resistance: >50% resistant individuals and reduced efficacy reported					
Busseola fusca	Maize	Cry1Ab	South Africa	8	van Rensburg, 2007; Kruger *et al.*, 2011; Van den Berg *et al.*, 2013
Diabrotica virgifera virgifera	Maize	Cry3Bb	USA	7	Gassmann *et al.*, 2011, 2012
D. v. virgifera[e]	Maize	mCry3A	USA	4	Gassmann *et al.*, 2014
H. zea	Cotton	Cry1Ac	USA	6	Luttrell *et al.*, 2004; Ali *et al.*, 2006; Tabashnik *et al.*, 2008a,b
P. gossypiella	Cotton	Cry1Ac	India	6[f]	Monsanto, 2010; Dhurua and Gujar, 2011; Sumerford *et al.*, 2013
Spodoptera frugiperda	Maize	Cry1F	Brazil	2	Farias *et al.*, 2014; Monnerat *et al.* (Chapter 3)
S. frugiperda	Maize	Cry1F	USA	3	Storer *et al.*, 2010, 2012

[a]*D. v. virgifera* is a beetle (coleopteran); the other eight species are lepidopterans.
[b]Years elapsed between the first year of commercial planting in the region studied and: (i) for the eight cases with either practical resistance or >50% resistant individuals and reduced efficacy expected, the first year of field sampling that yielded evidence of resistance, or (ii) for all other cases, the most recent year of monitoring data reviewed here.
[c]Reduced efficacy is expected, but has not been reported.
[d]May reflect some cross-resistance caused by selection with Cry1Ac.
[e]Registered October 2006, grown 2007, first resistance reported from populations sampled in 2011; reflects cross-resistance from selection with mCry3A.
[f]Excludes years when Bt cotton was grown illegally in India before it was commercialized in 2002. Resistance was first detected in samples collected in 2008, 6 years after commercial planting. If illegal planting started in 2000, the total years elapsed would be 8 (Tabashnik *et al.*, 2013).

Table 1.2. Susceptibility to Bt crops: 12 cases in seven pest species from four countries with monitoring data showing no evidence of resistance. (Adapted from Tabashnik *et al.*, 2013.)

Pest[a]	Crop	Bt toxin	Country	Years[b]	Reference/s
Diatraea grandiosella	Maize	Cry1Ab	USA	6	Huang *et al.*, 2007
Diabrotica virgifera virgifera	Maize	Cry34/35Ab	USA	5	Gassmann *et al.*, 2014
Helicoverpa punctigera	Cotton	Cry1Ac	Australia	10	Downes *et al.*, 2009
Heliothis virescens	Cotton	Cry1Ac	USA	11	Ali *et al.*, 2006; Blanco *et al.*, 2009
		Cry1Ac	Mexico	11	Blanco *et al.*, 2009
		Cry2Ab	USA	2	Ali and Luttrell, 2007
Ostrinia nubilalis	Maize	Cry1Ab	Spain	4	Farinós *et al.*, 2004
		Cry1Ab	USA	15	Siegfried *et al.*, 2007; Siegfried and Hellmich, 2012
		Cry1F	USA	8	Siegfried *et al.*, 2014
Pectinophora gossypiella	Cotton	Cry1Ac	USA	13	Tabashnik *et al.*, 2010
		Cry2Ab	USA	5	Tabashnik *et al.*, 2010
Sesamia nonagroides	Maize	Cry1Ab	Spain	7	Andreadis *et al.*, 2007; Farinós *et al.*, 2011

[a]*D. v. virgifera* is a beetle (coleopteran); the other six species are lepidopterans.
[b]Years elapsed between the first year of commercial planting in the region studied and the most recent year of monitoring data reviewed here.

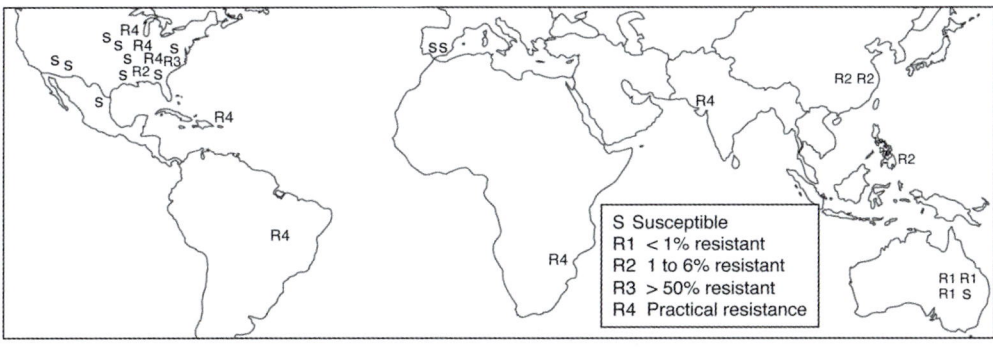

Fig. 1.2. Global status of field-evolved resistance to Bt crops. Each one-letter or letter-number code represents one of 27 cases evaluating the status of field-evolved resistance to one Bt toxin in populations of one pest species from one country (see Tables 1.1 and 1.2). (Adapted from Tabashnik *et al.*, 2013.)

with that in non-cropping areas. Downes *et al.* (2010) also detected an elevenfold increase from 2004/05 to 2008/09 in the frequency of resistance to Cry2Ab in populations exposed to this toxin. However, they estimated that the percentage of resistant individuals – a maximum of 0.2% – was too low to reduce the efficacy of Bt cotton. Furthermore, from 2008/09 to 2010/11 the frequency of resistance to Cry2Ab did not increase (Downes and Mahon 2012a). This shows that the statistically significant yet small rises in resistance allele frequency characteristic of incipient resistance do not necessarily indicate that further increases in resistance are imminent.

Zhang *et al.* (2011) used the phrase 'early warning' of resistance to describe the statistically significant increase in the percentage of individuals with resistance to Bt toxin Cry1Ac in *H. armigera* from northern China. Their 2010 survey showed that survival at a diagnostic concentration of Cry1Ac was significantly higher for 13 field populations from northern China, where exposure to Bt cotton producing Cry1Ac was extensive, relative to two field populations from northwestern China where exposure to Bt cotton was limited. For the populations from northern China surveyed in 2010, the mean survival at the diagnostic concentration was 1.3% (range: 0 to 2.6%) compared with 0% for the populations from northwestern China and a susceptible laboratory strain (Zhang *et al.*, 2011). Results of screening in 2009 and 2011 also support the conclusion that exposure to Bt cotton increased the frequency of *H. armigera* resistance to Cry1Ac in northern China, with up to 5.4% resistant individuals in a population (Zhang *et al.*, 2012; Jin *et al.*, 2013).

Three other cases also fit the early warning category, with a statistically significant increase in resistance relative to susceptible populations and between 1 and 6% of resistant individuals: resistance to Cry1Ac in Bt cotton of *Pectinophora gossypiella* in China, and resistance to Cry1Ab in Bt maize of *Ostrinia furnacalis* in the Philippines and *Diatraea saccharalis* in the southern USA (Table 1.1). As with incipient resistance, the percentage of resistant individuals in this category is too low to substantially reduce the efficacy of Bt crops. However, field-evolved resistance with >1% resistant individuals detected warrants consideration of enhanced actions to manage resistance, such as increases in monitoring, refuge requirements and alternative methods of control. It remains to be seen what actions, if any, are taken in these four cases of early warning of resistance and how such actions would affect the trajectory of resistance. The absence of cases with >6 to <50% resistant individuals suggests that after the percentage of resistant individuals surpasses 6%, it quickly rises above 50%.

1.3.2 Practical resistance: >50% resistant individuals and reduced efficacy reported

In the seven cases of practical resistance to Bt crops, one or more pest populations had >50% resistant individuals and reduced efficacy of the Bt crop was reported (Table 1.1). These seven cases entail practical resistance to Bt maize in three pests, *Busseola fusca*, *Diabrotica virgifera virgifera* and *Spodoptera frugiperda*; and practical resistance to Bt cotton in two pests, *Helicoverpa zea* and *P. gossypiella*.

Practical resistance to Bt maize

In the US territory of Puerto Rico, *S. frugiperda* (fall armyworm) evolved resistance to Bt maize producing Cry1F in 3 years, which is one the fastest documented cases of practical resistance to a Bt crop (Table 1.1), and the first case of resistance leading to withdrawal of a Bt crop from the marketplace (Storer *et al.*, 2010, 2012). In 2011, some 4 years after Dow AgroSciences and Pioneer Hi-Bred International voluntarily stopped selling Cry1F maize in Puerto Rico, field populations remained highly resistant (Storer *et al.*, 2012). Despite the opportunity to learn from this failure in Puerto Rico, a similar pattern has been repeated in Brazil, as described by Monnerat *et al.* (Chapter 3). After only a few years of commercial cultivation of Bt maize producing Cry1F, practical resistance of *S. frugiperda* to Cry1F is widespread in Brazil (Farias *et al.*, 2014). The limited gene flow between populations in Puerto Rico and Brazil (Nagoshi *et al.*, 2010) implies that this resistance arose independently in Brazil. It will be intriguing to compare the genetic basis of resistance between the strains of *S. frugiperda* from these two regions.

Practical resistance of *B. fusca* (maize stem borer) to Bt maize producing Cry1Ab occurred in South Africa in 8 years (van Rensburg 2007; Tabashnik *et al.*, 2009a; Van den Berg *et al.*, 2013; Van den Berg and Campagne, Chapter 4), with striking parallels to *S. frugiperda* resistance to Cry1F maize in Puerto Rico. In both cases, proactive resistance monitoring was not conducted and observations of reduced efficacy in the field preceded the documentation of resistance with bioassays (Storer *et al.*, 2012; Van den Berg *et al.*, 2013). In South Africa, however, Cry1Ab maize was not withdrawn from sales, with >1 million ha planted in 2013 (James, 2013). Continued planting of Cry1Ab maize has yielded widespread resistance and hundreds of reports of product failure during the 2010/11 and 2011/12 seasons (Kruger *et al.*, 2009; Van den Berg *et al.*, 2013). Monsanto, the company that developed the predominant type of Cry1Ab maize grown in South Africa, compensated growers for their insecticide sprays on this Bt maize (Kruger *et al.*, 2009). Large-scale planting of two-toxin Bt maize producing Cry1A.105 (similar to Cry1Ab; Tabashnik *et al.*, 2009a) and Cry2Ab began during the 2012/13 season in South Africa (Van den Berg *et al.*, 2013).

Field and laboratory data indicate practical resistance to Bt maize producing Cry3Bb or mCry3A in some populations of *D. v. virgifera* (Western corn rootworm) from Iowa (Gassmann *et al.*, 2011, 2012, 2014; Gassmann, 2012). Survival in bioassays on Bt maize plants producing either Cry3Bb or mCry3A was significantly higher for nine strains derived in 2011 from 'problem' fields with severe damage to Bt maize producing either of these two toxins compared with survival in eight 'control' laboratory strains that had not been exposed to Bt toxins (Gassmann *et al.*, 2014).

Survival on Cry3Bb maize and mCry3A maize was correlated across the 17 strains of *D. v. virgifera* examined by Gassmann *et al.* (2014). This correlation could reflect cross-resistance, correlated exposure to the two toxins (populations exposed more to Cry3Bb were also exposed more to mCry3A), or both. Previous exposure to Cry3Bb maize was not correlated with exposure to mCry3A maize across the 17 strains ($r = -0.15$, $t = -0.58$, $df = 15$, $P = 0.57$), which leaves cross-resistance as the best explanation for the observed correlation in survival between Cry3Bb maize and mCry3A maize across strains. This cross-resistance is not surprising (Tabashnik and Gould, 2012), because the two toxins have high amino acid

sequence similarity overall (80%) and particularly for domain II (83%), which determines specificity and is associated with cross-resistance in other insects (Tabashnik *et al.*, 1996; Carrière *et al.*, 2015). To put this in perspective, the amino acid sequence similarity between Cry1Aa and Cry1Ac is 80% overall and 81% for domain II, with strong cross-resistance typically seen between these two toxins (Carrière *et al.*, 2015).

Practical resistance to Bt cotton

Despite strong evidence of practical resistance to Bt cotton in *P. gossypiella* from western India and *H. zea* in the southeastern USA (Table 1.1), these cases have been controversial (Moar *et al.*, 2008; Tabashnik *et al.*, 2008b, 2013, 2014; Luttrell and Jackson, 2012). Resistance of *P. gossypiella* (pink bollworm) to Bt cotton producing Cry1Ac was first detected with laboratory bioassays of the offspring of insects collected from non-Bt cotton fields in 2008 in the state of Gujarat in western India (Dhurua and Gujar, 2011). India ranks second in cotton production, behind only China, and Gujarat accounted for a third of India's cotton production in 2009/10, which is equivalent to about half of the annual cotton production in the USA during 2009 and 2010 (Tabashnik *et al.*, 2014).

Several studies have confirmed widespread resistance of *P. gossypiella* to Cry1Ac in Gujarat, which was associated with an unusually high abundance of larvae on Cry1Ac cotton and moths caught in pheromone traps (Monsanto, 2010; Sumerford *et al.*, 2013; Tabashnik *et al.*, 2013; Fabrick *et al.*, 2014). Fabrick *et al.* (2014) discovered severe disruptions in the transcripts encoding cadherin (a protein that binds Cry1Ac in the midgut of susceptible larvae) associated with resistance to Cry1Ac in the resistant insects from Gujarat and in insects sampled from Bt cotton plants in the neighbouring state of Madhya Pradesh (see Fabrick and Wu, Chapter 7).

Farmers in India have switched to cotton hybrids producing two Bt toxins (Cry1Ac and Cry2Ab), which are effective against pink bollworm larvae resistant to Cry1Ac

(Tabashnik *et al.*, 2002, 2009b). In 2013, they planted two-toxin Bt cotton on 10 million ha, representing 94% of India's cotton (James, 2013). Dhurua and Gujar (2011) concluded that their bioassay results with seed powder from two-toxin cotton incorporated in an artificial diet 'imply slightly reduced susceptibility' to Cry2Ab in a population from Gujarat that was sampled in 2008. In bioassays with plants, however, larvae from a laboratory-selected strain of pink bollworm from Arizona with >200-fold resistance to Cry2Ab did not survive on two-toxin Bt cotton (Tabashnik *et al.*, 2009b).

As with *P. gossypiella* in India, the extensive documentation of the practical resistance of *H. zea* (cotton bollworm) to Cry1Ac in the USA includes reports of decreased efficacy of Bt cotton in the field. Luttrell *et al.* (2004) reported the first evidence of field-evolved resistance to a Bt crop, including data showing that two populations of *H. zea* that inflicted 'unacceptable levels of boll damage' in problem fields of Bt cotton in 2002 were resistant to Cry1Ac in laboratory bioassays (Tabashnik *et al.*, 2008b). The compelling evidence confirming this case of practical resistance includes >50% survival at a diagnostic concentration of Cry1Ac for four strains derived from the field in 2003, a significant association between larval survival on Bt cotton leaves and decreased susceptibility to Cry1Ac in bioassays, as well as Cry1Ac resistance ratios >100 for eight strains derived during 2003 to 2006 from field sources other than Bt crops (Ali *et al.*, 2006; Luttrell and Ali, 2007; Tabashnik *et al.*, 2008b). These resistance ratios were calculated by dividing the concentration of Cry1Ac killing 50% of larvae (LC_{50}) for each field-derived strain by the LC_{50} of a susceptible laboratory strain called UALab (Ali *et al.*, 2006). This robust strain, which was maintained at the University of Arkansas, provides an excellent standard for comparison because the mean LC_{50} of Cry1Ac was virtually identical for UALab and two susceptible laboratory strains from North Carolina State University and the US Department of Agriculture (Ali *et al.*, 2006; Tabashnik *et al.*, 2008b).

One important difference between the two cases of practical resistance to Cry1Ac cotton is that the initial efficacy of this crop was greater against *P. gossypiella* than *H. zea*. For example, the efficacy of Cry1Ac cotton ranged from 77 to 96% against *P. gossypiella* in India from 2002 to 2005 (Bambawale *et al.*, 2004, 2010), but was only 44% against *H. zea* in North Carolina in 2001 (Jackson *et al.*, 2004). This difference in efficacy occurred before field-evolved resistance was detected in either pest; it reflects the 70-fold greater inherent susceptibility to Cry1Ac of pink bollworm relative to *H. zea* (Sivasupramaniam *et al.*, 2008). Thus, the reduction in efficacy associated with resistance to Cry1Ac cotton is almost certainly larger for pink bollworm in India than for *H. zea* in the USA.

Similar to the transition to two-toxin Bt cotton in India described above, cotton farmers in the USA have completely replaced one-toxin Bt cotton producing Cry1Ac with two-toxin Bt cotton, predominantly plants producing Cry1Ac and Cry2Ab (Brévault *et al.*, 2013; Tabashnik *et al.*, 2013). Here, we confirm the classification of *H. zea* resistance to Cry2Ab as >50% resistant individuals detected, with reduced efficacy of the Bt crop expected (Tabashnik *et al.*, 2013). The percentage of *H. zea* populations tested that had >50% survival at a diagnostic concentration of Cry2Ab rose from 0% in 2002 to 50% in 2005, only 2 years after the commercialization of Bt cotton producing Cry2Ab and Cry1Ac (Ali and Lutrell, 2007; Tabashnik *et al.*, 2009a). Three populations sampled from non-Bt plants in Arkansas in 2005 had such low mortality in bioassays that LC_{50} values could not be calculated (Ali and Luttrell, 2007). The decreased susceptibility to Cry2Ab detected in 2005, when cotton producing this toxin was not common, suggests that resistance to Cry1Ac caused some cross-resistance to Cry2Ab, which is consistent with the genetic correlation between resistance to these two toxins (Jackson *et al.*, 2006; Ali and Luttrell, 2007; Tabashnik *et al.*, 2013).

In addition, data from field populations in Arkansas show that mortality caused by a diagnostic concentration of Cry2Ab decreased substantially in 2010 compared with the previous 4 years (Jackson *et al.*, 2011). This evidence of field-evolved resistance to Cry2Ab coincided with higher abundance of *H. zea* in the field and increased insecticide sprays targeting *H. zea* on Bt cotton in 2010 (Jackson *et al.*, 2011). In the USA from 1999 to 2011, the percentage of Bt cotton producing two toxins increased from 0 to 90%, while the sprays against *H. zea* on Bt cotton tripled (Williams, 2012; Tabashnik *et al.*, 2013).

1.4 Lessons from Global Patterns of Field-evolved Resistance

The number of cases of field-evolved resistance to Bt crops with practical implications for pest control has increased to seven, encompassing the resistance of some populations of five pest species to five Bt toxins in four countries on four continents (Table 1.1 and Fig. 1.2), compared with only one such case as of 2005 (Tabashnik *et al.*, 2008a, 2013). Factors contributing to this surge in documented cases of practical resistance to Bt crops include more extensive monitoring as well as increases in the area planted to Bt crops (Fig. 1.1), the number of pest populations exposed to Bt crops and the cumulative duration of exposure.

Retrospective analyses of global patterns show that, as predicted from theory, abundant refuges of non-Bt host plants, recessive inheritance of resistance and a low initial frequency of resistance can delay the evolution of resistance to Bt crops (Gould 1998; Tabashnik *et al.*, 2008a, 2013; Carrière *et al.*, 2010). Nearly all of the seven cases of practical resistance (Table 1.1) entail a scarcity of refuges. For example, it appears that little or no planting of refuges spurred the rapid evolution of resistance to Bt crops in South Africa, Puerto Rico, India and Brazil (Storer *et al.*, 2012; Tabashnik *et al.*, 2013; Van den Berg *et al.*, 2013; Farias *et al.*, 2014; Monnerat *et al.*, Chapter 3).

Modelling results and empirical evidence suggest that a 'one size fits all' approach to refuge requirements is not optimal. In particular, under the ideal conditions of

recessive inheritance of resistance and low initial resistance allele frequency, refuges that account for a relatively small percentage of a pest's host plants may be sufficient to delay resistance for many years. Conversely, to substantially delay pest adaptation, much larger refuges are needed when resistance is not recessive or resistance alleles are not rare (Tabashnik *et al.*, 2008a, 2013; Carrière *et al.*, 2010). For example, in the Midwestern USA, the required 20% refuge of non-Bt maize has sustained the efficacy of Cry1Ab against *O. nubilalis* for more than a decade, generating billions of dollars in benefits for growers (Siegfried *et al.*, 2007; Hutchison *et al.*, 2010), but has not substantially slowed resistance to Cry3Bb in *D. v. virgifera* (Gassmann *et al.*, 2011, 2014; Tabashnik and Gould, 2012). Selection experiments and extensive screening of field populations show that alleles conferring resistance to Cry1Ab are exceedingly rare in *O. nubilalis* (Siegfried *et al.*, 2007). By contrast, non-recessive alleles conferring resistance to Cry3Bb are not rare in *D. v. virgifera*, yielding rapid evolution of resistance in the laboratory, greenhouse and field (Tabashnik and Gould, 2012).

Although most of the available data on field-evolved resistance to Bt crops involve responses to first-generation plants that each produce a single toxin, Bt plant pyramids producing two or three toxins active against the same pest have become increasingly prevalent during the past decade (Carrière *et al.*, 2015); see also Chapters 13 (Jakka *et al.*) and 16 (Huang). We hypothesize that, as observed with single-toxin Bt crops, pyramids will be extremely effective in delaying resistance under ideal conditions, but decidedly less successful under suboptimal conditions (Brévault *et al.*, 2013;, Carrière *et al.*, 2015). For example, two-toxin pyramids are expected be especially durable when they are used proactively – when resistance to both toxins is rare – as two-toxin cotton producing Cry1Ac and Cry2Ab has been used in Australia against *H. armigera* and *H. punctigera* (Downes *et al.*, 2012a, b). However, the advantages of pyramids may be diminished or lost when they are used remedially to counter resistance to one of the toxins in the pyramid, as in India, where cotton producing Cry1Ac and Cry2Ab is effectively a single-toxin crop against pink bollworm resistant to Cry1Ac (Tabashnik *et al.*, 2013). In general, the expected benefits of pyramids rely on 'redundant killing', which means that resistance to only one toxin in a pyramid does not increase survival on the pyramid (Brévault *et al.*, 2013; Carrière *et al.*, 2015). The expected benefits of pyramids are undermined by any factors that reduce redundant killing, including the survival of some susceptible individuals on pyramids or cross-resistance between the toxins in pyramids (Brévault *et al.*, 2013; Carrière *et al.*, 2015).

Since 2007, the US Environmental Protection Agency (US EPA) has greatly reduced refuge requirements for Bt crops despite four cases of practical resistance to Bt crops in the USA (Table 1.1), the observed association between the limited planting of refuges and rapid evolution of resistance, and recommendations from public sector scientists to maintain or increase refuge requirements (US EPA, 2002, 2013a; Alyokhin, 2011; Tabashnik and Gould, 2012). These reductions in refuge areas have been spurred, in part, by overly optimistic predictions about the durability of pyramids from computer models simulating ideal conditions (Alyokhin, 2011; Brévault *et al.*, 2013). Currently, in the USA, refuges of non-Bt maize can be as little as 5% of the total area planted to maize (US EPA, 2013b). Refuges of non-Bt cotton are no longer required for Bt cotton in most of the country, primarily because non-Bt host plants other than cotton are considered 'natural' refuges (US EPA, 2007). However, this approach largely ignores variation in host plant distribution that can yield a scarcity of refuges near Bt crops, thereby favouring the emergence and spread of resistance (Sisterson *et al.*, 2005; O'Rourke *et al.*, 2010; Onstad and Carrière, 2014). We hope that the US EPA and other regulatory agencies worldwide will heed one of the lessons learned from the failures and successes during the first 18 years of Bt crops: when

conditions are not optimal, large refuges are needed to sustain the efficacy of Bt crops and maximize their benefits.

Acknowledgement

This work was supported by USDA Biotechnology Risk Assessment Grant 2011-33522-30729.

References

Alcantara, E., Estrada, A., Alpuerto, V. and Head, G. (2011) Monitoring Cry1Ab susceptibility in Asian corn borer (Lepidoptera: Crambidae) on Bt corn in the Philippines. *Crop Protection* 30, 554–559.

Ali, M.I. and Luttrell, R.G. (2007) Susceptibility of bollworm and tobacco budworm (Noctuidae) to Cry2Ab2 insecticidal protein. *Journal of Economic Entomology* 100, 921–931.

Ali, M.I., Luttrell, R.G. and Young, S.Y. III (2006) Susceptibilities of *Helicoverpa zea* and *Heliothis virescens* (Lepidoptera: Noctuidae) populations to Cry1Ac insecticidal protein. *Journal of Economic Entomology* 99, 164–175.

Alyokhin, A. (2011) Scant evidence supports EPA's pyramided Bt corn refuge size of 5%. *Nature Biotechnology* 529, 577–578.

Andreadis, S.S., Álvarez-Alfageme, F., Sánchez-Ramos, L., Stodola, T.J., Andow, D.A. *et al.* (2007) Frequency of resistance to *Bacillus thuringiensis* toxin Cry1Ab in Greek and Spanish population of *Sesamia nonagrioides* (Lepidoptera: Noctuidae). *Journal of Economic Entomology* 100, 195–201.

Bambawale, O.M., Singh, A., Sharma, O.P., Bhosle, B.B., Lavekar, R.C. *et al.* (2004) Performance of Bt cotton (MECH-162) under integrated pest management in farmers' participatory field trial in Nanded district, central India. *Current Science* 86, 1628–1633.

Bambawale, O.M., Tanwar, R.K., Sharma, O.P., Bhosle, B.B., Lavekar, R.C. *et al.* (2010) Impact of refugia and integrated pest management on the performance of transgenic (*Bacillus thuringiensis*) cotton (*Gossypium hirsutum*). *Indian Journal of Agricultural Sciences* 80, 730–736.

Blanco, C.A., Andow, D.A., Gould, F., Abel, C.A., Sumerford, D.V., *et al.* (2009) *Bacillus thuringiensis* Cry1Ac resistance frequency in tobacco budworm (Lepidoptera: Noctuidae). *Journal of Economic Entomology* 102, 381–387.

Brévault T., Heuberger, S., Zhang, M., Ellers-Kirk, C., Ni, X. *et al.* (2013) Potential shortfall of pyramided Bt cotton for resistance management. *Proceedings of the National Academy of Sciences of the United States of America* 110, 5806–5811.

Carrière, Y., Crowder, D.W. and Tabashnik, B.E. (2010) Evolutionary ecology of insect adaptation to Bt crops. *Evolutionary Applications* 3, 561–573.

Carrière, Y., Crickmore, N. and Tabashnik, B.E. (2015) Optimizing pyramided transgenic crops for sustainable pest management. *Nature Biotechnology* 33, 161–168.

Dhurua, S. and Gujar, G.T. (2011) Field-evolved resistance to Bt toxin Cry1Ac in the pink bollworm, *Pectinophora gossypiella* (Saunders) (Lepidoptera: Gelechiidae), from India. *Pest Management Science* 67, 898–903.

Downes, S. and Mahon, R. (2012a) Evolution, ecology and management of resistance in *Helicoverpa* spp. to Bt cotton in Australia. *Journal of Invertebrate Pathology* 110, 281–286.

Downes, S. and Mahon, R. (2012b) Successes and challenges of managing resistance in *Helicoverpa armigera* to Bt cotton in Australia. *GM Crops and Food* 3, 228–234.

Downes, S., Parker, T.L. and Mahon, R.J. (2009) Frequency of alleles conferring resistance to the *Bacillus thuringiensis* toxins Cry1Ac and Cry2Ab in Australian populations of *Helicoverpa punctigera* (Lepidoptera: Noctuidae) from 2002 to 2006. *Journal of Economic Entomology* 102, 733–742.

Downes, S., Parker, T. and Mahon, R. (2010) Incipient resistance of *Helicoverpa punctigera* to the Cry2Ab Bt toxin in Bollgard II® cotton. *PLoS ONE* 5(9): e12567.

Edgerton, M.D., Fridgen, J., Anderson, J.R. Jr, Ahigrim, J., Criswell, M. *et al.* (2012) Transgenic insect resistance traits increase corn yield and yield stability. *Nature Biotechnology* 30, 493–496.

Fabrick, J.A., Ponnuraj, J., Singh, A., Tanwar, R.K., Unnithan, G.C. *et al.* (2014) Alternative splicing and highly variable cadherin transcripts associated with field-evolved resistance of pink bollworm to Bt cotton in India. *PLoS ONE* 9(5): e97900.

Farias, J.R., Andow, D.A., Horikoshi, R.J., Sorgatto, R.J., Fresia, P. *et al.* (2014) Field-evolved resistance to Cry1F maize by *Spodoptera frugiperda* (Lepidoptera: Noctuidae) in Brazil. *Crop Protection* 64, 150–158.

Farinós, G.P., de la Poza, M., Hernández-Crespo, P., Ortego, F. and Castañera, P. (2004) Resistance monitoring of field populations of the corn borers *Sesamia nonagrioides* and *Ostrinia nubilalis* after 5 years of Bt maize cultivation in Spain. *Entomologia Experimentalis et Applicata* 110, 23–30.

Farinós, G.P., Andreadis, S.S., de la Poza, M., Mironidis, G.K., Ortego, F. *et al.* (2011) Comparative assessment of the field-susceptibility of *Sesamia nonagrioides* to the Cry1Ab toxin in areas with different adoption rates of Bt maize and in Bt-free areas. *Crop Protection* 30, 902–906.

Gassmann, A.J. (2012) Field-evolved resistance to Bt maize by western corn rootworm: predictions from the laboratory and effects in the field. *Journal of Invertebrate Pathology* 110, 287–293.

Gassmann, A.J., Petzold-Maxwell, J.L., Keweshan, R.S. and Dunbar, M.W. (2011) Field-evolved resistance to Bt maize by western corn rootworm. *PLoS ONE* 6(7): e22629.

Gassmann, A.J., Petzold-Maxwell, J.L., Keweshan, R.S. and Dunbar, M.W. (2012) Western corn rootworm and Bt maize: challenges of pest resistance in the field. *GM Crops and Food* 3, 235–244.

Gassmann, A.J., Petzold-Maxwell, J.L., Clifton, E.H., Dunbar, M.W. and Hoffmann, A.M. (2014) Field-evolved resistance by western corn rootworm to multiple *Bacillus thuringiensis* toxins in transgenic maize. *Proceedings of the National Academy of Sciences of the United States of America* 111, 5141–5146.

Gould, F. (1998) Sustainability of transgenic insecticidal cultivars: integrating pest genetics and ecology. *Annual Review of Entomology* 43, 701–726.

Huang, F., Leonard, B.R., Cook, D.R., Lee, D.R., Andow, D.A. *et al.* (2007) Frequency of alleles conferring resistance to *Bacillus thuringiensis* maize in Louisiana populations of southwestern corn borer (Lepidoptera: Crambidae). *Entomologia Experimentalis et Applicata* 122, 53–58.

Huang, F., Ghimire, M.N., Leonard, B.R., Daves, C., Levy, R. *et al.* (2012) Extended monitoring of resistance to *Bacillus thuringiensis* Cry1Ab maize in *Diatraea saccharalis* (Lepidoptera: Crambidae). *GM Crops and Food* 3, 245–254.

Hutchison, W.D., Burkness, E.C., Mitchell, P.D., Moon, R.D., Leslie, T.W. *et al.* (2010) Areawide suppression of European corn borer with Bt maize reaps savings to non-Bt maize growers. *Science* 330, 222–225.

Jackson, R.E., Bradley, J.R. Jr, Van Duyn, J.W. and Gould, F. (2004) Comparative production of *Helicoverpa zea* (Lepidoptera: Noctuidae) from transgenic cotton expressing either one or two *Bacillus thuringiensis* proteins with and without insecticide oversprays. *Journal of Economic Entomology* 97, 1719–1725.

Jackson, R.E., Gould, F., Bradley, J.R. Jr and Van Duyn, J. (2006) Genetic variation for resistance to *Bacillus thuringiensis* toxins in *Helicoverpa zea* (Lepidoptera: Noctuidae) in eastern North Carolina. *Journal of Economic Entomology* 99, 1790–1797.

Jackson, R.E., Catchot, A., Gore, J. and Stewart, S.D. (2011) Increased survival of bollworms on Bollgard II cotton compared to lab-based colony. In: Boyd, S., Huffman, M. and Robertson, B. (eds) *Proceedings, 2011 Beltwide Cotton Conferences, 4–7 January 2011, Atlanta, Georgia*. National Cotton Council of America, Memphis, Tennessee, pp. 893–894.

James, C. (2013) *Global Status of Commercialized Biotech/GM Crops: 2013*. ISAAA Brief 46, International Service for the Acquisition of Agri-Biotech Applications, Ithaca, New York.

Jin, L., Wei, Y., Zhang, L., Yang, Y., Tabashnik, B.E. and Wu, Y. (2013) Dominant resistance to Bt cotton and minor cross-resistance to Bt toxin Cry2Ab in cotton bollworm from China. *Evolutionary Applications* 6, 1222–1235.

Kruger, M.J., van Rensburg, J.B.J. and Van den Berg, J. (2009) Perspective on the development of stem borer resistance to Bt maize and refuge compliance at the Vaalharts irrigation scheme in South Africa. *Crop Protection* 28, 684–689.

Kruger, M.J., van Rensburg, J.B.J. and Van den Berg, J. (2011) Resistance to Bt maize in *Busseola fusca* (Lepidoptera: Noctuidae) from Vaalharts, South Africa. *Environmental Entomology* 40, 477–483.

Lu, Y., Wu, K., Jiang, Y., Guo, Y. and Desneux, N. (2012) Widespread adoption of Bt cotton and insecticide decrease promotes biocontrol services. *Nature* 487, 362–365.

Luttrell, R.G. and Ali, M.I. (2007) Exploring selection for Bt resistance in heliothines: results of laboratory and field studies. In: Richter, D.A. (ed.) *Proceedings, 2007 Beltwide Cotton Conferences, 9–12 January 2007, New Orleans, Louisiana*. National Cotton Council of America, Memphis, Tennessee, pp. 1073–1086.

Luttrell, R.G. and Jackson, R.E. (2012) *Helicoverpa zea* and Bt cotton in the United States. *GM Crops and Food* 3, 213–227.

Luttrell, R.G., Ali, I., Allen, K.C., Young, S.Y. III, Szalanski, A. *et al.* (2004) Resistance to Bt in Arkansas populations of cotton bollworm. In: Richter, D.A. (ed.) *Proceedings, 2004 Beltwide Cotton Conferences, 5–9 January 2004, San Antonio, Texas*. National Cotton Council of

America, Memphis, Tennessee, pp. 1373–1383.

Moar, W., Roush, R., Shelton, A., Ferré, J., MacIntosh, S. *et al.* (2008) Field-evolved resistance to Bt toxins. *Nature Biotechnology* 26, 1072–1074.

Mendelsohn, M., Kough, J., Vaituzis, Z. and Matthews, K. (2003) Are Bt crops safe? *Nature Biotechnology* 21, 1003–1009.

Monsanto (2010) Pink Bollworm Resistance to GM Cotton in India. Available at: http://www.monsanto.com/newsviews/Pages/india-pink-bollworm.aspx (accessed 5 May 2010).

Nagoshi, R.N., Meagher, R.L. and Jenkins, D.A. (2010) Puerto Rico fall armyworm has only limited interactions with those from Brazil or Texas but could have substantial exchanges with Florida populations. *Journal of Economic Entomology* 103, 360–367.

National Research Council (2010) *The Impact of Genetically Engineered Crops on Farm Sustainability in the United States*. National Academies Press, Washington, DC.

Onstad, D.W. and Carrière, Y. (2014) The role of landscapes in insect resistance management. In: Onstad, D.W. (ed.) *Insect Resistance Management: Biology, Economics and Prediction*, 2nd edn. Academic Press (imprint of Elsevier), London, pp. 327–372.

O'Rourke, M.E., Sappington, T.W. and Fleischer, S.J. (2010) Managing resistance to Bt crops in a genetically variable insect herbivore, *Ostrinia nubilalis*. *Ecological Applications* 20, 1228–1236.

Pardo-López, L., Soberón, M. and Bravo, A. (2013) *Bacillus thuringiensis* insecticidal 3-domain Cry toxins: mode of action, insect resistance and consequences for crop protection. *FEMS Microbiology Reviews* 37, 3–22.

Sanahuja, G., Banakar, R., Twyman, R., Capell, T. and Christou, P. (2011) *Bacillus thuringiensis*: a century of research, development and commercial applications. *Plant Biotechnology Journal* 9, 283–300.

Siegfried, B.D. and Hellmich, R.L. (2012) Understanding successful resistance management: the European corn borer and Bt corn. *GM Crops and Food* 184–193.

Siegfried, B.D., Spencer, T., Crespo, A.L., Storer, N.P., Head, G.P. *et al.* (2007) Ten years of Bt resistance monitoring in the European corn borer: what we know, what we don't know, and what we can do better. *American Entomologist* 53, 208–214.

Siegfried, B.D., Rangasamy, M., Wang, H., Spencer, T., Haridas, V.V. *et al.* (2014) Estimating the frequency of Cry1F resistance in field populations of the European corn borer

(Lepidoptera: Crambidae). *Pest Management Science* 70, 725–733.

Sisterson, M.S., Carrière, Y., Dennehy, T.J. and Tabashnik, B.E. (2005) Evolution of resistance to transgenic crops: interactions between insect movement and field distribution. *Journal of Economic Entomology* 98, 1751–1762.

Sivasupramaniam, S., Moar, W.J., Ruschke, L.G., Osborn, J.A., Jiang, C. *et al.* (2008) Toxicity and characterization of cotton expressing *Bacillus thuringiensis* Cry1Ac and Cry2Ab2 proteins for control of lepidopteran pests. *Journal of Economic Entomology* 101, 546–554.

Storer, N.P., Babcock, J.M., Schlenz, M., Meade, T., Thompson, G.D. *et al.* (2010) Discovery and characterization of field resistance to Bt maize: *Spodoptera frugiperda* (Lepidoptera: Noctuidae) in Puerto Rico. *Journal of Economic Entomology* 103, 1031–1038.

Storer, N.P., Kubiszak, M.E., King, J.E, Thompson, G.D. and Santos, A.C. (2012) Status of resistance to Bt maize in *Spodoptera frugiperda*: lessons from Puerto Rico. *Journal of Invertebrate Pathology* 110, 294–300.

Sumerford, D.V., Head, G.P., Shelton, A., Greenplate, J. and Moar, W. (2013) Field-evolved resistance: assessing the problem and moving forward. *Journal of Economic Entomology* 106, 1525–1534.

Tabashnik, B.E. and Gould, F. (2012) Delaying corn rootworm resistance to Bt corn. *Journal of Economic Entomology* 105, 767–776.

Tabashnik, B.E., Malvar, T., Liu, Y.B., Borthakur, D., Shin, B.S. *et al.* (1996) Cross-resistance of diamondback moth indicates altered interactions with domain II of *Bacillus thuringiensis* toxins. *Applied and Environmental Microbiology* 62, 2839–2844.

Tabashnik, B.E., Dennehy, T.J., Sims, M.A., Larkin, K., Head, G.P. *et al.* (2002) Control of resistant pink bollworm by transgenic cotton with *Bacillus thuringiensis* toxin Cry2Ab. *Applied and Environmental Microbiology* 68, 3790–3794.

Tabashnik, B.E., Gassmann, A.J., Crowder, D.W. and Carrière, Y. (2008a) Insect resistance to Bt crops: evidence versus theory. *Nature Biotechnology* 26, 199–202.

Tabashnik, B.E., Gassmann, A.J., Crowder, D.W. and Carrière, Y. (2008b) Field-evolved resistance to Bt toxins. [Reply to Moar *et al.*, 2008.] *Nature Biotechnology* 26, 1074–1076.

Tabashnik, B.E., van Rensburg, J.B.J. and Carrière, Y. (2009a) Field-evolved insect resistance to Bt crops: definition, theory, and data. *Journal of Economic Entomology* 102, 2011–2025.

Tabashnik, B.E., Unnithan, G.C., Masson, L., Crowder, D.W., Li, X. *et al.* (2009b) Asymmetrical cross-resistance between *Bacillus thuringiensis* toxins Cry1Ac and Cry2Ab in pink bollworm. *Proceedings of the National Academy of Sciences of the United States of America* 106, 11889–11894.

Tabashnik, B.E., Sisterson, M.S., Ellsworth, P.C., Dennehy, T.J., Antilla, N. *et al.* (2010) Suppressing resistance to Bt cotton with sterile insect releases. *Nature Biotechnology* 28, 1304–1307.

Tabashnik, B.E., Brévault, T. and Carrière, Y. (2013) Insect resistance to Bt crops: lessons from the first billion acres. *Nature Biotechnology* 31, 510–521.

Tabashnik, B.E., Mota-Sanchez, D., Whalon, M.E., Hollingworth, R.M. and Carrière, Y. (2014) Defining terms for proactive management of resistance to Bt crops and pesticides. *Journal of Economic Entomology* 107, 496–507.

USDA ERS (2013) Adoption of Genetically Engineered Crops in the U.S. USDA Economic Research Service, Washington, DC. Available at: http://www.ers.usda.gov/datafiles/Adoption_of_Genetically_Engineered_Crops_in_the_US/alltables.xls (accessed 9 October 2014).

US EPA (2002) *Final Meeting Minutes. FIFRA [Federal Insecticide, Fungicide, and Rodenticide Act] Scientific Advisory Panel Meeting Held August 27–29, 2002. Corn Rootworm Plant-Incorporated Protectant Nontarget Insect and Insect Resistance Management Issues. SAP Meeting Minutes No. 2002-05.* Environmental Protection Agency, Washington, DC. Available at: http://www.epa.gov/scipoly/sap/meetings/2002/august/august2002final.pdf (accessed 9 October 2014).

US EPA (2007) Pesticide News Story: EPA Approves Natural Refuge for Insect Resistance Management in Bollgard II Cotton. US Environmental Protection Agency, Washington, DC. Available at: http://www.epa.gov/oppfead1/cb/csb_page/updates/2007/bollgard-cotton.htm (accessed 9 October 2014).

US EPA (2013a) December 4–6, 2013: Scientific Uncertainties Associated with Corn Rootworm Resistance Monitoring for Bt corn Plant Incorporated Protectants (PIPs). US Environmental Protection Agency, Washington, DC. Available at: http://www.epa.gov/scipoly/sap/meetings/2013/120413meeting.html (accessed 9 October 2014).

US EPA (2013b) Current & Previously Registered Section 3 PIP Registrations. US Environmental Protection Agency, Washington, DC. Available at: http://www.epa.gov/pesticides/biopesticides/pips/pip_list.htm (accessed 9 October 2014).

Van den Berg, J., Hilbeck, A. and Bøhn, T. (2013) Pest resistance to Cry1Ab Bt maize: field resistance, contributing factors and lessons from South Africa. *Crop Protection* 54, 154–160.

van Rensburg, J.B.J. (2007) First report of field resistance by stem borer, *Busseola fusca* (Fuller) to Bt-transgenic maize. *South African Journal of Plant and Soil* 24, 147–151.

Wan, P., Huang, Y., Wu, H., Huang, M., Cong, S. *et al.* (2012) Increased frequency of pink bollworm resistance to Bt toxin Cry1Ac in China. *PLoS ONE* 7(1): e29975.

Williams, M.R. (2012) Cotton insect loss estimate. In: Boyd, S., Huffman, M. and Robertson, B. (eds) Proceedings of the 2012 Beltwide Cotton Conferences, Orlando, Florida, *3–6 January 2012*. National Cotton Council of America, Memphis, Tennessee, pp. 1001–1012.

Zhang, H., Yin, W., Zhao, J., Jin, L., Yang, Y. *et al.* (2011) Early warning of cotton bollworm resistance associated with intensive planting of Bt cotton in China. *PLoS ONE* 6(8): e22874.

Zhang, H., Tian, W., Zhao, J., Jin, L., Yang, J. *et al.* (2012) Diverse genetic basis of field-evolved resistance to Bt cotton in cotton bollworm from China. *Proceedings of the National Academy of Sciences of the United States of America* 109, 10275–10280.

2 Status of Resistance to Bt Cotton in China: Cotton Bollworm and Pink Bollworm

Yulin Gao,[1]* Chenxi Liu,[1] Yidong Wu[2] and Kongming Wu[1]

[1]*State Key Laboratory for Biology of Plant Disease and Insect Pests, Institute of Plant Protection, Chinese Academy of Agricultural Sciences, Beijing, People's Republic of China;* [2]*Department of Entomology, Nanjing Agricultural University, Nanjing, People's Republic of China*

Summary

Transgenic cotton that expresses a gene derived from the bacterium *Bacillus thuringiensis* (Bt) has been deployed for combating the cotton pests *Helicoverpa armigera* and *Pectinophora gossypiella* since 1997 in China. The pest management tactics associated with Bt cotton have resulted in a drastic reduction in insecticide use. However, the evolution of resistance by the pests threatens the continued success of Bt cotton. The development of resistance to Bt is of great concern, and there is a vast body of research in this area aimed at ensuring the continued success of Bt cotton. Here, we review studies on the evolution of Bt resistance in these two bollworms, focusing on the commercial release of Bt cotton varieties in China, and the biochemical and molecular basis of Bt resistance. We also discuss resistance management strategies, and monitoring programmes implemented in China and other countries.

2.1 Introduction

The cotton bollworm, *Helicoverpa armigera*, is one of the most important insect pests in the cotton-growing regions of China. *H. armigera* completes three to four generations annually in the Yellow River Region (YRR) and the Northwestern Region (NR), and four to six generations in the Changjiang River Region (CRR) (Wu and Guo, 2005). Rainfall during cotton growth is an important climatic factor that influences the regional population dynamics of the pest during a season. Especially in the CRR, high levels of rainfall early in the season drastically inhibit the population development of *H. armigera*, which usually reaches outbreak status in later generations that occur during the drought season. The populations of *H. armigera* in all of China can be divided into four regional groups: the tropical, subtropical, temperate and Xinjiang genotypes. Their adaptive zones are, respectively, in: southern China; the middle and lower CRR, which includes Sichuan, Hunan, Hubei and Zhejiang provinces; the YRR, which includes Henan, Hebei and Shandong provinces; and the Xinjiang Uygur Autonomous Region and Gansu Province (Fig. 2.1) (Wu and Guo, 2005).

The pink bollworm, *Pectinophora gossypiella*, a cotton pest probably native to Indo-Pakistan, invaded China at the beginning of the 20th century (Liu *et al.*, 2008).

---

* Corresponding author. E-mail address: ylgao@ippcaas.cn

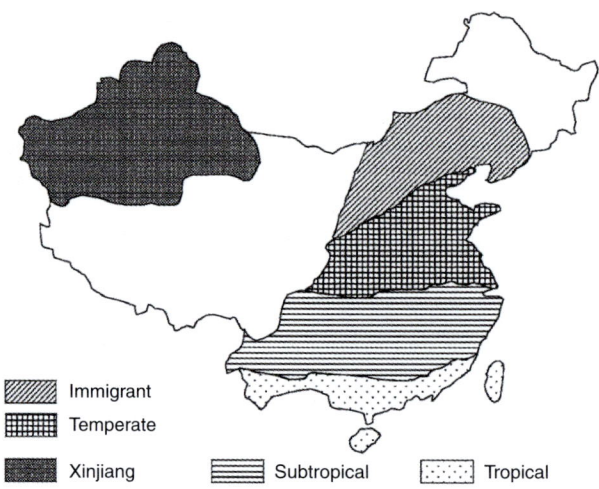

Fig. 2.1. Ecological zones of the four different genotypes (tropical, subtropical, temperate and Xinjiang) and the immigrant zone of temperate genotypes of the cotton bollworm, *Helicoverpa armigera*, in China.

Chinese *P. gossypiella* has been assumed to be the result of indiscriminate introductions from Pakistan and America via the transport of cotton seeds. Although pink bollworm is not the primary pest targeted by the use of Bt cotton throughout China, it is a major pest in the Yangtze River Valley of China (Wan *et al.*, 2004, 2012; Wu and Guo, 2005), where millions of resource-poor farmers plant more than a million hectares of cotton each year. The pink bollworm completes three to four generations each year. The use of insecticides is often futile, because the entire larval development of *P. gossypiella* is completed in the internal part of the infested bolls (Wu and Guo, 2005).

2.2 Commercial Release of Bt Cotton Varieties in China

China is one of the first countries to have grown commercialized genetically modified (GM) crops. Bt cotton was commercially released in 1997 and was rapidly adopted by farmers thereafter (Wu and Guo, 2005; Liu *et al.*, 2010). The use of Bt cotton over the total cotton areas exceeded 65% after 2004. Nearly all farmers planted Bt cotton in northern China and the Yangtze River basin after the middle 2000s (James, 2013). Bt

cotton is well reported as a successful case of biotechnology adoption in China. While smallholders in most GM crop growing countries are adopting GM technologies from multinational companies (MNCs), China's public sector has generated its own impressive GM technology; Bt cotton is one of the most cited examples of research and development progress of GM technology in the country. In 1997, two varieties of Bt cotton with different sources of Bt genes could be obtained by Chinese smallholders in certain provinces: one variety patented by the Chinese Academy of Agricultural Sciences (CAAS) is competitive with the one (NC33B) integrating the Monsanto Cry1Ac gene developed by the Monsanto company itself (Huang *et al.*, 2002; Wu and Guo, 2005). The Ministry of Agriculture (MOA) simultaneously approved these two varieties for commercialization: the one owned by CAAS was allowed to be cultivated in Shanxi, Anhui, Shandong and Hubei provinces, while the one developed by Monsanto was grown in Hebei Province. Table 2.1 indicates that in 1999, one and two varieties were allowed to be cultivated in Jiangsu Province (the YRR production zone) and Xinjiang Province, respectively. Since 2004, four varieties adapted to the agronomic conditions were commercially released in the

Table 2.1. Approval of commercial releases of Bt cotton in China by start year and by province.

Province	Cotton production zone	Start year	Variety	Affiliation
Anhui	Huang-Huai-Hai	1997	Bt cotton	Biotechnology Research Institute, CAAS[a]
Shanxi	Huang-Huai-Hai	1997	Bt cotton	Biotechnology Research Institute, CAAS
Shandong	Huang-Huai-Hai	1997	Bt cotton	Biotechnology Research Institute, CAAS
Hubei	Huang-Huai-Hai	1997	Bt cotton	Biotechnology Research Institute, CAAS
Hebei	Huang-Huai-Hai	1997	NC33B	Monsanto
Henan	Huang-Huai-Hai	1999	GK12, GK95-1	Biotechnology Research Institute, CAAS
Liaoning	Huang-Huai-Hai	1999	GK95-1	Biotechnology Research Institute, CAAS
Jiangsu	Yellow River Region	1999	GK-12	Biotechnology Research Institute, CAAS
Xinjiang	Xinjiang	1999	GK-12, GK95-1	Biotechnology Research Institute, CAAS
Shaanxi	Huang-Huai-Hai	2004	GKz1, GKz2	Biotechnology Research Institute, CAAS
Jiangxi	Yellow River Region	2004	DP410B	Monsanto
			GKz18	Biotechnology Research Institute, CAAS
Hunan	Yellow River Region	2004	DP410B	Monsanto
			GKz17	Biotechnology Research Institute, CAAS
Sichuan	Yellow River Region	2004	DP410B	Monsanto
			GKz34	Biotechnology Research Institute, CAAS
Zhejiang	Yellow River Region	2004	GKz18	Biotechnology Research Institute, CAAS

[a]Chinese Academy of Agricultural Sciences.

Yangtze River production zone (Wu and Guo, 2005; James, 2013).

The national area of cotton planted in China in 2012, at 4.9 million ha, was significantly lower than that planted in 2011 (5.5 million ha), but the adoption rate of Bt cotton increased to 80% in 2012, thus offsetting the decrease in the total area of cotton. The area of Bt cotton planted in 2012, 3.9 million ha, was approximately the same as that in 2011, when the adoption rate was only 71.5% (James, 2013).

2.3 Biochemical and Molecular Basis of Resistance

Midgut digestive proteinases have critical roles in the activation of Bt protoxin in the insect midgut. Therefore, a reduction or loss of proteases that convert Bt protoxin to the active toxin is a general mechanism of Bt resistance in resistant insects. A serine protease from the midgut of *H. armigera* resistant to the Bt toxin Cry1Ac displayed significantly reduced expression, resulting in improper processing of the protoxin (Rajagopal *et al.*, 2009). Accordingly, it

was proposed that esterase sequestration could be a mechanism for Bt resistance. Furthermore, esterases can also bind to and thus detoxify Cry1Ac before it reaches its target site. Consistent with this, Gunning *et al.* (2005) reported that resistant *H. armigera* strains showed greatly increased esterase titres compared with susceptible ones. In addition, alterations in other non-binding site proteinases (glutathione-*S*-tranferases, chymotrypsin and total midgut protease) have been found in three groups of *H. armigera* strains with different resistance levels and from different geographic origins and selection pressures (Cao *et al.*, 2013).

The specific binding of Bt toxins to midgut receptors is a critical event in Bt toxins achieving their toxic form. Loss or reduced expression of these receptors could result in reduced toxicity to insects. To date, this is the most common mechanism by which insects have become resistant to toxins (Ferré and van Rie, 2002). Cry1A resistance has been linked to gene mutations or reduced expression of insect midgut receptors such as cadherin-like proteins, aminopeptidase N (APN) and alkaline phosphatase (ALP). Understanding the molecular mechanisms of

Bt resistance is critical for developing efficient DNA-based monitoring of the frequency of toxin resistance. Such sensitive monitoring methods will enable early detection of resistance, allowing for early resistance warnings and proactive resistance management. Cadherin proteins are important receptors of the Bt toxin Cry1Ac in the midgut brush border membrane vesicles of lepidopteran pests. Mutations of *H. armigera* cadherin alleles (r_1–r_{15}) are linked with resistance to Cry1Ac in a laboratory-selected strain and in field-selected populations from northern China that were exposed intensively to Bt cotton producing Cry1Ac (Xu *et al.*, 2005; Yang *et al.*, 2007; Zhao *et al.*, 2010; Zhang *et al.*, 2012a; and Fabrick and Wu, Chapter 7). Alterations in *P. gossypiella* cadherin (r_1–r_{12D}) have been found in laboratory-selected and field-selected populations (Morin *et al.*, 2003; Fabrick *et al.*, 2011, 2014). A glycosylphosphatidylinositol (GPI)-anchored APN in the midgut epithelium of *H. armigera* larvae was characterized as a Cry1A toxin receptor, and Zhang *et al.* (2009) identified a deletion mutation of APN (HaAPN1) in laboratory-selected Cry1Ac-resistant strains of *H. armigera* by using cDNA-amplified fragment length polymorphism analysis. Similarly, a GPI-anchored ALP in the midgut epithelium of *H. armigera* larvae was also characterized as a Cry1A toxin receptor (Ning *et al.*, 2010). Reduced expression of ALP is a common mechanism of resistance to Bt toxins in Lepidoptera (Jurat-Fuentes *et al.*, 2011).

2.4 Resistance Monitoring of Cotton Bollworm to Bt-Cry1Ac Cotton

Resistance monitoring is generally accepted as an integral tool for managing resistance. In China, the susceptibility of *H. armigera* field populations to the Bt insecticidal protein Cry1Ac was monitored using traditional bioassays from 1997 to 2006 (Wu, 2007). The monitoring data indicated that the field populations have retained their susceptibility to Cry1Ac protein, and that *H. armigera* populations are not showing

a shift towards resistance. It has been suggested that *H. armigera* resistance genes are still rare (Li *et al.*, 2007; Gao *et al.*, 2009), and the existing refugia in maize, soybean, peanut and other vegetable crops may be a major factor in maintaining the susceptibility of *H. armigera* to Bt cotton, even after several years of large-scale commercial cultivation. In addition, gene flow derived from the migration of cotton bollworm over a large area may be an important factor in delaying the evolution of Bt resistance (Wu, 2007; Gao *et al.*, 2010).

Although a resistance monitoring system using traditional bioassays is already in place, it is not sensitive enough to detect frequencies of resistance as low as 0.005. Therefore, the relative average development rating (RADR) screening technique specific to bollworm was used from 2002 to 2013 in northern China to monitor Bt resistance in Xiajin County of Shandong Province and Anci County of Hebei Province. The dynamics of the RADR for each year in the F_1 generation populations during 2002–2013 in both regions are summarized in Fig. 2.2 (Li *et al.*, 2007; Gao *et al.*, 2009; An *et al.*, 2014). Although these results show significant decreases in cotton bollworm susceptibility to Cry1Ac during 2002–2013, the rate of evolution of resistance to Bt cotton in field populations was slow under the strong selection pressure in northern China after more than 16 years of commercialization. One main factor delaying the development of resistance in northern China may be the high percentage of host plants of crops other than cotton that do not produce Bt toxins and may act as natural refuges for *H. armigera*. In addition, *H. armigera* adults, as a typical migrant pest of economic crops, showed a strong movement ability, as demonstrated by damage to different hosts and by long-distance migrations between provinces, both of which occur frequently in northern China. So the gene flow derived from migration activities of the bollworm moths over a large area may be a major factor contributing to maintain the efficacy of transgenic cotton.

Bioassays have been used as a primary tool to evaluate Bt resistance, but this

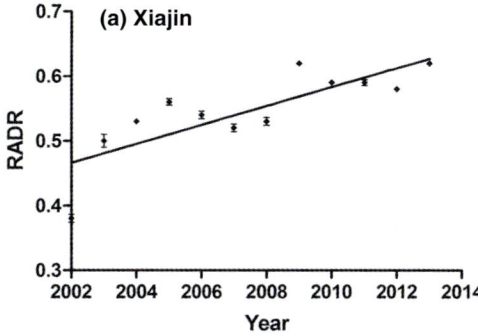

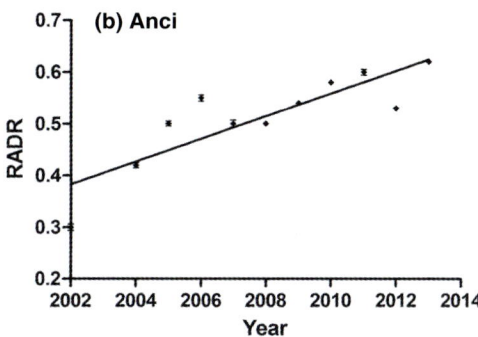

Fig. 2.2. Dynamics of the relative average development rating (RADR) for each year in the F_1 generation in populations of the cotton bollworm, *Helicoverpa armigera*, in (a) Xiajin County, Shandong Province and (b) Anci County, Hebei Province, northern China, during 2002–2013. (From Li *et al.*, 2007; Gao *et al.*, 2009; An *et al.*, 2014.)

approach may need huge manpower and material resources for sampling insects from the field and rearing their progeny in the laboratory. Moreover, when resistance is recessive, bioassays fail to distinguish between heterozygotes (*rs*) and homozygous susceptible individuals (*ss*), so it is hard to detect the extremely rare *rr* individuals in field by this technique. Compared with bioassays, DNA screens can save labour and time, for these can be done on all life stages of appropriately preserved insects. Further, DNA screens are feasible and efficient for detecting resistance alleles at low frequencies. None the less, these two monitoring approaches are complementary and

mutually confirmatory. A prerequisite of DNA diagnostics is prior identification of the mutations with major effects on resistance. A number of target site mutations conferring insecticide resistance have been identified and exploited to develop DNA-based diagnostics, which have been widely used to examine the spread and frequency of resistance mutations (ffrench-Constant 2007; Bass *et al.*, 2011; Puinean *et al.*, 2013). DNA-based screening has become applicable to Bt resistance since cadherin mutations were confirmed to cause Cry1Ac resistance in three lepidopteran pests of cotton (Gahan *et al.*, 2001; Morin *et al.*, 2003; Xu *et al.*, 2005; Zhang *et al.*, 2013).

By using the family screen bioassays in F_1 and F_2 screens, a total of 15 resistance alleles of *HaCad* (r_1–r_{15}) were identified in field populations of *H. armigera* from northern China (Yang *et al.*, 2007; Zhao *et al.*, 2010; Zhang *et al.*, 2012a,b, 2013). However, the mutational diversity of the cadherin gene may complicate DNA screening for Bt resistance allele frequency in the field.

Two methods have been used to estimate genotype frequencies: an indirect method based on the Hardy–Weinberg principle (which states that allele and genotype frequencies in a population will remain constant from generation to generation in the absence of other evolutionary influences) (Stern, 1943); and a direct method based on the genotypes of resistant individuals detected with the F_1 and F_2 screens. For populations from northern China, results from the F_1 and F_2 screens analysed by both the indirect and direct methods show that even though non-recessive alleles accounted for at most 25% of all resistance alleles detected, individuals with at least one non-recessive resistance allele (*Rx*) accounted for 59–94% of resistant individuals (Fig. 2.3). Based on the F_1 screen data for northern China, the estimated frequency of $r_c r_c$ (any two cadherin alleles) was 0.0053 from the indirect method and 0.018 from the direct method, compared with 0.027 for *Rx* from both methods. Results from the F_2 screen for northern China show a similar pattern, with an estimated frequency of 0.0015 for $r_c r_c$ and

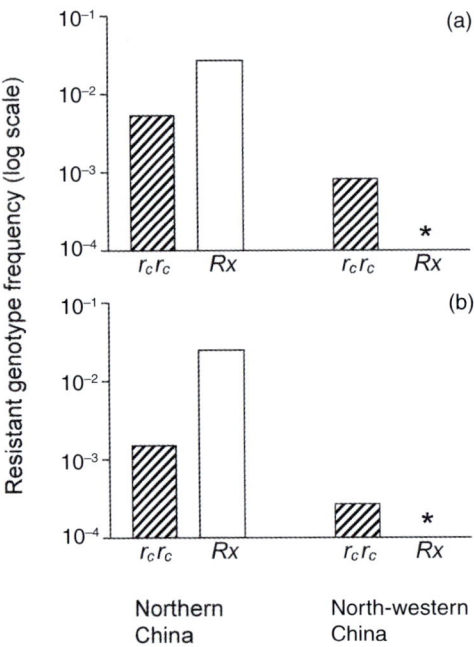

Fig. 2.3. Resistant genotype frequencies in the cotton bollworm, *Helicoverpa armigera* from northern and north-western China estimated from allele frequencies using the indirect method based on the Hardy–Weinberg principle (Zhang *et al.*, 2012a). The two classes of resistant genotypes depicted are $r_c r_c$ (any two recessive cadherin alleles) and Rx (one non-recessive resistance allele at any locus and any other allele at the same locus). (a) F_1 screen of field-collected males; (b) F_2 screen of field-mated females. Asterisks indicate that the frequency of Rx in north-western China was zero.

0.025 for Rx from the indirect method, and 0.0055 for $r_c r_c$ and 0.012 for Rx from the direct method (Zhang *et al.*, 2012a).

In addition, unlike most of the cadherin mutants conferring recessive resistance, an allele (r_{15}) with a 55 amino acid deletion in the intracellular domain of cadherin (HaCad) was identified to cause non-recessive resistance to Cry1Ac in *H. armigera* (Zhang *et al.*, 2012b). Screening the r_{15} allele in field populations of *H. armigera* showed that populations from northern China had a higher frequency of r_{15} than those from north-western China (Xinjiang) (Zhang *et al.*, 2013).

2.5 Resistance Monitoring of Pink Bollworm to Bt-Cry1Ac Cotton

In six provinces of the Yangtze River Valley, the percentage of total cotton planted with Bt cotton increased from 9% in 2000 to 52% in 2004, 84% in 2006, and 92–94% in 2008–2010 (Wan *et al.*, 2012). Resistance monitoring data show that the susceptibility of pink bollworm to Cry1Ac decreased significantly in 2008–2010 compared with 2005–2007 (Wan *et al.*, 2012). Results from laboratory diet bioassays of 51 field-derived strains of pink bollworm from 16 sites in the Yangtze River Valley show that susceptibility to Cry1Ac was significantly lower in 2008–2010 compared with 2005–2007 (Fig. 2.4) (Tabashnik *et al.*, 2012; Wan *et al.*, 2012).

Although the data reviewed above show decreased susceptibility to Cry1Ac in 2008–2010 compared with 2005–2007, resistance to Cry1Ac did not increase from 2009 to 2010, even though Bt cotton accounted for 94% of the cotton planted in the Yangtze River Valley in both years. These results may reflect stochastic factors that can affect the frequency of resistance and the results of resistance monitoring, particularly with relatively small sample sizes over short time intervals when resistance is close to the limit of detection (Wan *et al.*, 2012). The finding that the survival of pink bollworm larvae on Bt cotton in the Yangtze River Valley was as high as 11% during October 2001 and 2002, 6 and 7 years before field-evolved resistance was first detected in 2008, implies that Bt cotton did not kill all or nearly all susceptible larvae. So it can be inferred that even small decreases in susceptibility to Cry1Ac could reduce the efficacy of Bt cotton in the field. Based on the field data described above and our bioassay results (Wan *et al.*, 2012), the magnitude of resistance documented here reduces the efficacy of Cry1Ac-producing Bt cotton against pink bollworm in the field, at least during some part of the growing season (Tabashnik *et al.*, 2012; Wan *et al.*, 2012).

In the USA, samples of *P. gossypiella* – 5571 insects in all – were collected from 59 cotton fields in Arizona, California and

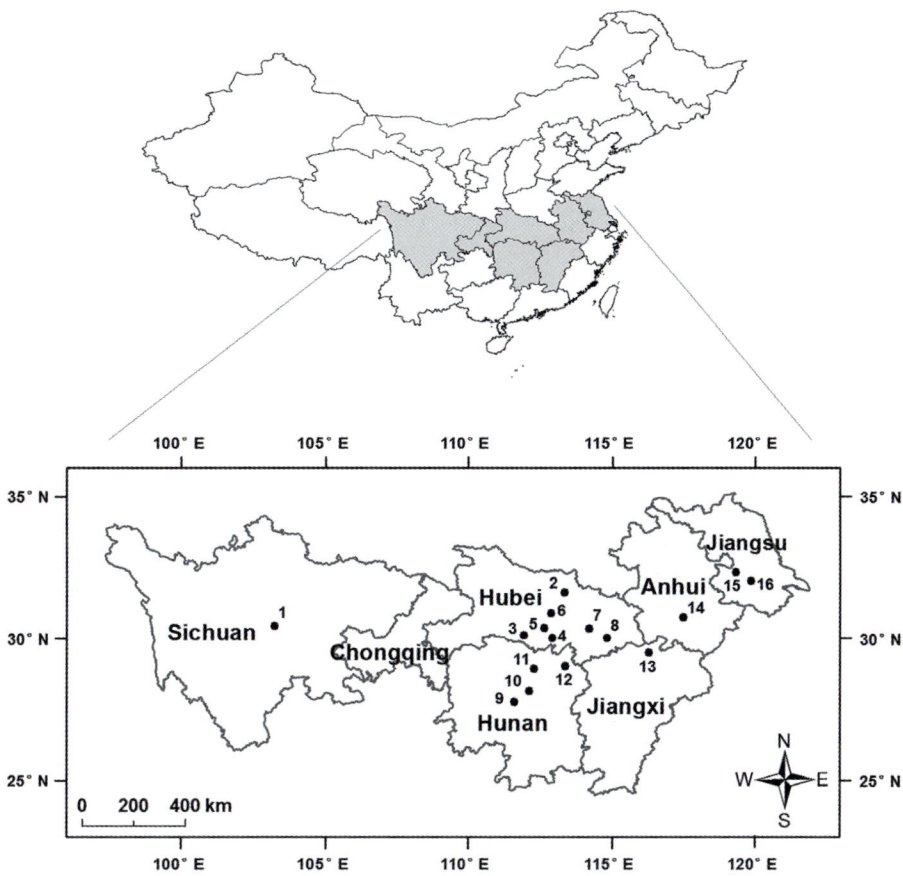

Fig. 2.4. Pink bollworm, *Pectinophora gossypiella*, resistance to Bt cotton monitoring sites in the Yangtze River Valley, China, during 2005–2010. (Wan *et al.*, 2012.)

Texas during 2001–2005, and were screened with a DNA-based diagnostic PCR; no resistance alleles of cadherin were detected (Tabashnik *et al.*, 2006). However, DNA sequencing of pink bollworm derived from resistant and susceptible field populations in India revealed eight novel, severely disrupted cadherin alleles associated with resistance to Cry1Ac. For these eight alleles, analysis of cDNA revealed a total of 19 transcript isoforms, each containing a premature stop codon, a deletion of at least 99 base pairs, or both. Seven of the eight disrupted alleles each produced two or more different transcript isoforms, which implicates alternative splicing of mRNA. This represents the first example of alternative splicing associated with field-evolved resistance that reduced the efficacy of a Bt crop (see Fabrick and Wu, Chapter 7).

2.6 Resistance Management Strategies in China

The greatest threat to the continued efficacy of Bt crops against their target pests is the evolution of resistance (Wu *et al.*, 2002; Burd *et al.*, 2003). Several deployment tactics designed to delay resistance have been proposed. The most promising resistance management strategy entails the use of plants with a high dose of toxin in combination with the maintenance of refuge crops that help to maintain Bt-susceptible insects within the overall pest population (Wu and Guo, 2005). The outcome of China's experiment with 'natural' refuges of non-Bt host plants other than cotton shows different results (Zhang *et al.*, 2011). Although widespread control failures have not been reported after 14 years of commercialization of Bt cotton producing Cry1Ac, non-Bt cotton accounted for more than 20% of the total cotton planted in northern China until 2003 (Wu *et al.*, 2008). Thus, field-evolved resistance of cotton bollworm to Cry1Ac in northern China was detected within 8 years after Bt cotton exceeded 80% of the total area of cotton planted. The first evidence of pink bollworm resistance to Bt cotton was detected in 2008, 8 years after Bt cotton was introduced in the Yangtze River Valley, though non-Bt cotton accounted for more than 37% of the total cotton planted in this region until 2006, when the amount dropped to 16% (Wan *et al.*, 2012). As a result, the first evidence of field-evolved resistance to Cry1Ac in the region was detected only 2 years after Bt cotton exceeded 80% of the total area of cotton planted.

The field outcomes with Bt cotton in China can be compared with those in India, the USA and Australia. The worst outcome has occurred with pink bollworm in the state of Gujarat in western India, where widespread control failures of Bt cotton associated with resistance to Cry1Ac have been reported (Bagla, 2010; Dhurua and Gujar, 2011). In India, non-Bt cotton refuges have been required, but compliance has been low (Stone, 2004). We do not know why pink bollworm resistance to Cry1Ac apparently evolved faster in India than in China, even though non-Bt cotton refuges were not required in China.

In the south-western USA, the susceptibility of pink bollworm to Cry1Ac has been sustained for 16 years (Tabashnik *et al.*, 2010, 2012). This success has been achieved first by abundant refuges (1996–2005) and second with a multi-tactic eradication programme that included mass releases of sterile moths and reduced refuge abundance (2006–2011) (Tabashnik *et al.*, 2005, 2010, 2012). In contrast, field-evolved resistance of *H. zea* to Cry1Ac and Cry2Ab produced by Bt cotton has been reported in the south-eastern USA (Tabashnik, 2008; Tabashnik *et al.*, 2009). Factors favouring this resistance appear to be a failure to meet the 'high-dose' criterion – 13 years of growing Bt cotton producing only Cry1Ac, and 7 years of overlap between Bt cotton plants producing a single toxin (Cry1Ac) and two toxins (Cry1Ac and Cry2Ab) (Tabashnik *et al.*, 2009).

In Australia, susceptibility to Cry1Ac has been sustained in cotton bollworm for more than a decade (Downes *et al.*, 2010). This has been achieved with a requirement of 70% refuges of non-Bt cotton for Bt cotton producing only Cry1Ac and complete replacement of this single-toxin Bt cotton with two-toxin Bt cotton during the 2004/05 growing season (Downes *et al.*, 2010).

Monitoring of cotton bollworm and pink bollworm in China has provided a warning that may be early enough to spur proactive measures to limit the consequences of resistance to Cry1Ac. One option is to switch to Bt cotton producing both Cry2Ab and Cry1Ac, which is already approved for small-scale trials in China. For countering resistance to Cry1Ac, cotton producing two or more toxins that are distinct from each other and from Cry1Ac could be more durable than cotton producing Cry1Ac and Cry2Ab, particularly if cross-resistance is limited or nil among Cry1Ac and the other two toxins. Other options for countering resistance to Cry1Ac are to increase the planting of non-Bt cotton and integrate other control tactics with Bt cotton for pest management (Wan *et al.*, 2012). For example, sterile moth releases and other tactics have been used in combination with Bt cotton to suppress pink bollworm in the USA (Tabashnik *et al.*, 2010, 2012). This approach has been implemented in Arizona since 2006 and pink bollworm has remained susceptible to Cry1Ac, even when the percentage of total cotton planted with Bt cotton exceeded 96% statewide.

2.7 Conclusion

Bt cotton has been widely planted in China since 1997, and this has resulted in efficient control of bollworm populations and helped to suppress pests, decrease the usage of insecticide sprays and promote biological control. However, the evolution of resistance by these pests can diminish these benefits and threaten the continued success of Bt cotton. Monitoring data from China have provided an early warning of resistance to Bt cotton in the cotton bollworm and pink bollworm. To delay pest resistance and guarantee the continued success of Bt crops in China, non-Bt-cotton refuges and pyramided plants would be the efficient ways to delay insect resistance. We advocate incorporating Bt crops as one tool in integrated pest management (IPM). IPM uses the best available combination of pest control tactics, including transgenic and conventionally bred host plant resistance, biological control, crop rotation and the judicious application of insecticide sprays. IPM can extend the efficacy of Bt crops while promoting sustainable agriculture that limits pest damage, optimizes returns to growers and preserves environmental quality.

References

An, J.J., Gao, Y., Lei, C., Gould, F. and Wu, K. (2014) Monitoring cotton bollworm resistance to Cry1Ac in two counties of northern China during 2009–2013. *Pest Management Science* Early view, doi 10.1002/ps.3807.

Bagla, P. (2010) Hardy cotton-munching pests are latest blow to GM crops. *Science* 327, 1439.

Bass, C., Puinean, A.M., Andrews, M., Cutler, P., Daniels, M. *et al.* (2011) Mutation of a nicotinic acetylcholine receptor subunit is associated with resistance to neonicotinoid insecticides in the aphid *Myzus persicae. BMC Neuroscience* 12:51.

Burd, A.D., Gould, F., Bradley, J.R., Vanduyn, J.W. and Moar, W.J. (2003) Estimated frequency of non-recessive Bt resistance genes in bollworm, *Helicoverpa zea* (Bolddie) (Lepidoptera: Noctuidae) in eastern North Carolina. *Journal of Economic Entomology* 96, 137–142.

Cao, G., Zhang, L., Liang, G., Li, X. and Wu, K.

(2013) Involvement of nonbinding site proteinases in the development of resistance of *Helicoverpa armigera* (Lepidoptera: Noctuidae) to Cry1Ac. *Journal of Economic Entomology* 106, 2514–2521.

Dhurua, S. and Gujar, G.T. (2011) Field-evolved resistance to Bt toxin Cry1Ac in the pink bollworm, *Pectinophora gossypiella* (Saunders) (Lepidoptera: Gelechiidae), from India. *Pest Management Science* 67, 898–903.

Downes, S., Mahon, R.J., Rossiter, L., Kauter, G., Leven, T. *et al.* (2010) Adaptive management of pest resistance by *Helicoverpa* species (Noctuidae) in Australia to the Cry2Ab Bt toxin in Bollgard II® cotton. *Evolutionary Applications* 3, 574–584.

Fabrick, J.A., Mathew, L.G., Tabashnik, B.E. and Li, X. (2011) Insertion of an intact CR1 retrotransposon in a cadherin gene linked with Bt resistance in the pink bollworm, *Pectinophora gossypiella. Insect Molecular Biology* 20, 651–665.

Fabrick, J.A., Ponnuraj, J., Singh, A., Tanwar, R.K., Unnithan, G.C. *et al.* (2014) Alternative splicing and highly variable cadherin transcripts associated with field-evolved resistance of pink bollworm to Bt cotton in India. *PLoS ONE* 9(5): e97900.

Ferré, J. and van Rie, J. (2002) Biochemistry and genetics of insect resistance to *Bacillus thuringiensis. Annual Review of Entomology* 47, 501–533.

ffrench-Constant, R.H. (2007) Which came first insecticides or resistance? *Trends in Genetics* 23, 1–4.

Gahan, L.J., Gould, F. and Heckel, D.G. (2001) Identification of a gene associated with Bt resistance in *Heliothis virescens. Science* 293, 857–860.

Gao, Y., Wu, K. and Gould, F. (2009) Frequency of Bt resistance alleles in *H. armigera* during 2006–2008 in northern China. *Environmental Entomology* 38, 1336–1342.

Gao, Y., Feng, H. and Wu, K. (2010) Regulation of the seasonal population patterns of *Helicoverpa armigera* moths by Bt cotton planting. *Transgenic Research* 19, 557–562.

Gunning, R.V., Dang, H.T., Kemp, F.C. and Moores, G.D. (2005) New resistance mechanism in *Helicoverpa armigera* threatens transgenic crops expressing *Bacillus thuringiensis* Cry1Ac toxin. *Applied and Environmental Microbiology* 71, 2558–2563.

Huang, J., Hu, R., Rozelle, S., Qiao, F. and Pray, C.E. (2002) Transgenic varieties and productivity of smallholder cotton farmers in China. *The Australian Journal of Agricultural and Resource Economics* 46, 1–21.

James, C. (2013) *Global Status of Commercialized Biotech/GM Crops: 2013*. ISAAA Brief No. 46, International Service for the Acquisition of Agri-Biotech Applications, Ithaca, New York.

Jurat-Fuentes, J.L., Karumbaiah, L., Jakka, S.R.K., Ning, C., Liu, C. *et al.* (2011) Reduced levels of membrane-bound alkaline phosphatase are common to lepidopteran strains resistant to Cry toxin from *Bacillus thuringiensis*. *PLoS ONE* 6(3): e17606.

Li, G., Wu, K., Gould, F., Wang, J., Mao, J. *et al.* (2007) Increasing tolerance to Cry1Ac cotton from cotton bollworm was confirmed in Bt cotton farming area of China. *Ecological Entomology* 32, 366–375.

Liu, C., Li, Y., Gao, Y., Ning, C. and Wu, K. (2010) Cotton bollworm resistance to Bt transgenic cotton: a case analysis. *Science in China Series C: Life Sciences* 53, 934–941.

Liu, Y., Wu, K. and Guo, Y. (2008) Population structure and introduction history of the pink bollworm, *Pectinophora gossypiella*, in China. *Entomologia Experimentalis et Applicata* 130, 160–172.

Morin, S., Biggs, R.W., Sisterson, M.S., Shriver, L., Ellers-Kirk, C. *et al.* (2003) Three cadherin alleles associated with resistance to *Bacillus thuringiensis* in pink bollworm. *Proceedings of the National Academy of Sciences of the United States of America* 100, 5004–5009.

Ning. C., Wu, K., Liu, C., Gao, Y. and Jurat-Fuentes, J.L. (2010) Characterization of a Cry1Ac toxin-binding alkaline phosphatase in the midgut from *Helicoverpa armigera* (Hübner) larvae. *Journal of Insect Physiology* 56, 666–672.

Puinean, A.M., Lansdell, S.J., Collins, T., Bielza, P. and Millar, N.S. (2013) A nicotinic acetylcholine receptor transmembrane point mutation (G275E) associated with resistance to spinosad in *Frankliniella occidentalis*. *Journal of Neurochemistry* 124, 590–601.

Rajagopal, R., Arora, N., Sivakumar, S., Rao, N.G.V., Nimbalkar, S.A. *et al.* (2009) Resistance of *Helicoverpa armigera* to Cry1Ac toxin from *Bacillus thuringiensis* is due to improper processing of the protoxin. *Biochemical Journal* 419, 309–316.

Stern, C. (1943) The Hardy–Weinberg law. *Science* 97, 137–138.

Stone, G.D. (2004) Biotechnology and the political ecology of information in India. *Human Organization* 63, 127–140.

Tabashnik, B.E. (2008) Delaying insect resistance to transgenic crops. *Proceedings of the National Academy of Sciences of the United States of America* 105, 19029–19030.

Tabashnik, B.E., Dennehy, T.J. and Carrière, Y. (2005) Delayed resistance to transgenic cotton in pink bollworm. *Proceedings of the National Academy of Sciences of the United States of America* 102, 15389–15393.

Tabashnik, B.E., Fabrick, J.A., Henderson, S., Biggs, R.W., Yafuso, C.M. *et al.* (2006) DNA screening reveals pink bollworm resistance to Bt cotton remains rare after a decade of exposure. *Journal of Economic Entomology* 99, 1525–1530.

Tabashnik, B.E., van Rensburg, J. and Carrière, Y. (2009) Field-evolved insect resistance to Bt crops: definition, theory, and data. *Journal of Economic Entomology* 102, 2011–2025.

Tabashnik, B.E., Sisterson, M.S., Ellsworth, P.C., Dennehy, T.J., Antilla, L. *et al.* (2010) Suppressing resistance to Bt cotton with sterile insect releases. *Nature Biotechnology* 28, 1304–1307.

Tabashnik, B.E., Wu, K. and Wu, Y. (2012) Early detection of field-evolved resistance to Bt cotton in China: cotton bollworm and pink bollworm. *Journal of Invertebrate Pathology* 110, 301–306.

Wan, P., Wu, K., Huang, M. and Wu, J. (2004) Seasonal pattern of infestation by pink bollworm *Pectinophora gossypiella* (Saunders) in field plots of Bt transgenic cotton in the Yangtze River Valley of China. *Crop Protection* 23, 463–467.

Wan, P., Huang, Y., Wu, H., Huang, M., Cong, S. *et al.* (2012) Increased frequency of pink bollworm resistance to Bt toxin Cry1Ac in China. *PLoS ONE* 7(1): e29975.

Wu, K. (2007) Monitoring and management strategy for *Helicoverpa armigera* resistance to Bt cotton in China. *Journal of Invertebrate Pathology* 95, 220–223.

Wu, K. and Guo, Y. (2005) The evolution of cotton pest management practices in China. *Annual Review of Entomology* 50, 31–52.

Wu, K., Guo, Y. and Gao, S. (2002) Evaluation of the natural refuge function for *Helicoverpa armigera* (Lepidoptera: Noctuidae) within *Bacillus thuringiensis* transgenic cotton growing areas in northern China. *Journal of Economic Entomology* 95, 832–837.

Wu, K., Lu, Y., Feng, H., Jiang, Y. and Zhao, J. (2008) Suppression of cotton bollworm in multiple crops in China in areas with Bt toxin-containing cotton. *Science* 321, 1676–1678.

Xu, X., Yu, L. and Wu, Y. (2005) Disruption of a cadherin gene associated with resistance to Cry1Ac δ-endotoxin of *Bacillus thuringiensis* in *Helicoverpa armigera*. *Applied and Environmental Microbiology* 71, 948–954.

Yang, Y., Chen, H., Wu, Y. and Wu, S. (2007) Mutated cadherin alleles from a field population of *Helicoverpa armigera* confer resistance to *Bacillus thuringiensis* toxin Cry1Ac. *Applied and Environmental Microbiology* 73, 6939–6944.

Zhang, S., Cheng, H., Gao, Y., Wang, G., Liang, G. *et al.* (2009) Mutation of an aminopeptidase N gene is associated with *Helicoverpa armigera* resistance to *Bacillus thuringiensis* Cry1Ac toxin. *Insect Biochemistry and Molecular Biology* 39, 421–429.

Zhang, H., Yin, W., Zhao, J., Jin, L., Yang, Y. *et al.* (2011) Early warning of cotton bollworm resistance associated with intensive planting of Bt cotton in China. *PLoS ONE* 6(8): e22874.

Zhang, H., Tian, W., Zhao, J., Jin, L., Yang J. *et al.* (2012a) Diverse genetic basis of field-evolved resistance to Bt cotton in cotton bollworm from China. *Proceedings of the National Academy of Sciences of the United States of America* 109, 10275–10280.

Zhang, H., Wu, S., Yang, Y., Tabashnik, B.E. and Wu, Y. (2012b) Non-recessive Bt toxin resistance conferred by an intracellular cadherin mutation in field-selected populations of cotton bollworm. *PLoS ONE* 7(12): e53418.

Zhang, H., Tian, W., Yang, F., Yang, Y. and Wu, Y. (2013) DNA-based screening for an intracellular cadherin mutation conferring non-recessive Cry1Ac resistance in field populations of *Helicoverpa armigera*. *Pesticide Biochemistry and Physiology* 107, 148–152.

Zhao, J., Jin, L., Yang, Y. and Wu, Y. (2010) Diverse cadherin mutations conferring resistance to *Bacillus thuringiensis* toxin Cry1Ac in *Helicoverpa armigera*. *Insect Biochemistry and Molecular Biology* 40, 113–118.

3 Insect Resistance to Bt toxins in Brazil and Latin America

Rose Monnerat,[1]* Erica Martins,[2] Paulo Queiroz,[2] Lilian Praça,[1] Carlos Marcelo Soares[2]

[1]*Embrapa Recursos Genéticos e Biotecnologia (CENARGEN), Brasília, Distrito Federal, Brazil;* [2]*Instituto Mato-Grossense do Algodão, Cuiabá, Mato Grosso, Brazil*

Summary

Bioinsecticides based on *Bacillus thuringiensis* (Bt) and transgenic plants expressing its toxins have been widely used to control insects in Latin America. In Brazil, during the season of 2013/14, 4 million ha were treated with Bt biopesticides, and the area of transgenic crops under cultivation reached 40.3 million ha. Countries such as Honduras, Guatemala, Costa Rica, Nicaragua and Brazil reported cases of resistance of *Plutella xylostella* to Bt biopesticides. Insect resistance to toxins expressed in Bt crops was reported in Puerto Rico and Brazil, and reports of low efficacy were also communicated from several regions of Brazil and Argentina. Faced to this scenario, it is important that companies providing technologies, extension services, farmers and the scientific community and governments of countries either using Bt as bioinsecticides or as Bt crops unite in their efforts to establish guidelines that will enable greater durability of Bt technology.

3.1 Insect Control with Chemical Insecticides

Climatic conditions in Latin America predispose to the occurrence of a large number of agricultural pests (CEPAL/FAO/ IICA, 2014). To minimize losses caused by these organisms, large amounts of chemicals are being used in agriculture in this region. Their indiscriminate use, in addition to the non-specific harm they cause to living organisms and to the environment, has resulted in the selection of pest populations that are increasingly resistant to the different active ingredients applied in the field (Baek *et al.*, 2005; Cruz *et al.*, 2013). The application costs of these products also represent a large investment that be up to 30% of the total cost, thus burdening production or making it unfeasible in some cases (Silva-Filho and Falco, 2000; Castelo Branco and Medeiros, 2001; Castelo Branco *et al.*, 2001).

Much of the pesticide produced worldwide is used in Latin America, with Brazil accounting for 86% of the use in this region and 20% of the total world usage. The reason for this excessive use of pesticides is associated with several factors, including: an increase in the amount of transgenic soybean resistant to glyphosate; the evolution of resistance in weeds, fungi and insects; and/or an increase in the prevalence of crop diseases such as soybean rust (Pignati and Machado, 2011). In 18 years, the amount of pesticides used on agricultural fields in Brazil has more than doubled from 70 kg ha^{-1} in 1992 to >150 kg ha^{-1} in 2010 (IBGE, 2012). In 2008, the Brazilian market

---

* Corresponding author. E-mail address: rose.monnerat@embrapa.br

for pesticides surpassed that of the USA and assumed the position of the world's largest market for pesticides (Carneiro *et al.*, 2012).

3.2 Use of *Bacillus thuringiensis* in Insect Control

The importance of controlling pests and increasing the awareness of the population of the direct and indirect effects of chemical pesticides on human health and the environment as a whole has demanded the use of new ways of controlling insects, which are less expensive and less harmful to the ecosystem. One alternative is the use of *Bacillus thuringiensis* (Bt). This bacterium can be used as the wild type form, as the active ingredient in biopesticide formulations or as a source of genes for expression in plants – Bt plants.

Products based on Bt began to be used in biological control in 1938, when a rather rudimentary formulation based on this bacterium, Sporeína, was produced in France (Weiser, 1986). At the beginning of the 1950s, many countries, including Russia, Czechoslovakia, France, Germany and the USA began producing industrially based biological Bt insecticides (Weiser, 1986). In Latin America, biopesticides are used in combating a wide range of pests in various crops. The first product that was used dates back to the early 1950s in Argentina, where it was used for control of the alfalfa caterpillar *Colias lesbia pyrrothea* (Sosa-Gómez and Moscardi, 1991). However, products formulated with Bt as the active ingredient have had limited use in Argentina due the large-scale use of conventional insecticides (Polanczyk *et al.*, 2012).

In Brazil, the first studies with Bt pesticides began in the 1960s and continued until the 1990s; three imported commercial formulations were available during this period (Polanczyk *et al.*, 2012). Nevertheless, by 2001, only 150,000 ha had been treated with Bt (Souza, 2001). There are many causes for the low adoption of the technology, but these especially included the following: effectiveness only on immature stages of insects; low residual efficacy caused by inactivation of the spore/crystal complex by UV radiation; slow action compared with conventional chemical insecticides; and a narrow spectrum of action. Coupled with these factors, competition with chemical products, which are usually available at a lower cost, pushed the use of formulated Bt biopesticides into the background, in addition to the establishment of extensive plantations of genetically modified (GM) plants expressing Bt toxins, such as cotton and maize and, more recently, soybeans.

Despite this situation, in the period 2011/12, many farmers in Brazil started to adopt the use of a tank mixture of Bt and chemical insecticides to control Lepidoptera in conventional crops; according to reports from these producers, the mixing increased the effectiveness of the application. With the adoption of this technique, the area treated with Bt-based bioproducts reached close to 2 million ha. A further increase in the area treated with Bt was observed after the finding of crop fields of cotton, maize and soybean infected with *Helicoverpa armigera* in 2013 (Specht *et al.*, 2013); this allowed the registration on an emergency basis of various chemical and biological agents, among them some Bt bioproducts. In the season of 2013/14, some 4 million ha were treated with Bt biopesticides, with a significant portion supplied by homemade brews, made by farmers on their own properties, and named 'homemade-Bt'. The use of homemade-Bt was principally due to the incapacity of registered Bt formulations to satisfy market demand. As the homemade product is produced without standardization, concern about its role in the selection of resistant insect populations has increased among technicians and researchers.

With the exception of Paraguay, whose agricultural situation is similar to that of Brazil and Argentina, with large farms that grow (GM) Bt cotton, Bt maize and Bt soybean, the other countries of Latin America, such as Chile, Colombia, Bolivia, Peru, Mexico, Costa Rica, Venezuela and Ecuador, make use of Bt products mainly in horticultural crops. Cuba has developed a proprietary Bt production system in the face of the economic embargo conducted against it by the USA (Polanczyk *et al.*, 2012).

3.3 Insect Resistance to Formulated Products Based on Bt toxins

Because different strains of *B. thuringiensis* produce different toxins with different modes of action, there was little expectation of the selection of resistant insects. However, in 1990, it was reported that populations of *Plutella xylostella* in Hawaii that had been exposed to continuous applications of commercial products showed a level of resistance 25–33 times higher than that of control populations (Tabashnik *et al.*, 1990; Shelton *et al.*, 1993).

In Latin America, the first studies of resistance to Bt biopesticides were conducted in Honduras, Guatemala, Costa Rica, Nicaragua and Brazil (Perez and Shelton, 1997; Perez *et al.*, 1997; Castelo Branco and Gatehouse, 2001). Values for the medium lethal concentration (LC_{50}) of *B. t.* subspp. *kurstaki* (Btk) and *aizawai* (Bta) against field populations of *P. xylostella* in Guatemala, Honduras and Costa Rica suggested that Central American populations of this pest in had developed resistance to Btk, as they demonstrated LC_{50} values 4.3 to 77.2 times higher than those obtained with a susceptible control population (Perez and Shelton, 1997). Studies with a resistant population coming from Honduras suggested that the inclusion of an area of refuge could be a strategy for the management of resistance (Perez *et al.*, 1997). The resistance of *P. xylostella* to *B. thuringiensis* toxins in Costa Rica was also reported by Cartín *et al.* (1999). In Brazil, the first reported case occurred with *P. xylostella* in the region of the Federal District in the midwest region of the country in 1996 and 1997, where it was observed that the LC_{50} of the commercial product based on *B. t. kurstaki* was 36 times higher than that of the susceptible population (Castelo Branco and Gatehouse, 2001). Subsequently, resistance was also documented in crops located in the south-eastern

and southern regions of Brazil (Ribeiro, 2010; Zago *et al.*, 2014). Ribeiro (2010) compared two field populations, one resistant and the other susceptible, and attributed the resistance found to various mechanisms, ranging from alterations in membrane receptors to processes involved in the activation of the toxin. Zago *et al.* (2014) conducted experiments with ten field populations collected in cabbage crops. Most populations showed levels of resistance 2–54 times higher than expected to Bt-based products, primarily to Xentari®WDG. It was observed that most females exposed to the Dipel®WP product oviposited on untreated surfaces and there was a correlation between reduced oviposition on treated surfaces that suggested a behavioural response among the colonies that developed resistance to Dipel biopesticide (Zago *et al.*, 2014).

In Mexico, Díaz-Gomez *et al.* (2000) evaluated the susceptibility of five populations of *P. xylostella* collected from commercial cruciferous crops at three locations to commercial formulations of Bt and found that bioassays confirmed a variation in the levels of susceptibility of up to 12 times lower to Dipel (Btk), seven times lower to Xentari (Bta) and five times lower to Agree (Btk and Bta). Comparisons with other studies conducted by this group indicated that a variation of 12 times in the level of susceptibility to Dipel resulted in significant differences in the control achieved in the field. The authors suggested that high rates of resistance indicated that the populations are becoming less heterogeneous and more tolerant to toxins used in the fields due to an increased frequency of resistant genotypes. The mechanism of resistance of these populations has not been elucidated. Nevertheless, in a resistant *P. xylostella* population from Hawaii, it was shown that the resistance was autosomal recessive and apparently controlled by one or a few loci (Tabashnik *et al.*, 1990, 1993). Later work identified the resistance gene allele, which mapped to an ANBCC2 transporter gene that is a putative receptor in the gut of susceptible insects (Baxter *et al.*, 2011; Tabashnik *et al.*, 2011).

3.4 Use of GM Plants Expressing Insecticidal Proteins from Bt

GM plants expressing toxic proteins from Bt have been used in the field since 1996 and are important tools for controlling insects and reducing the use of chemical products (Tabashnik *et al.*, 2013). In Latin America, according to the International Service for the Acquisition of Agri-biotech Applications (ISAAA), the area cultivated with GM seeds in 2013 reached approximately 71 million ha, almost half of the global area of transgenic crops, which is 175.2 million ha (James, 2013). According to data from the same institution, the area cultivated with GM seeds in Brazil in 2013 reached 40.3 million ha, making it the second largest country cultivating transgenic plants in the world and the first largest in Latin America, where it accounted for 57% of the planting in the region (James, 2013). The second Latin American country where GM plants are most planted is Argentina, with an area of 24.4 million ha, followed by Paraguay, Uruguay, Bolivia, Mexico and Colombia. GM plants are also found in Chile, Cuba, Costa Rica and Puerto Rico. Table 3.1 shows the Latin American countries in which GM crops are planted, along with the acreage and crops planted (James, 2013).

Commercial release of GM crops in Latin American countries occurred according to the legislation of each country (Borsani *et al.*, 2010). The first country to release GM plants for commercial use in Latin America was Argentina, where the commercial release of Bt cotton and Bt maize was approved in 1998, and that of Bt soybean in 2013 (Trigo and Cap, 2003; ArgenBio, 2014). In Brazil, the first commercially released GM crop for insect control was Bt cotton, in 2005. Later, in 2007, there were several transformation events (an event is a specific genetic modification in a specific species) of maize expressing Bt toxins that were released for sale in Brazil, and in 2010, Bt soybean was also approved (CTNBio, 2014). In Paraguay, the third country to cultivate GM crops in Latin America, the first authorization for planting GM crops occurred in 2012 (INBio, 2014).

Generally, the plants used in Latin America express the Cry1-3 Bt proteins (Cry1Ab, Cry1Ac, Cry1A.105, Cry1F, Cry2Ab2, Cry2Ae, Cry3Bb1, Cry34Ab1 and Cry35Ab1) (Table 3.2) (CTNBio, 2014). The main targets of these toxins in the transgenic events that have been released are one coleopteran (*Diabrotica* spp.) and ten insect lepidopterans: *Spodoptera frugiperda* (fall armyworm); *Alabama argillacea* (cotton leafworm); *Heliothis virescens* and *Helicoverpa zea* (corn earworm); *Pectinophora gossypiella* (pink bollworm); *Anticarsia gemmatalis* (woolly pyrol moth); *Chrysodeixis includens*

Table 3.1. Countries and acreage planted with genetically modified (GM) plants in Latin America. (Data from James, 2013.)

Country	Area (million ha)	GM crop
Brazil	40.3	Soybean, maize, cotton
Argentina	24.4	Soybean, maize, cotton
Paraguay	3.6	Soybean, maize, cotton
Uruguay	1.5	Soybean, maize
Bolivia	1.0	Soybean
Mexico	0.1	Cotton
Colombia	<0.1	Cotton, maize
Chile	<0.1	Maize, soybean, canola
Honduras	<0.1	Maize
Cuba	<0.1	Maize
Costa Rica	<0.1	Cotton, soybean

(soybean looper moth); *Diatraea saccharalis*; *Agrotis ipsilon*; and *Elasmopalpus lignosellus* (Table 3.2). Since the detection of *H. armigera* in Brazil (Specht *et al.*, 2013), the technology has also been used to control this pest. The variability in the susceptibility to Cry toxins of some of these insects has been reported by many authors (Waquil *et al.*, 2004; Monnerat *et al.*, 2006; Polania *et al.*, 2009; Blanco *et al.*, 2010; Rios-Diez *et al.*, 2012; Bernardi *et al.*, 2014a,b).

In general, transgenic crops have been well accepted by Latin American farmers owing to the high efficacy in controlling target pests In the 2013/14 season, almost 100% of the agricultural area in Brazil was planted with transgenic crops.

3.5 Insect Resistance to Bt Crops in Latin America

To date, insect resistance to Bt crops has been reported in Puerto Rico (Storer *et al.*, 2010) and Brazil (Farias, 2014; Monnerat *et al.*, 2015). Reports of low efficacy have also been communicated from several regions of Brazil and Argentina (Benintende, personal communication).

In Puerto Rico, resistance was detected in *S. frugiperda* found in crops of Bt maize expressing Cry1F protein (Storer *et al.*, 2010). Cry1F maize has been grown experimentally in Puerto Rico since 1996 and on a commercial scale since 2003. Observations of crop damage in Bt maize have been reported by farmers since 2006. These warning reports led to the company holding the technology to investigate the situation. High resistance of *S. frugiperda* to

Cry1F toxin was found and it was shown that resistance is autosomal, highly recessive and without maternal effects. The authors found no indication of strong cross-resistance to Cry1Ab and Cry1Ac toxins, and suggested that resistance may be due to changes in receptor-binding molecules (Storer *et al.*, 2010). Relevant factors to be considered in understanding the reasons for the development of resistance in this region are that Puerto Rico is an island, *S. frugiperda* is less susceptible to Cry1F than other maize pests and the climatic conditions on the island (which went through a dry spell). Together, these factors probably induced the migration of *S. frugiperda* to irrigated areas where Cry1F maize was being cultivated. This unusual combination of biological, geographical and operational factors was probably the cause of the evolution of resistance to Cry1F in Puerto Rico (Storer *et al.*, 2010). Upon confirmation of the resistance that had developed, the company providing the technology immediately stopped marketing Cry1 maize in Puerto Rico. Further studies monitoring insect populations in Puerto Rico and southern areas of USA showed high levels of Cry1F resistance in Puerto Rico, while populations collected from the southern US mainland continued to show full susceptibility to Cry1F and TC1507 maize (Storer *et al.*, 2012). Studies performed by Vélez (2013) and Vélez *et al.* (2013) indicated no significant Cry1F cross-resistance to Cry1Aa, Cry1Ba and Cry2Aa, and no cross-resistance with Vip3Aa; in contrast, significant cross-resistance was observed for Cry1Ab and Cry1Ac. An F_1 screen study performed to measure the frequency of

Table 3.2. Bt toxins present in transgenic events approved by CTNBio in Brazil (2012/13 harvest). (From CTNBio, 2014.)

Transgenic event	Bt toxins	Targets
Cotton	Cry1Ab, Cry1Ac, Cry2Ab, Cry 2Ae, Cry1F	Lepidoptera
Maize	Cry1Ab, Cry2Ab, Cry1A.105, Cry1F, VIP3Aa	Lepidoptera
	Cry3Bb, Cry34 Ab1, Cry35Ab1	Coleoptera
Soybean	Cry1Ac	Lepidoptera

Cry1F resistant alleles in 2010 and 2011 showed a frequency of resistant alleles of 0.13 and 0.02 in populations from Florida and Texas, respectively, indicating that resistant alleles could be found in US populations.

In Argentina, there was unexpected damage to Cry1F maize fields caused by *D. saccharalis* in the north-eastern region of San Luis. At this location, pest pressure was exceptionally high because it is an area with high temperatures, high amounts of sunshine and mild winters, with a long frost-free period and high moisture due to irrigation. Refuges were not used in the areas planted with Bt maize and neither were there rotations with susceptible crops established in the area, which represents an environment in which more generations of the pest develop compared with other areas (ASA, 2013; Edelstein, 2013; INTA, 2013). When the situation was noticed, the case was analysed, and the Argentinean authorities were informed of the discovery and are currently monitoring the region, together with the Asociación Semilleros Argentinos (ASA). At the same time, a mitigation and control plan was launched and it was proposed that maize production be stopped in this region during the 2013/14 season.

Recently, in Brazil, severe attacks of *S. frugiperda* were recorded in the main GM crops of the cerrado (Cruz *et al.*, 2013). *S. frugiperda* is one of the most important pests of maize, cotton and vegetables in Brazil (Soares and Vieira, 1998; Cruz *et al.*, 1999; Montesbravo, 2001); it causes damage from emergence to maturity of the plants, i.e. at all stages of development of the various crops (Cruz *et al.*, 1997; Fundação MT, 2001; Gallo *et al.*, 2002). It is important to note that in the Brazilian cerrado region, annual crops of maize, soybeans and cotton begin in October and extend through to June (Embrapa, 2013). Additionally, there is the implementation of the 'green bridge' that is formed by the additional cultivation of sorghum, millet and beans (Embrapa, 2013). Some of the GM seeds used in this region are not certified, so it is unclear what amount of toxin is being produced by the Bt plants that are being used. Also, the refuge areas, which should constitute 5–20% of the planted area (Leite *et al.*, 2011) are being implemented either at less than the recommended level or are not being implemented at all (Grigolli and Lourenção, 2013).

Since 2013, maize farmers in Cabeceiras de Goiás, in the Brazilian cerrado, have reported that over the 4 years that Bt maize expressing Cry1F toxin has being grown, the effectiveness of the control of *S. frugiperda* has decreased, thus forcing them to use chemicals to reduce the damage caused by this pest. The selection pressure to which this insect is being exposed is extremely high, as this is a polyphagous insect that feeds throughout the year, and all transgenic events express the same toxins. Moreover, the weather conditions favour short and continuous life cycles of the pest. Experiments conducted with populations collected at this site allowed the identification of individuals with significant tolerance to Cry1F toxin (Monnerat *et al.*, 2015). The population resistant to Cry1F showed cross-resistance to the Cry1Aa, Cry1Ab and Cry1Ac toxins that are present in other transgenic events being used in Brazil, but no cross-resistance to Cry2Aa and Cry2Ab toxins. However, susceptibility to Cry2A toxins is relatively low, implying that these toxins are not a good alternative for the control of *S. frugiperda*. Heterologous competition assays between Cry1F and Cry1A toxins showed that these toxins share binding sites in *S. frugiperda* and it is therefore not recommended that plants pyramided with these toxins be used to control *S. frugiperda*, or that other GM crops expressing these proteins are used in the same region,. Other important data generated in this study showed that the susceptibility of this population to Cry1F toxin did not change after ten generations without exposure to Cry1F toxin, indicating that the colony is now positioned as a pest with a stable phenotype of tolerance to Cry1F toxin (Monnerat *et al.*, 2015).

These results confirmed the hypotheses proposed by Hernández-Rodríguez *et al.* (2013), which are based on the data obtained in the laboratory, in which it was shown that Cry1A.105, Cry1Ab, Cry1Ac and Cry1Fa competed with high affinity for the same binding sites but that Cry2Ab and Cry2Ae did not compete for the binding sites of Cry1 proteins in *Ostrinia nubilatis* and *S. frugiperda*. According to the results from these authors, the development of cross-resistance among Cry1Ab, Cry1Ac, Cry1A.105 and Cry1Fa proteins is possible in these two insect species if the mutations that have occurred affect the shared binding sites of these toxins. Conversely, cross-resistance between these proteins and Cry2A proteins is unlikely.

A parallel study also showed high levels of resistance to Cry1F in *S. frugiperda* populations in other regions of Brazil, including São Paulo, Santa Catarina, Rio Grande do Sul, Bahia, Mato Grosso, Mato Grosso do Sul and Paraná (Farias, 2014). It has been shown that this resistance phenotype is due to a single recessive gene locus, and that the frequency of this resistance allele in the field has increased significantly in recent years (Farias, 2014).

Additional studies, not yet completed, conducted with caterpillars collected in fields of Bt maize and Bt cotton expressing Cry1 and Cry2A toxins, either individually or pyramided, indicated that resistance in *S. frugiperda* is occurring with high frequency in the states of Mato Grosso, Bahia and Goiás in the Brazilian cerrado region (personal communication).

3.6 Conclusion

The adoption of GM plants expressing toxins with larvicidal activity obtained from *B. thuringiensis* has shown high efficacy in controlling some of the main caterpillar pests of soybean, maize and cotton (Tabashnik *et al.*, 2013). This technology has increased yields in some regions, reduced the use of chemical insecticides, thereby preserving biodiversity and improving the quality of life of farmers. However, the evolution of pest resistance and consequent resumption of the use of chemicals may cancel these benefits. The evolution of resistance is a reality and can happen in any biological system. The difference may be in taking decisions that may delay the evolution of resistance, as has been reported in India (Dhurua and Gujar, 2011), China (Zhang *et al.*, 2011) and the USA (Tabashnik *et al.*, 2009).

The best alternative is to continue applying strategies to manage the evolution of resistance and to enhance research focused on developing new methods to control insect pests and reduce their negative impacts on production. It is essential that a well-designed plan is implemented, including the monitoring of insect resistance and the mandatory use of refuges because these strategies have helped to preserve the effectiveness of the technology and delay the selection of insects resistant to Bt toxins in various regions (Tabashnik *et al.*, 2013).

It is important that all participants, including the companies providing technologies, extension services, farmers, the scientific community and the governments of producing countries unite in their efforts to establish guidelines that will enable a greater durability of Bt technology. In Latin America, we must seek a sustainable agrosystem with high productivity that will avoid the failure of this technology.

References

ArgenBio (2014) Cultivos resistentes a insectos o Bt. Consejo Argentino para la Información y el Desarollo de la Biotecnología, Buenos Aires. Available at: http://argenbio.org/index.php?action=novedades&note=261 (accessed 27 May 2014).

ASA (2013) Daño no esperado de *Diatraea saccharalis* en maíces Bt en San Luis. Asociación Semilleros Argentinos, Buenos Aires. Available at: http://www.asa.org.ar/vertext.asp?id=3793 (accessed 19 September 2013).

Baek, J.H., Kim, J.I., Lee, D.W., Chung, B.K., Miyata, T. *et al.* (2005) Identification and characterization of ace1-type acetylcholinesterase likely associated with organophosphate resistance in *Plutella xylostella*. *Pesticide Biochemistry and Physiology* 81, 164–175.

Baxter, S.W., Badenes-Perez, F.R., Morrison, A., Vogel, H, Crickmore, N. *et al.* (2011) Parallel evolution of Bt toxin resistance in Lepidoptera. *Genetics* 189, 675–679.

Bernardi, O., Sorgatto, R., Barbosa, A., Domingues, F., Dourado, P. *et al.* (2014a) Low susceptibility of *Spodoptera cosmioides*, *Spodoptera eridania* and *Spodoptera frugiperda* (Lepidoptera: Noctuidae) to genetically-modified soybean expressing Cry1Ac protein. *Crop Protection* 58, 33–40.

Bernardi, O., Amado, D., Sousa, R., Segatti, F., Fatoretto, J. *et al.* (2014b) Baseline susceptibility and monitoring of Brazilian populations of *Spodoptera frugiperda* (Lepidoptera: Noctuidae) and *Diatraea saccharalis* (Lepidoptera: Crambidae) to Vip3Aa20 insecticidal protein. *Journal of Economic Entomology* 107, 781–790.

Blanco, C., Portilla, M., Jurat-Fuentes, J.L., Sanchez, J., Viteri, D. *et al.* (2010) Susceptibility of isofamilies of *Spodoptera frugiperda* (Lepidoptera: Noctuidae) to Cry1Ac and Cry1F proteins of *Bacillus thuringiensis*. *Southwestern Entomologist* 35, 409–415.

Borsani, O., Castiglioni, E., Chiappe, M., Ferenczi, A., García, F. *et al.* (2010) Biotecnología moderna, cultivares transgénicos y proceso de adopción en Uruguay. In: García Préchec, F., Ernst, O., Arbeletche, P., Pérez Bidegain, M., Pritsch, C. *et al.* (eds) *Intensificación Agrícola, Oportunidades y Amenazas para un País Productivo y Natural*. Colección Art. 2, Fondo Universitario para Contribuir a la Comprensión Pública de Temas de Interés General/Comisión Sectorial de Investigación Científica (CSIC), Montevideo, pp. 29–66.

Carneiro, F.F., Pignati, W., Rigotto, R.M., Augusto, L.G.S., Rizollo, A. *et al.* (2012) Dossiê ABRASCO – Um Alerta sobre os Impactos dos Agrotóxicos na Saúde, 1ª Parte. Associação Brasileira de Pós-Graduação em Saúde Coletiva, Rio de Janeiro.

Cartín, L., Carazo , R., Lobo, S., Monge, V. and Araya R. (1999) Resistance by *Plutella xylostella* to *Bacillus thuringiensis* in Costa Rica. *Jornal Manejo Integrado de Plagas* 54, 31–36.

Castelo Branco, M. and Gatehouse, A.G. (2001) A survey of insecticide susceptibility in *Plutella xylostella* (L.) (Lepidoptera: Yponomeutidae) in the Federal District, Brazil. *Neotropical Entomology (Londrina)* 30, 327–332.

Castelo Branco, M. and Medeiros, M.A. (2001) Impacto de inseticidas sobre parasitóides de traça-das-crucíferas em repolho no Distrito Federal. *Pesquisa Agropecuária Brasileira* 36, 7–13.

Castelo Branco, M., França, F.H., Medeiros, M.A. and Leal, J.G.T. (2001) Uso de inseticidas para o controle da traça-do-tomateiro e traça-das-crucíferas: um estudo de caso. *Horticultura Brasileira* 9, 60–63.

CEPAL/FAO/IICA (2014) *Perspectivas de la Agricultura y del Desarrollo Rural en las Américas: Una Mirada Hacia América Latina y el Caribe, 2013*. UN Comisión Económica para América Latina y el Caribe, Santiago/ Organización de las Naciones Unidas para la Alimentación y la Agricultura, Santiago/Instituto Interamericano de Cooperación para la Agricultura IICA, San José, Costa Rica. FAO, Santiago.

Cruz, I., Valicente, F.H., Santos, J.P. dos, Waquil, J.M. and Viana, P.A. (1997) *Manual de Identificação de Pragas da Cultura do Milho*. Centro Nacional de Pesquisa de Milho e Sorgo (Embrapa-CNPMS), Sete Lagoas, Brazil.

Cruz, I., Viana, P.A. and Waquil, J.M. (1999) *Manejo das Pragas Iniciais de Milho o Tratamento de Sementes com Inseticidas Sistêmicos*. Circular Técnica, 31, Centro Nacional de Pesquisa de Milho e Sorgo (Embrapa-CNPMS), Sete Lagoas, Brazil.

Cruz, I., Valicente, F.H., Viana, P.A. and Mendes, S.M. (2013) *Risco Potencial das Pragas de Milho e de Sorgo*. Circular Técnica 150, Centro Nacional de Pesquisa de Milho e Sorgo (Embrapa-CNPMS), Sete Lagoas, Brazil.

CTNBio (2014) CTNBio. Commissão Técnica Nacional de Biossegurança, Brasília. Available at: http://www.ctnbio.gov.br/index.php/content/view/2.html (accessed 13 October 2014).

Dhurua, S. and Gujar, G.T. (2011) Field-evolved resistance to Bt toxin Cry1Ac in the pink bollworm, *Pectinophora gossypiella* (Saunders) (Lepidoptera: Gelechiidae), from India. *Pest Management Science* 67, 898–903.

Díaz-Gomez, O., Rodríguez, J.C., Shelton, A.M., Lagunes-T., A. and Bujanos-M., R. (2000) Susceptibility of *Plutella xylostella* (L.) (Lepidoptera: Plutellidae) populations in Mexico to commercial formulations of *Bacillus thuringiensis*. *Journal of Economic Entomology* 93, 963–970.

Edelstein, J.D. (ed.) (2013) *Informe de la Red de Monitoreo Sistemático de Insectos Diciembre/2013*. Red de Monitoreo Sistemático de Insectos 1(3), Instituto Nacional de Tecnología Agropecuária, Buenos Aires. Available at: http://inta.gob.ar/documentos/red-de-monitoreo-sistematico-de-insectos-diciembre-2013 (accessed 13 October 2014).

Embrapa (2013) *Ações Emergenciais Propostas pela Embrapa para o Manejo Integrado de*

Helicoverpa spp. *em Áreas Agrícolas*. Empresa Brasileira de Pesquisa Agropecuária, Brasília. Available at: https://www.embrapa.br/docu ments/10180/1602515/A%C3%A7%C3%B5es+ emergenciais+propostas+pela+Embrapa+- +Documento+oficial/3a569ce1-c132-4bfa-8314- bc993ce8b920 (accessed 13 October 2014).

Farias, J.R. (2014) Resistance risk assessment of *Spodoptera frugiperda* (J.E. Smith) (Lepidoptera: Noctuidae) to Cry1F protein from *Bacillus thuringiensis* Berliner in Brazil. PhD thesis, College of Agriculture, University of São Paulo "Luis de Queiroz", São Paulo, Brazil.

Fundação MT (2001) *Boletim de Pesquisa de Algodão*. Boletim 4, Fundação MT (Mato Grosso) de Apoio à Pesquisa Agropecuária de Mato Grosso, Rondonópolis, Brazil.

Gallo, D. (*in memoriam*), Nakano, O., Neto, S.S., Carvalho, R.P.L., Baptista, G.C. *et al.* (2002) *Entomologia Agrícola, Biblioteca de Ciências Agrárias Luiz de Queiroz, Volume 10*. Fundação de Estudos Agrários Luiz de Queiroz (FEALQ), Piracicaba, Brazil.

Grigolli, J.F.J. and Lourenção, A.L.F. (2013) Alta Infestação de lagartas na cultura do milho Bt. Resultados da Pesquisa 2, Fundação MS (Mato Grosso do Sul) para Pesquisa e Difusão de Tecnologias Agropecuárias, Maracajú, Brazil.

Hernández-Rodríguez, C.S., Hernández-Martínez, P., Van Rie, J., Escriche, B. and Ferré, J. (2013) Shared midgut binding sites for Cry1A.105, Cry1Aa, Cry1Ab, Cry1Ac and Cry1Fa proteins from *Bacillus thuringiensis* in two important corn pests, *Ostrinia nubilatis* and *Spodoptera frugiperda*. *PLoSONE* 8(7), e68164.

IBGE (2012) *Indicadores de Desenvolvimento Sustentável – Brasil 2012*. Estudos e Pesquisas, Informação Geográfica 9, Instituto Brasileiro de Geografia e Estatística, Rio de Janeiro.

INBIO (2014) El Instituto de Biotecnología Agrícola – INBIO, Ciancio, Asunción, Paraguay. Available at: http://www.inbio.org.py/cultivos_aplicaciones/ eventos_aprobados (accessed 27 May 2014).

INTA (2013) El INTA alerta sobre la presencia de una nueva plaga. Instituto Nacional de Tecnología Agropecuaria, Buenos Aires. Available at: http://intainforma.inta.gov.ar/?p= 19702 (accessed 15 November 2013).

James, C. (2013) *Global Status of Commercialized Biotech/GM Crops: 2013*. ISAAA Brief No. 46, International Service for the Acquisition of Agri- Biotech Applications, Ithaca, New York.

Leite, N., Mendes, S., Waquil, J. and Pereira, E. (2011) *O Milho Bt no Brasil: a Situação e a Evolução da Resistência de Insetos*. Série Documentos 133, Embrapa Milho e Sorgo (CNPMS), Sete Lagoas, Brazil.

Monnerat, R.G., Queiroz, P., Orduz, S., Benitende, G., Cozzi, J. *et al.* (2006) Genetic variability in *Spodoptera frugiperda* Smith populations in Latin America is associated to variations in susceptibility to *Bacillus thuringiensis* Cry toxins. *Applied and Environmental Microbiology* 72, 7029–7035.

Monnerat, R., Martins, E., Macedo, C., Queiroz, P., Praça, L. *et al.* (2015) Evidence of field-evolved resistance of *Spodoptera frugiperda* to Bt corn expressing Cry1F in Brazil that is still sensitive to modified Bt toxins. Submitted *PLoS ONE*.

Montesbravo, E.P. (2001) Control biológico de *Spodoptera frugiperda* (J.E. Smith) en maíz. Departamento de Manejo de Plagas, Instituto de Investigaciones de Sanidad Vegetal (INISAV), Playa Ciudad de la Habana, Cuba. Available at: http://www.aguascalientes.gob.mx/ codagea/produce/SPODOPTE.htm (accessed 13 October 2014).

Perez, C.J. and Shelton, A.M. (1997) Resistance of *Plutella xylostella* (Lepidoptera: Plutellidae) to *Bacillus thuringiensis* Berliner in Central America. *Journal of Economic Entomology* 90, 87–93.

Perez, C.J., Shelton, A.M. and Roush, R.T. (1997) Managing diamondback moth *Plutella xylostella* (Lepidoptera: Plutellidae) resistance to foliar applications of *Bacillus thuringiensis*: testing strategies in field cages. *Journal of Economic Entomology*, 90, 1462–1470.

Pignati, W.A. and Machado, J.M.H. (2011) O agronegócio e seus impactos na saúde dos trabalhadores e da população do estado de Mato Grosso. In: Minayo, C., Machado, J.M.H. and Pen, P.G.L. (orgs) *Saúde do Trabalhador na Sociedade Brasileira Contemporânea*. Fundação Oswaldo Cruz (FIOCRUZ), Rio de Janeiro, pp. 245–272.

Polanczyk, R., Valicente, F.H. and Barreto, M. (2012) Utilização de *Bacillus thuringiensis* no controle de pragas agrícolas na América Latina. In: Alves, S.B. and Lopes, R.B. (eds) *Controle Microbiano de Pragas na America Latina*, 3rd edn. Fundação de Estudos Agrários Luiz de Queiroz (FEALQ), Piracicaba, Brasil, pp. 111– 136.

Polania, I.Z., Maldonado, H.A.A., Cruz, R.M. and Sanchez, J.L.D. (2009) *Spodoptera frugiperda*: respuesta de distintas poblaciones a la toxina Cry1Ab. *Revista Colombiana de Entomologia* 35, 34–41.

Ribeiro, L.M.S. (2010) Respostas imunológicas e mecânicas em população suscetível e re- sistente de *Plutella xylostella* (L.) (Lepidoptera: Plutellidae) frente a formulações comerciais à base de *Bacillus thuringiensis* Berliner. MSc thesis, Universidade Federal do Pernambuco, Pernambuco, Brazil.

Rios-Diez, J.D., Siegfried, B. and Saldamando-Benjumea, C.I. (2012) Susceptibility of *Spodoptera frugiperda* (Lepidoptera: Noctuidae) strains from central Colombia to Cry1Ab and Cry1Ac entotoxins of *Bacillus thuringiensis*. *Southwestern Entomologist* 37, 281–293.

Shelton, A.M, Robertson, J.L., Tang, J.D., Perez, C., Eigenbrode, S.D. *et al.* (1993) Resistance of diamondback moth (Lepidoptera: Plutellidae) to *Bacillus thuringiensis* subspecies in the field. *Journal of Economic Entomology* 86, 697–705.

Silva-Filho, M.C. and Falco, M.C. (2000) Interação planta–inseto: adaptação dos insetos aos inibidores de proteinase produzidos pelas plantas. *Biotecnologia Ciência e Desenvolvimento* 2, 38–42.

Soares, J.J. and Vieira, R.M. (1998) *Spodoptera frugiperda* ameaça a cotonicultura brasileira. Comunicado Técnico, Centro Nacional de Pesquisa do Algodão (Embrapa-CNPA), Campina Grande, Brazil.

Sosa-Gómez, D. and Moscardi, F. (1991) Microbial control and insect pathology in Argentina, *Ciencia e Cultura* 43, 375–379.

Souza, M.L. (2001).Utilização de microrganismos na agricultura. *Biotecnologia Ciencia e Desenvolvimento* 21, 28–31.

Specht, A., Sosa-Gómez, D., Paula-Moraes, S. and Yano, S. (2013) Identificação morfológica e molecular de *Helicoverpa armigera* (Lepidoptera: Noctuidae) e ampliação de seu registro de ocorrência no Brasil. *Pesquisa Agropecuária Brasileira* 48, 689–692.

Storer, N.P., Babcock, J.M., Schlenz, M., Meade, T., Thompson, G.D. *et al.* (2010) Discovery and Characterization of Field Resistance to Bt Maize: *Spodoptera frugiperda* (Lepidoptera: Noctuidae) in Puerto Rico. *Journal of Economic Entomology* 103, 1031–1038.

Storer, N.P., Kubiszak, M.E., King, J.E., Thompson, G.D. and Santos, A.C. (2012) Status of resistance to Bt maize in *Spodoptera frugiperda*: lessons from Puerto Rico. *Journal of Invertebrate Pathology*, 110, 294–300.

Tabashnik, B.E., Cushing, N.L., Finson, N. and Johson, M.W. (1990) Field development of resistance to *Bacillus thuringiensis* in diamondback moth (Lepidoptera: Plutellidae). *Journal of Economic Entomology* 83, 1671–1676.

Tabashnik B.E., Finson N., Johnson M.W. and Moar W.J. (1993) Resistance to toxins from *Bacillus thuringiensis* subs. *aizawai* in the diamondback moth (Lepidoptera: Plutellidae). *Applied and Environmental Microbiology* 59, 1332–1335.

Tabashnik, B.E., van Rensburg, J.B.J. and Carrière, Y. (2009) Field-evolved insect resistance to Bt crops: definition, theory, and data. *Journal of Economic Entomology* 102, 2011–2025.

Tabashnik, B.E., Huang, F., Ghimire, M.N., Leonard, B.R., Siegfried, B.D. *et al.* (2011) Efficacy of genetically modified Bt toxins against insects with different mechanisms of resistance. *Nature Biotechnology* 29, 1128–1131.

Tabashnik, B.E., Brévault, T. and Carrière, Y. (2013) Insect resistance to Bt crops: lessons from the first billion acres. *Nature Biotechnology* 31, 510–521.

Trigo, E.J. and Cap, E.L. (2003) The impact of the introduction of transgenic crops in Argentinean agriculture. *AgBioForum* 6, 87–94.

Vélez, A.M. (2013) Characterization of resistance to the Cry1F toxin from *Bacillus thuringiensis* in resistant fall armyworm, *Spodoptera frugiperda* (J.E. Smith) (Lepidoptera: Noctuidae) from Puerto Rico. PhD thesis, University of Nebraska-Lincoln, Lincoln, Nebraska.

Vélez, A.M., Spencer, T.A., Alves, A.P., Moellenbeck, D., Meagher, R.L. *et al.* (2013) Inheritance of Cry1F resistance, cross-resistance and frequency of resistant alleles in *Spodoptera frugiperda* (Lepidoptera: Noctuidae). *Bulletin of Entomological Research* 103, 700–713.

Waquil, J., Vilella, F., Siegfried, B. and Foster, J. (2004) Atividade biológica das toxinas do Bt, Cry1Ab e Cry1F em *Spodoptera frugiperda* (Smith) (Lepidoptera: Noctuidae). *Revista Brasileira de Milho e Sorgo* 3, 161–171.

Weiser, J. (1986) Impact of *Bacillus thuringiensis* on applied entomology in eastern Europe and in Soviet Union. In: Krieg, A. and Huger, A.M. (eds) *Mitteilungen aus der Biologischen Bundesanstalt für Land und Forstwirtschaft Berlin-Dahlem Heft 233*. Paul Parey, Berlin, pp. 37–50.

Zago, H.B., Herbert, A.A.S., Pereira, E.J.G., Picanço, M.C. and Barros, R. (2014) Resistance and behavioural response of *Plutella xylostella* (Lepidoptera: Plutellidae) populations to *Bacillus thuringiensis* formulations. *Pest Management Science* 70, 488–495.

Zhang, H., Yin, W., Zhao, J., Jin, L., Yang, Y. *et al.* (2011) Early warning of cotton bollworm resistance associated with intensive planting of Bt cotton in China. *PLoS ONE* 6(8): e22874.

4

Resistance of *Busseola fusca* to Cry1Ab Bt Maize Plants in South Africa and Challenges to Insect Resistance Management in Africa

Johnnie Van den Berg[1]* and Pascal Campagne[2]

[1]*Unit for Environmental Sciences and Management, North-West University, Potchefstroom, South Africa;* [2]*Institute of Integrative Biology, University of Liverpool, Liverpool, UK*

Summary

The evolution of resistance to *Bacillus thuringiensis* (Bt) maize by the African stem borer, *Busseola fusca*, in South Africa highlighted the importance of the development of appropriate integrated resistance management (IRM) strategies for stem borers in Africa. Landscape heterogeneity is characteristic of African agroecosystems. This heterogeneity, in addition to between-field and within-field spatial mosaics resulting from variable gene expression in Bt maize, will provide challenges to managing resistance evolution of the lepidopteran stem borers that attack maize. Adding to this landscape heterogeneity is the cultivation of open-pollinated maize varieties (OPVs) and bimodal rainfall patterns that allow two maize cropping seasons each year in many subtropical and tropical areas. The role that these factors, as well as aspects such as low-dose expression events, refuge compliance, the genetic bases of resistance, pest behaviour, host plant range and farming practices, may play in the evolution of stem borers to Bt maize in Africa are addressed in this chapter.

4.1 Importance of Maize in Africa: From Small- to Large-scale Farming Systems

Maize is the most important cereal crop in Africa and sustains more than 300 million of Africa's most vulnerable people (La Rovere *et al.*, 2010). Maize farming within the African context is mostly done on small plots of land with low productivity. Increasing the productivity of these small-scale farming systems would ensure both food security in various contexts and an increased contribution to the economies of African countries (Byerlee and Heisey, 1996).

Production systems for maize in Africa are highly diverse. The crop is grown in a wide range of agroecologies, which may vary between unimodal and bimodal rainfall regions, forest zones, lowland tropics and highland temperate regions, each with its unique challenges. The sizes of maize fields also vary greatly. Numerous small fields are characteristic of small-scale maize farming systems in Africa. For example, Aheto *et al.* (2013) reported 58 (average size = 0.81 ha) and 97 (average size = 0.41 ha) maize fields

* Corresponding author. E-mail address: Johnnie.VanDenBerg@nwu.ac.za

km^{-2} in Ghana and Zambia, respectively. The use of fertilizers, herbicides and insecticides in these fields is minimal and the planting of open pollinated varieties (OPVs) is common. In some areas of the continent, large-scale commercial agriculture is practised, both under rainfed and irrigated conditions. In these mechanized systems, the use of fertilizers, herbicides and insecticides is high, and the maize is produced as a cash crop rather than for local consumption, as in most of the small-scale farming systems. In South Africa, for example, most of the maize (2.5 million ha) is cultivated in large-scale high-input farming systems in which the size of the maize fields can be several hundreds of hectares. There are, however, also large numbers of small-scale farmers with field sizes similar to those reported in the rest of Africa (Van den Berg, 2013).

The average maize yields in several African countries, where maize is a highly important staple food crop, are below 1 t ha^{-1}, which compares with a world average of 4.7 t ha^{-1} in 2005 and an average of 9.4 t ha^{-1} in the USA (Worku *et al.*, 2012). These low yields are ascribed to poor soil fertility, droughts, a high incidence of insect pests, diseases and weeds and limited access to fertilizer and improved maize seed.

4.2 Stem Borer Pests of Maize in Africa: Distribution and Pest Status

Stem borers seriously limit potentially attainable maize yields by infesting the crop from the seedling stage to maturity. Many species belonging to the families Pyralidae and Noctuidae have been found to attack maize in various parts of Africa. The most important of these are *Busseola fusca* (Fuller) (Lepidoptera: Noctuidae), *Chilo partellus* (Swinhoe) (Lepidoptera: Crambidae), *Sesamia calamistis* Hampson (Lepidoptera: Noctuidae) and *Eldana saccharina* Walker (Lepidoptera: Pyralidae) (Kfir *et al.*, 2002). Of these species, *B. fusca* and *S. calamistis* occur virtually throughout sub-Saharan Africa. *C. partellus* occurs mostly in the eastern and southern parts of the continent, while another species of *Chilo*, *C. orichalociliellus* (Strand), seem to be confined

to coastal low altitude areas of East and southern Africa. In addition to the above-mentioned stem borer pest species, several noctuids, *B. segeta* (Bowden), *B. phaia* (Bowden) and *Pirateolea piscator* (Fletcher) have also recently been reported from maize in East Africa (Le Ru *et al.*, 2006a; Ong'amo *et al.*, 2013).

Estimates of crop losses vary greatly in different regions and agroecological zones. In Kenya alone, losses due to *B. fusca* damage fluctuate around 14% on average (De Groote, 2002), but can be as high as 73% (Mailafiya *et al.*, 2009). Kfir (1998) reported maize yield losses due to *B. fusca* to be between 10 and 60%. A review of *C. partellus* yield loss studies indicated losses between 4 and 73% depending on cultivar, planting date, pest population density and phenological stage of the crop at the time of infestation (Seshu-Reddy and Walker, 1990). In West Africa, stem borers are responsible for losses ranging between 25 and 55% in maize (Ndemah *et al.*, 2007).

While *Bacillus thuringiensis* (Bt) maize targets all stem borer species that attack the crop, the main species targeted differs between regions. In East and southern Africa, *B. fusca* and *C. partellus* are the main target species, while in Egypt they are *Sesamia cretica* (Lederer) (Lepidoptera: Noctuidae) and *Ostrinia nubilalis* (Hübner) (Lepidoptera: Pyralidae) (El-Shazlya *et al.*, 2013). The major stem borers of maize in West Africa include *B. fusca*, *S. calamistis* and *E. saccharina* (Bosque-Pèrez and Schulthess, 1998).

4.3 Commercial Release of Bt Maize in Africa

South Africa and Egypt are the only African countries that currently cultivate Bt maize, with cultivation having commenced in 1997 (Gouse *et al.*, 2005) and 2008 (El-Shazlya *et al.*, 2013), respectively. While Bt maize was quickly adopted in South Africa, the area of its cultivation in Egypt reportedly grew from 700 ha in 2008 to only 2000 ha in 2010 (James, 2010). However, recent reports do not list Egypt as a genetically modified (GM) maize-producing country

(James, 2013). The adoption rates of GM maize among commercial farmers in South Africa were high and in certain areas the market penetration of the Bt trait is nearly 100%. Today, Bt maize is planted on approximately 1.8 million ha in South Africa, making it the eighth largest producer of GM crops in the world (James, 2013). This high adoption rate has been ascribed to significant benefits of GM maize, such as the convenience of target pest management (Kruger et al., 2009) and economic benefits. Gouse et al. (2005) reported that despite paying more for seed, adopters of GM maize enjoyed increased income over non-GM maize adopters through savings on pesticides and protection from yield losses due to target pest species. In order for these benefits to be realized in the future, insect resistance management (IRM) strategies need to be effectively implemented.

4.4 Resistance Evolution to Bt Cry1Ab Maize

Until 2006, when event Bt11 was commercially released in South Africa, all Bt maize hybrids contained event MON810 (an event is a specific genetic modification in a specific species). From the 2012/13 growing season onwards, the pyramided event MON89034, which expresses two different Cry (crystal) proteins, Cry1A.105 and Cry2Ab2, was planted on a large scale in South Africa.

The first evaluation of the efficacy of several Bt maize events against B. fusca was conducted between 1994 and 1997 (van Rensburg, 1999). Under field conditions, the ranges of survival of B. fusca after 14 days of feeding on MON810 hybrids were 0.5–0.9% after early infestations and 1.3–2.4% after late infestations (van Rensburg, 2001). Although B. fusca survival was higher than the high-dose requirement (see Section 4.5.2), the efficacy of the event MON810 was considered sufficient to protect maize from stem borer damage. Bt maize hybrids have since been shown to provide effective control of C. partellus and to provide partial to very good control of S. calamistis and

B. fusca (van Rensburg, 1999; Van den Berg and Van Wyk, 2007; Tende et al., 2010).

Damage caused by B. fusca to Bt maize was observed at a number of localities during the first harvest season (1999) after the first commercial plantings of Bt maize in South Africa (van Rensburg, 2001). The first official report of field resistance to Bt maize in B. fusca was subsequently made in 2006, when van Rensburg (2007) showed that significant numbers of the F_1 generation of diapause larvae collected on Bt maize in the Christiana area (27° 57′ S, 25° 05′ E) in the Northern Cape Province survived on Bt maize. A year after this first official report of resistance, other cases of resistance were reported and confirmed in the Vaalharts area, approximately 50 km from the initial site (Kruger et al., 2011, 2012, 2014). Results from an extensive farmer survey conducted during 2010 indicated the presence of resistant populations in the maize production region and also showed that borer damage to Bt maize had been observed over a number of cropping seasons between 2003 and 2008 (Kruger et al., 2011). During these surveys, farmers indicated that stem borers were generally effectively controlled with Bt maize, but in several districts it was indicated that stem borer damage to Bt maize was already prevalent from the 2003/04 season onward (Kruger et al., 2011). A conservative estimate is that approximately 250 cases of product failure have been reported annually for the 2010/11 and 2011/2012 growing seasons (Van den Berg et al., 2013).

Farmer's observations on the levels of stem borer infestation in Bt maize fields over time (Kruger et al., 2012), show that infestation levels decrease with distance away from the initial site where resistance was reported (Fig. 4.1). These infestation levels are also below 10% in most cases, which is also below the economic threshold value of 10% for the application of chemical control measures. Despite the low stem borer infestation levels observed further than 100 km away from the original site (up to 2011), the numbers of farmers reporting the presence of stem borer damage in Bt maize fields were high (Fig. 4.2). Today, B. fusca populations with resistance to

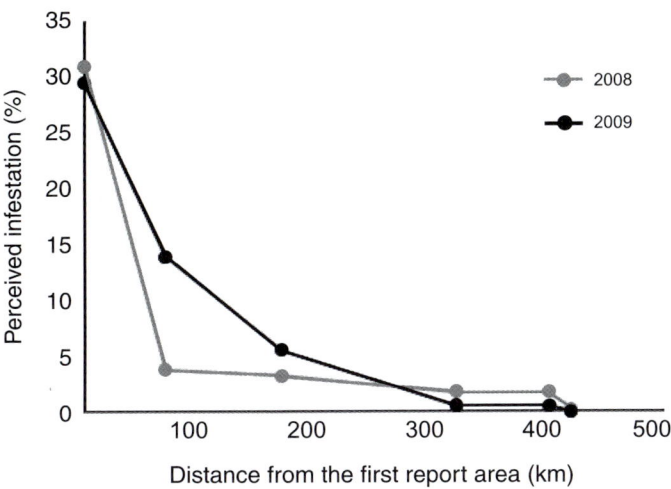

Fig. 4.1. Average infestation level of Bt maize perceived by farmers as a function of the distance from the site in South Africa where resistance of *Busseola fusca* was first reported. (Based on Kruger *et al.*, 2012.)

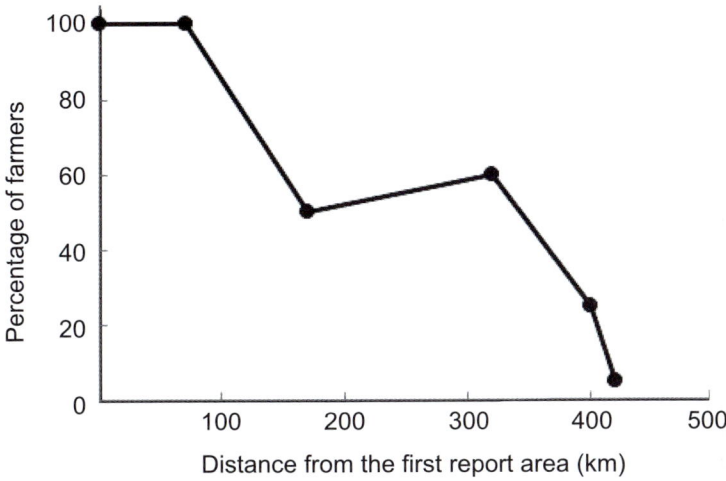

Fig. 4.2. Proportion of farmers who observed signs of *Busseola fusca* infestation on *Bt* maize in 2009, as a function of the distance from the site in South Africa where resistance of *Busseola fusca* was first reported. (Based on Kruger *et al.*, 2012.)

Cry1Ab-expressing maize occur throughout the maize production region. Although this is the case, the nature of the distribution is such that product failure is not reported on all farms, but rather on particular farms throughout certain geographical regions (Van den Berg *et al.*, 2013).

4.5 The High-dose Refuge Strategy

The current success of Bt maize in sustaining the susceptibility of some of the major pests is ascribed to the implementation of the 'high dose/refuge' IRM strategy (Tabashnik *et al.*, 2013). This is also the IRM strategy deployed in South Africa. The principle of this strategy is to promote survival of susceptible insects in refuges of non-Bt plants and random mating between the few resistant insects that survive on Bt plants with the many susceptible insects originating from the non-Bt refuge maintained nearby (Gould, 1998).

This strategy has certain requirements and is based on several assumptions and the interactions between these. These assumptions, as summarized by Roush (1994), Gould (1998), the US Environmental Protection Agency (US EPA, 1998, 2001), Ferré and Van Rie (2002) and Glaser and Matten (2003), are that:

- Non-transgenic refugia sustain a susceptible pest population.
- The crop expresses a high toxin concentration and resistance is recessive.
- Resistance genes are initially rare.
- There is random mating between resistant and susceptible adults.

Although the evolution of resistance by *B. fusca* has largely been ascribed to non-compliance with refuge requirements (Kruger *et al.*, 2011, 2012), other important factors have contributed. Indeed, IRM strategies cannot only be based on assumptions of single traits with simple effects (Gassmann *et al.*, 2009). The success of the high-dose refuge strategy in Africa may depend on how IRM strategies are adapted to a variety of farming contexts. The above-mentioned assumptions, as well as other possible contributing factors, such as pest biology and behaviour, are therefore revisited below.

4.5.1 Non-transgenic refugia

Refuges are habitats in which the target pest is not under selection pressure from the toxin, and which provide a sustainable habitat where pest development can take place and where a susceptible pest population is sustained. The planting of non-Bt maize is the most common form of providing a refuge. As noted above, it is widely accepted that non-compliance with refuge require-

ments contributed towards resistance evolution in *B. fusca*. Throughout the maize production region of South Africa, compliance with refuge requirements was low for the first 8 years after the release of Bt maize (Kruger *et al.*, 2009, 2011). Planting dedicated areas of non-Bt maize in close proximity to the crop is the prescribed strategy in South Africa and is generally considered the appropriate strategy to adopt in large-scale farming systems. However, unstructured refugia in the form of non-Bt maize fields of non-adopters of the technology, which will be a common occurrence in Africa, would also suffice as refugia, provided that certain measures are put in place so as not to allow a too high a proportion of Bt maize compared with non-Bt maize in a particular geographical area.

Wild host plants and alternative host plants have been suggested as a refuge strategy for stem borers in Africa (Mulaa *et al.*, 2011). Until recently, *B. fusca* was considered to be a species feeding not only on maize and cultivated and wild sorghums but also on many wild grasses. Wild host plants, mostly thick-stemmed grasses, were therefore considered to host sufficient numbers of *B. fusca* larvae to serve as a refuge. In contradiction of this, extensive field surveys in East and southern African over the past decade have shown that there is a low diversity of wild hosts for this stem borer (Le Ru *et al.*, 2006a,b; Ong'amo *et al.*, 2006; Moolman *et al.*, 2013), and that these wild hosts do not suffice as a refuge for stem borers in Africa.

4.5.2 High dose and functionally recessive resistance

The high-dose component of the high-dose/refuge strategy requires that the dose should theoretically be 25 times the dose needed to kill 99% of the susceptible individuals (Roush, 1994; Gould, 1998; US EPA, 1998, 2001; Glaser and Matten, 2003). This high-dose requirement is challenged by the deployment of Bt maize against a range of pests throughout the world, because a high dose for one pest may not be a high dose for another.

Although data have not been reported on the dominance of *B. fusca* resistance to Cry1Ab, pre-commercialization field data implied that the Cry1Ab maize did not kill 99% of larvae (van Rensburg, 1999). According to a review by Tabashnik *et al.* (2009), available pre-commercialization field data showed that the high-dose standard was not met by event MON810.

The diversity of the genetic bases of resistance and their inheritance is an important aspect in resistance management. A previous study (Campagne *et al.*, 2013) showed that resistance to Cry 1Ab in *B. fusca* was not recessively inherited, contrary to the important assumptions of the 'high dose/refuge' resistance management strategy. The study was carried out with maize stems under laboratory conditions, which did not directly reflect survival on plants in the field. Nevertheless, the results were consistent with those of a previous study showing the resistance of larvae originating from the Vaalharts area in whole plant bioassays conducted in greenhouses (van Rensburg, 1999) and are further supported by farmers' observations (Kruger *et al.*, 2011). Functionally non-recessive resistance results in a higher proportion of resistant phenotypes in a population in comparison with a recessively inherited resistance, all other things being equal. It is, therefore, expected to lead to rapid evolution of resistance in a pest population and to a drastic reduction in the efficiency of the refuge strategy.

When resistance is not recessive, refuges need to be much more abundant to delay resistance evolution effectively (Tabashnik and Gould, 2012; Brévault *et al.*, 2013). Because the high-dose standard was not met by event MON810, non-recessive in-heritance of resistance appears to have hastened the evolution of *B. fusca* resistance to Cry 1Ab in Bt maize (Tabashnik *et al.*, 2009). Simple population genetics models (Wright, 1942) may be used to assess the effects of non-recessive inheritance of resistance traits on the rate at which resistance evolves.

4.5.3 Effects of selection and dominance on the evolution of resistance

Let us consider resistance due to a single locus, where R denotes the resistance allele and S is the susceptible allele in a landscape where Bt plants and non-Bt plants would be randomly distributed. Considering a monogenic resistance, the relative fitness of different genotypes under the selection operated by Bt toxins may be described by the following equations:

$$RR, s_1 = 1 + s \qquad (1)$$

$$RS, s_2 = 1 + hs \qquad (2)$$

$$SS, 1 \qquad (3)$$

where s is the selection coefficient ($s > 0$, in this case) and h is the dominance level ($0 < h < 1$) of resistance. Assuming no effect of spatial structure, the selection coefficient s may be modelled as a function of the proportion of the Bt crop (β) in the system.

$$s = \frac{1}{1-\beta} - 1 \qquad (4)$$

The change in the frequency of the resistance allele (p) from one to the next generation (i.e. t to $t + 1$) is described by the equation of Wright (1942), as a function of dominance and selection:

$$p_{t+1} = p_t \frac{1 + s_1 p_t + s_2 (1 - p_t)}{1 + s_1 p_t^2 + 2s_2 p_t (1 - p_t)} \qquad (5)$$

Dominance of a trait has a strong effect on the dynamics of change of allele frequency across generations, as suggested by the latter model. Across generations, non-recessive resistance is characterized by a strong increase in allele frequency, especially when the frequency of the resistance allele is low in comparison with a strictly recessive case. In contrast, the frequency increase of a recessive resistance allele will result in a fast elimination of the susceptible allele in comparison with non-recessive cases ($h > 0$) (Fig. 4.3) (Felsenstein, 2007).

As a consequence, non-recessive re-sistance alleles are expected to invade the

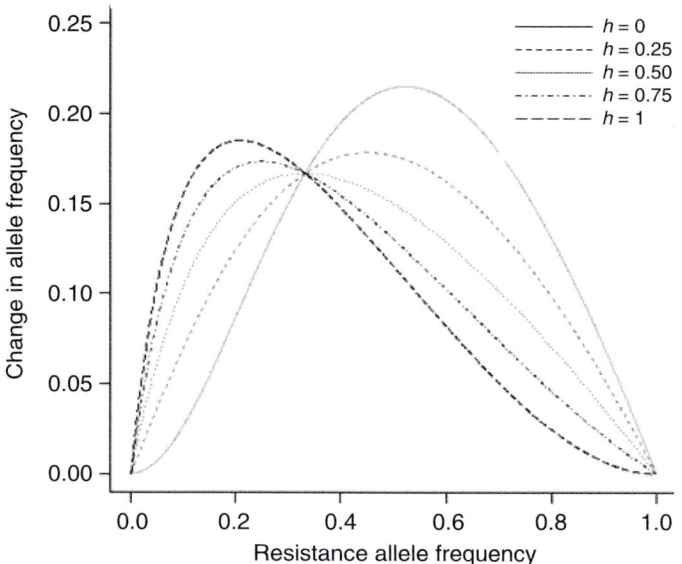

Fig. 4.3. Change in allele frequency as a function of frequency the resistance allele, in various cases: resistance allele is fully recessive ($h = 0$) to fully dominant ($h = 1$). The proportion of Bt crop in the model is $\beta = 0.8$.

pest population much more quickly than recessive resistance alleles (Tabashnik *et al.*, 2008). The time required for a change of gene frequency (from an initial frequency of p_0 to a much higher frequency p_T) cannot be analytically derived from the Wright model. However, approximations using a time continuous model may be obtained for the strictly recessive and dominant cases, T_{res} and T_{dom}, respectively (see Felsenstein, 2007):

$$T_{res} = \frac{1}{s}\left[-\frac{1}{p_T} + ln\left(\frac{p_T}{1-p_T}\right) + \frac{1}{p_0} - \right.$$
$$\left. ln\left(\frac{p_0}{1-p_0}\right) + 2s\,ln\left(\frac{1-p_T}{1-p_0}\right)\right] \quad (6)$$

$$T_{dom} = \frac{1}{s}\left[-\frac{1}{1-p_T} + ln\left(\frac{p_T}{1-p_T}\right) + \right.$$
$$\left. \frac{1}{1-p_0} - ln\left(\frac{p_0}{1-p_0}\right)\right] \quad (7)$$

Setting p_T at 0.25, for the sake of simplicity, and p_0 at 10^{-x} (with $x > 3$), the previous equations (6) and (7) may numerically be reduced to:

$$T_{res} \approx 10^x.\left(\frac{1}{\beta}-1\right) \quad (8)$$

and

$$T_{dom} \approx \left[x\,ln(10)-0.765\right]\left(\frac{1}{\beta}-1\right) \quad (9)$$

Although both of these equations depend on the initial frequency of the resistance allele, they exhibit strong contrasts. Dominant resistance is expected to evolve with a time lapse that is proportional to the logarithm of the initial frequency of the resistance allele, while in the recessive case, the corresponding time would be inversely proportional to the initial resistance allele frequency. For example, the time required for a change in resistance allele frequency from $p_0 = 10^{-4}$ to $p_T = 0.25$ would be:

$$T_{res} \approx 10000\left(\frac{1}{\beta}-1\right) \quad (10)$$

and

$$T_{dom} \approx 8.45\left(\frac{1}{\beta} - 1\right) \tag{11}$$

in the recessive and dominant cases, respectively. In other words, irrespective of the proportion of Bt crop in the system (which determines the selection coefficient), the model predicts that a fully dominant resistance would be acquired 1000 times faster than a recessive resistance.

Intermediate cases of dominance may be numerically explored using Eqn (6) (see Table 4.1). In line with previous results (e.g. Tabashnik *et al.*, 2008) the Wright (1942) model predicts that non-recessive resistance will drastically increase the rate of resistance evolution compared with the recessive case.

In a context where the genetic bases of resistance to Bt crops are not necessarily uniform (Campagne *et al.*, 2013; Jin *et al.*, 2013), possible high-dose failures in maize cultivated in small-scale African farming systems should be anticipated. The use of the technology by farmers in various agroecosystems and intrinsically non-recessive resistance are indeed crucial components in the evolution of resistance.

4.5.4 Pest biology and larval migration

Because pest biology, feeding behaviour and movement between plants determines the level of exposure to Bt plants, these aspects should be considered in resistance management strategies. They are discussed below for *B. fusca*.

Larval behaviour of Busseola fusca

The high level of between-plant larval migration by *B. fusca* (Calatayud *et al.*, 2014) will provide challenges regarding resistance management. The movement of larvae between plants in the refuge area and in adjacent Bt maize blocks (or between plants in a seed mixture planting) pose a potential concern for resistance evolution.

B. fusca larvae migrate throughout all their larval stages. This migration commences immediately after egg hatch and ceases during the last instar, 4 to 5 weeks post hatching, when the larvae prepare pupal cells in which they become pupae, or go into diapause (van Rensburg *et al.*, 1987a). Migration does not cease after the larvae leave plant whorls to feed inside maize stems. The larvae migrate until the

Table 4.1. Time required for a change in resistance allele frequency, from p_0 to $p_T = 0.25$ as a function of different levels of allele dominance (*h*) and the proportion of Bt crop in the system (β). T_{res} and T_{dom} are the strictly recessive and dominant cases.

Proportion of Bt crop	Initial frequency	No. generations to $p_T = 0.25$						
β	p_0	$h = 0$	$h = 0.25$	$h = 0.5$	$h = 0.75$	$h = 1$	T_{res}	T_{dom}
0.50	1×10^{-6}	$>1 \times 10^6$	56	32	24	19	$>1 \times 10^6$	13
0.50	1×10^{-5}	100,017	46	26	20	16	100,006	11
0.50	1×10^{-4}	10,012	35	21	15	13	10,003	8
0.50	1×10^{-3}	1008	25	15	11	9	1001	6
0.75	1×10^{-6}	333,349	23	15	12	10	333,335.7	4
0.75	1×10^{-5}	33,346	19	12	10	8	33,335	4
0.75	1×10^{-4}	3343	15	10	8	7	3334	3
0.75	1×10^{-3}	340	11	7	6	5	333	2
0.90	1×10^{-6}	111,125	11	8	7	6	111,112	2
0.90	1×10^{-5}	11,122	9	7	6	5	11,111	1
0.90	1×10^{-4}	1119	7	5	5	4	1111.0	1
0.90	1×10^{-3}	117	5	4	4	3	111	1

6th instar, a behaviour which is density dependent (van Rensburg *et al.*, 1987a). Furthermore, a significant positive relationship has been observed between the plant stand (which is very high, especially in irrigated systems where *B. fusca* resistance in South Africa is problematic) and the number of larvae that migrate successfully and survive on maize plants (van Rensburg *et al.*, 1988).

A reduction in the numbers of susceptible individuals due to pre-feeding movement from non-Bt to Bt plants could increase the potential for heterozygous (*RS*) larvae to survive and adults to mate, which could, in turn, lead to an increase in the incidence of homozygous (*RR*) resistant individuals in the offspring (Murphy *et al.*, 2010). The development of resistance may also occur when more mature larger larvae move from non-Bt to Bt plants, and in consequence are exposed to sublethal dosages of Bt proteins (Murphy *et al.*, 2010). This may lead to an increased risk of resistance development over time. A target pest exhibiting a more sedentary behaviour might, in contrast, not be affected in this way. Such insects (for example aphids) would endure longer exposure to Bt proteins throughout their life cycle (Gould, 2000).

Accordingly, the larval behaviour of all stem borer species that will be exposed to Bt maize in Africa will have to be considered in the development of IRM strategies.

Adult behaviour of Busseola fusca

Variability in pest biology and behaviour between different geographical regions may affect the exposure of pests to Bt maize, thereby influencing the rate of resistance evolution. Guse *et al.* (2002) and Onstad *et al.* (2002) showed that different stem borer species exhibit different adult behaviour patterns, particularly in mating, oviposition and male moth dispersal, and indicated that these interactions of landscape and insect behaviour must be understood in order to develop suitable resistance management strategies.

Agronomic practices such as fertilizer use, cultivation and irrigation practices all influence the landscape. The behaviours of *Diatraea grandiosella* (Lepidoptera: Crambidae) and *O. nubilalis*, for example, differ in irrigated and non-irrigated maize fields (Guse *et al.*, 2002; Onstad *et al.*, 2002). *B. fusca* moths also prefer to mate and move within moister vegetation than drier vegetation. This behaviour was indicated by van Rensburg (2007) as a possible factor that could have resulted in higher pest numbers in Bt maize than in non-Bt maize in South Africa, and thus contributed to resistance evolution. Guse *et al.* (2002) reported that this type of adult behaviour would strongly influence the evolution of resistance to Bt maize and that practices that would increase oviposition in natural refuges would delay resistance evolution because they would increase the source potential of refuges and reduce the intensity of selection.

Whereas *B. fusca* has only three generations per season in temperate South Africa, this situation is different in subtropical and tropical regions. Flight patterns of *B. fusca* moths in areas where only one rainy season occurs show two to three distinct generations with zero moth activity during the winter period (van Rensburg *et al.*, 1987b). However, less discernible patterns are observed in areas where maize is cultivated throughout the year (van Rensburg, 1997). Bimodal rainfall patterns, in effect, result in two cropping seasons a year, with large-scale staggering of planting dates. This phenomenon adds to landscape heterogeneity and consequently provides longer periods of availability of maize suitable for borer infestation. The possible effects of landscape heterogeneity on *B. fusca* exposure to Bt maize have not been considered before.

Other factors that can contribute to resistance evolution in *B. fusca* are that moths need no sexual maturation time and mating may start a few hours after moth emergence (Calatayud *et al.*, 2014), hence increasing the likelihood of non-random mating. Males can also mate several times, although only once a night, and polyandry is not obligatory and not necessary (Unnithan and Paye, 1990).

4.6 Insect Resistance Management in the African Context

Landscape heterogeneity is characteristic of African agroecosystems. This heterogeneity, in addition to between-field and within-field spatial mosaics resulting from variable gene expression in Bt maize, will be important in IRM. Adding to this landscape heterogeneity is the cultivation of OPV maize varieties and bimodal rainfall patterns that allow two maize cropping seasons a year in many subtropical and tropical areas. The role that these factors may play in resistance evolution to Bt maize in Africa will not have previously been considered.

Spatial mosaics resulting from variable gene expression within a field could significantly affect resistance evolution (Onstad and Carrière, 2013). For example, plant-gene expression and pollen dispersal, as well as typical African agricultural practices such as planting OPVs and seed saving, can significantly contribute to resistance evolution. Onstad and Carrière (2013) indicated that soil moisture, soil nutrients, herbivory and topography vary over space and influence the growth of plants and the production of toxin in these plants. Hence, expression of a gene for Bt production may vary among plants from a single cultivar over a crop field, creating a spatial mosaic of toxin doses. This type of spatial mosaic would be especially important in cases where neighbouring farmers plant multiple cultivars and where individual farmers plant both hybrid maize and OPVs on small plots of land.

One of the biggest potential contributors to both in-field and landscape-level spatial mosaics in terms of Bt protein expression levels is the cultivation of OPVs. Maize has a high risk of gene flow through cross-pollination, particularly when landholdings are fragmented, varieties are planted continuously and farmers recycle, exchange or mix maize seeds (Smale and De Groote, 2003).

Pollen dispersal between Bt and non-Bt plants (OPVs) and subsequent cultivation of F_1 and F_2 OPV seed by small farmers is a serious threat to the sustainable use of Bt maize. Cross-pollination could transform external refuges, in this case neighbouring farms on which OPVs are planted, into seed mixtures in subsequent cropping seasons. When larvae move between plants in such fields, the adventitious presence of Bt plants in OPV fields can be expected to accelerate resistance evolution. Aheto *et al.* (2013) showed that the potential for gene flow, be it in the form of pollen flow between fields or seed exchange between farmers, has significant potential to spread transgenes across landscapes. The combination of low-dose expression resulting from the cross-pollination of OPVs with Bt hybrids provides unique challenges to IRM in African farming systems.

Africa at large is particularly vulnerable to potential unintended and undesirable spread of genetically modified organisms (GMOs), with a consecutive mixing with non-GM material. Maize seed is easy to store and transport, and through pollen flow, traits can easily be transferred between varieties (Smale and De Groote, 2003). Throughout Africa, the cultivation of OPVs, seed saving and seed exchange are common (Smale and Phiri, 1998; Smale and DeGroote, 2003; Aheto *et al.*, 2013). This is also the case in South Africa, despite a strong and regulated private seed sector and seed production and marketing system (Mphinyane and Terblanché, 2005; Van den Berg, 2013).

4.7 Conclusion

Small-scale farming systems in Africa provide unique challenges to IRM strategies. Lessons learned from the case of *B. fusca* have shown that its biology is poorly understood and that the assumptions on which the high-dose refuge strategy is based are not warranted for this pest. Other challenges are in the form of large variability in stem borer ecology between regions, farming practices and the cultivation of OPVs.

References

Aheto, D.W, Bøhn, T., Breckling, B., Van den Berg, J., Lim, L.C. and Wikmark, O. (2013) Implications of GM crops in subsistence-based agricultural systems in Africa. GM-crop cultivation – ecological effects on a landscape scale. *Theorie in der Ökologie* 17, 93–103.

Bosque-Pèrez, N.A. and Schulthess, F. (1998) Maize: West and Central Africa. In: Polaszek A. (ed.) *African Cereal Stem Borers: Economic Importance, Taxonomy, Natural Enemies and Control.* CAB International. Wallingford, UK, pp. 11–27.

Brévault, T., Heuberger, S., Zhang, M., Ellers-Kirk, C., Ni, X. *et al.* (2013) Potential shortfall of pyramided Bt cotton for resistance management. *Proceedings of the National Academy of Sciences of the United States of America* 110, 5806–5811.

Byerlee, D. and Heisey, P.W. (1996) Past and potential impacts of maize research in sub-Saharan Africa: a critical assessment. *Food Policy* 3, 255–277.

Calatayud, P.-A., Le Ru, B., Van den Berg, J. and Schulthess, F. (2014) Ecology of the African maize stalk borer, *Busseola fusca* (Lepidoptera: Noctuidae) with special reference to insect-plant interactions. *Insects* 5, 539–563.

Campagne, P., Kruger, M., Pasquet, R., Le Ru, B. and Van den Berg, J. (2013) Dominant inheritance of field-evolved resistance to Bt corn in *Busseola fusca.* *PLoS ONE* 8(7): e69675.

De Groote H. (2002) Maize yield losses from stemborers in Kenya. *Insect Science and its Application* 22, 89–96.

El-Shazlya, E.A., Ismailb, I.A., El Shabrawya, H.A., Abdel-Moniemb, A.S.H. and Abdel-Rahman, R.S. (2013) Transgenic maize hybrids as a tool to control *Sesamia cretica* Led. compared by conventional method of control on normal hybrids. *Archives of Phytopathology and Plant Protection* 46, 2304–2313.

Felsenstein, J. (2007) *Theoretical Evolutionary Genetics.* University of Washington, Seattle, Washington.

Ferré, J. and Van Rie, J. (2002) Biochemistry and genetics of insect resistance to *Bacillus thuringiensis. Annual Review of Entomology* 47, 501–533.

Gassmann, A.J., Carrière, Y. and Tabashnik, B.E. (2009) Fitness costs of insect resistance to *Bacillus thuringiensis. Annual Review of Entomology* 54, 147–163.

Glaser, J.A. and Matten, S.R. (2003) Sustainability of insect resistance management strategies for transgenic Bt corn. *Biotechnology Advances* 22, 45–69.

Gould, F. (1998) Sustainability of transgenic insecticidal cultivars: integrating pest genetics and ecology. *Annual Review of Entomology* 43, 701–726.

Gould, F. (2000) Testing Bt refuge strategies in the field. *Nature Biotechnology* 18, 266–267.

Gouse, M., Pray, C.E., Kirsten, J. and Schimmelpfennig, D. (2005) A GM subsistence crop in Africa: the case of Bt white maize in South Africa. *International Journal of Biotechnology* 7, 84–94.

Guse, C.A., Onstad, D.W., Buschman, L.L., Porter, P., Higgins, R.A. *et al.* (2002) Modeling the development of resistance by stalk-boring Lepidoptera (Crambidae) in areas with irrigated transgenic corn. *Environmental Entomology* 31, 676–685.

James, C. (2010) A global overview of biotech (GM) crops: adoption, impact and future prospects. *GM Crops and Food* 1, 8–12.

James, C. (2013) *Global Status of Commercialized Biotech/GM Crops: 2013.* ISAAA Brief No. 46, International Service for the Acquisition of Agri-Biotech Applications, Ithaca, New York.

Jin, L., Wei, Y., Zhang, L., Yang, Y., Tabashnik, B.E. *et al.* (2013) Dominant resistance to Bt cotton and minor cross-resistance to Bt toxin Cry2Ab in cotton bollworm from China. *Evolutionary Applications* 6, 1222–1235.

Kfir, R. (1998) Maize and grain sorghum: southern Africa. In: Polaszek A. (ed.) *African Cereal Stem Borers: Economic Importance, Taxonomy, Natural Enemies and Control.* CAB International, Wallingford, UK, pp. 29–38.

Kfir, R., Overholt, W.A., Khan, Z.R. and Polaszek, A. (2002) Biology and management of economically important lepidopteran cereal stem borers in Africa. *Annual Review of Entomology* 47, 701–731.

Kruger, M., van Rensburg, J.B.J. and Van den Berg, J. (2009) Perspective on the development of stem borer resistance to Bt maize and refuge compliance at the Vaalharts irrigation scheme in South Africa. *Crop Protection* 28, 684–689.

Kruger, M., van Rensburg, J.B.J. and Van den Berg, J. (2011) Resistance to Bt maize in *Busseola fusca* (Lepidoptera: Noctuidae) from Vaalharts, South Africa. *Environmental Entomology* 40, 477–483.

Kruger, M., van Rensburg, J.B.J. and Van den Berg, J. (2012) Reproductive biology of Bt-resistant and susceptible field-collected larvae of the maize stem borer, *Busseola fusca* (Lepidoptera: Noctuidae). *African Entomology* 20, 35–43.

Kruger, M., van Rensburg, J.B.J. and Van den Berg, J. (2014) No fitness costs associated with resistance of *Busseola fusca* (Lepidoptera: Noctuidae) to genetically modified Bt maize. *Crop Protection* 55, 1–6.

La Rovere, R., Kostandini, G., Abdoulaye, T., Dixon, J., Mwangi, W. *et al.* (2010) *Potential Impact of Investments in Drought Tolerant Maize in Africa.* International Maize and Wheat Improvement Center (CIMMYT), Addis Ababa.

Le Ru, B.P., Ong'amo, G.O., Moyal, P., Muchugu, E., Ngala, L. *et al.* (2006a) Geographic distribution and host plant ranges of East African noctuid stem borers. *Annales de la Société Entomologique de France* 42, 353–361.

Le Ru, B.P., Ong'amo, G.O., Moyal, P., Ngala, L., Musyoka, B. *et al.* (2006b) Diversity of lepidopteran stem borers on monocotyledonous plants in eastern Africa and the islands of Madagascar and Zanzibar revisited. *Bulletin of Entomological Research* 96, 555–563.

Mailafiya, D.M., Le Ru, B., Kairu, E.W., Dupas, S. and Calatayud, P.A. (2009) Species diversity of lepidopteran stem borer parasitoids in cultivated and natural habitats in Kenya. *Journal of Applied Entomology* 133, 416–429.

Moolman, H.J., Van den Berg, J., Conlong, D., Cugala, D., Siebert, S.J. *et al.* (2014) Species diversity and distribution of lepidopteran stem borers in South Africa and Mozambique. *Journal of Applied Entomology* 138, 152–166.

Mphinyane, M.S. and Terblanché, S.E. (2005) Personal and socio-economical variables affecting the adoption of maize production intervention program by dryland farmers in the Vuwani district, Limpopo Province. *South African Journal of Agricultural Extension* 35, 221–240.

Mulaa, M.M., Bergvinson, J.D., Mugo, S.N., Wanyama, J.M., Tende, R.M. *et al.* (2011). Evaluation of stem borer resistance manage-ment strategies for Bt maize in Kenya based on alternative host refugia. *African Journal of Biotechnology* 10, 4732–4739.

Murphy, A.F., Ginzel, M.D. and Krupke, C.H. (2010) Evaluating western corn rootworm (Coleoptera: Chrysomelidae): emergence and root damage in a seed mix refuge. *Journal of Entomology* 103, 147–157.

Ndemah, R., Schulthess, F., Le Ru, B. and Bame, I. (2007) Lepidopteran cereal stem borers and associated natural enemies on maize and wild grass hosts in Cameroon. *Journal of Applied Entomology* 131, 658–668.

Ong'amo, G.O., Le Ru, B.P., Dupas, S., Moyal, P., Calatayud, P.-A. *et al.* (2006) Distribution, pest status and agro-climatic preferences of lepidopteran stem borers of maize in Kenya. *Annales de la Société Entomologique de France* 42, 171–177.

Ong'amo, G.O., Le Ru, B., Calatayud, P.-A., Ogol, C.K.P.O. and Silvain J.-F. (2013). Composition of stem borer communities in selected vegetation mosaics in Kenya. *Arthropod Plant Interactions* 7, 267–275.

Onstad, D.W. and Carrière, Y. (2013) The role of landscapes in insect resistance management. In: Onstad, D.W. (ed.) *Insect Resistance Management: Biology, Economics and Prediction.* 2nd edn. Academic Press (imprint of Elsevier),.London/Waltham, Massachusetts/ San Diego, California, pp. 327–372.

Onstad, D.W., Guse, C.A., Porter, P., Buschman, L.L., Higgins, R.A. *et al.* (2002) Modeling the development of resistance by stalk-boring lepidopteran insects (Crambidae) in areas with transgenic corn and frequent insecticide use. *Journal of Economic Entomology* 95, 1033–1043.

Roush, R.T. (1994) Managing pests and their resistance to *Bacillus thuringiensis*: can transgenic crops be better than sprays? *Biocontrol Science and Technology* 4, 501–516.

Seshu Reddy, K.V. and Walker, P.T. (1990) A review of the yield losses in graminaceous crops caused by *Chilo partellus* spp. *Insect Science and its Application* 11, 563–569.

Smale, M. and De Groote, H. (2003) Diagnostic research to enable adoption of transgenic crop varieties by smallholder farmers in sub-Saharan Africa. *African Journal of Biotechnology* 2, 586–595.

Smale, M. and Phiri, A. (1998) *Institutional Change and Discontinuities in Farmers' Use of Hybrid Maize Seed and Fertilizer in Malawi. Findings from the 1996–7 CIMMYT/MoALD Survey.* Economics Working Paper 98-01, International Maize and Wheat Improvement Center (CIMMYT), Mexico City.

Tabashnik, B.E. and Gould, F. (2012) Delaying corn root worm resistance to Bt crops. *Journal of Economic Entomology* 105, 767–776.

Tabashnik, B.E., Gassmann, A.J., Crowder, D.W. and Carriére, Y. (2008) Insect resistance to Bt crops: evidence versus theory. *Nature Biotechnology* 26, 199–202.

Tabashnik, B.E., van Rensburg, J.B.J. and Carrière, Y. (2009) Field-evolved insect resistance to Bt crops: definition, theory, and data. *Journal of Economic Entomology* 102, 2011–2025.

Tabashnik, B.E., Brévault, T. and Carrière, Y. (2013) Insect resistance to Bt crops: lessons from the first billion acres. *Nature Biotechnology* 31, 510–521.

Tende, R.M., Mugo, S.N., Nderitu, J.H., Olubayo, F.M., Songa, J.M. *et al.* (2010) Evaluation of *Chilo partellus* and *Busseola fusca* susceptibility to δ-endotoxins in Bt maize. *Crop Protection* 29, 115–120.

US EPA (1998) *Final Report of the FIFRA Scientific Advisory Panel Subpanel on* Bacillus thuringiensis *(Bt) Plant-pesticides and Resistance Management. Meeting Held on February 9 and 10, 1998.* US Environmental Protection Agency, Washington, DC. Available at: http://www.epa. gov/scipoly/sap/meetings/1998/0298_mtg.htm (accessed 14 October 2014).

US EPA (2001) *Biopesticides Registration Action Document – Bacillus thuringiensis Plant-incorporated Protectants.* US Environmental Protection Agency, Washington, DC. Available at: http://www.epa.gov/oppbppd1/biopesticides/ pips/bt_brad.htm (accessed 22 November 2010).

Unnithan, G.C. and Paye, S.O. (1990) Factors involved in mating, longevity, fecundity and egg fertility in the maize stem-borer, *Busseola fusca* (Fuller) (Lep., Noctuidae). *Journal of Applied Entomology* 109, 295–301.

Van den Berg, J. (2013) Socio-economic factors affecting adoption of improved agricultural practices by scale farmers. *African Journal of Agricultural Research* 8, 4490–4500.

Van den Berg J. and Van Wyk, A. (2007) The effect of Bt maize on *Sesamia calamistis* (Lepidoptera: Noctuidae) in South Africa. *Entomologia Experimentalis et Applicata* 122, 45–51.

Van den Berg, J., Hilbeck, H. and Bøhn, T. (2013) Pest resistance to Cry 1Ab Bt maize: field resistance, contributing factors and lessons from South Africa. *Crop Protection* 54, 154–160.

van Rensburg, J.B.J. (1997) Seasonal moth flight activity of the maize stalk borer, *Busseola fusca* (Fuller) (Lepidoptera: Noctuidae) in small farming areas of South Africa. *Applied Plant Science* 11, 20–23.

van Rensburg, J.B.J. (1999) Evaluation of Bt-transgenic maize for resistance to the stem borers *Busseola fusca* (Fuller) and *Chilo partellus* (Swinhoe) in South Africa. *South African Journal of Plant and Soil* 16, 38–43.

van Rensburg, J.B.J. (2001) Larval mortality and injury patterns of the African stalk borer, *Busseola fusca* on various plant parts of Bt-transgenic maize. *South African Journal of Plant and Soil* 18, 62–69.

van Rensburg, J.B.J. (2007) First report of field resistance by the stem borer, *Busseola fusca* (Fuller) to Bt-transgenic maize. *South African Journal of Plant and Soil* 24, 147–151.

van Rensburg, J.B.J., Walters, M.C. and Giliomee, J.H. (1987a) Ecology of the maize stalk borer, *Busseola fusca* (Fuller) (Lepidoptera: Noctuidae). *Bulletin of Entomological Research* 77, 255–269.

van Rensburg, J.B.J., Walters, M.C. and Giliomee, J.H. (1987b) The influence of rainfall on seasonal abundance and flight activity of the maize stalk borer, *Busseola fusca* in South Africa. *South African Journal of Plant and Soil* 4, 183–187.

van Rensburg, J.B.J., Walters, M.C. and Giliomee, J.H. (1988) Plant population and cultivar effects caused by the maize stalk borer, *Busseola fusca* (Lepidoptera: Noctuidae). *South African Journal of Plant and Soil* 5, 215–218.

Worku, M., Twumasi-Afriyie, S., Wolde, L., Tadesse, B., Demisie, G. *et al.* (eds) (2012) *Meeting the Challenges of Global Climate Change and Food Security through Innovative Maize Research. Proceedings of the Third National Maize Workshop of Ethiopia, April 18–20, 2011, Addis Ababa, Ethiopia.* International Maize and Wheat Improvement Center (CIMMYT), Mexico City.

Wright, S. (1942) Statistical genetics and evolution. *Bulletin of the American Mathematical Society* 48, 223–246.

5 Resistance of Cabbage Loopers to Btk in a Greenhouse Setting: Occurrence, Spread and Management

Alida Janmaat,[1] Michelle Franklin[2] and Judith H. Myers[3]*

[1]Biology Department, University of the Fraser Valley, Abbotsford, British Columbia, Canada; [2]Institute for Sustainable Horticulture, Kwantlen Polytechnic University, Surrey, British Columbia, Canada; [3]Department of Zoology, University of British Columbia, Vancouver, Canada

Summary

Resistance to the microbial insecticide Btk occurred in a number of laboratory populations of Lepidoptera before it was discovered in field populations. Therefore, it is not surprising that cabbage loopers (*Trichoplusia ni*) in semi-contained vegetable greenhouses were among the first examples of Btk resistance in agricultural situations. We have studied the occurrence of Btk resistance in cabbage loopers in greenhouse populations and the movement of resistance among greenhouse populations of moths. Migratory populations of cabbage loopers remained highly susceptible to Btk sprays in fields in the vicinity of greenhouses. Complete clean-up of greenhouses to remove any overwintering moths is necessary to reduce selection for resistance. Cabbage loopers represent a model system for the study of Btk resistance in contained, seasonal environments.

5.1 Introduction

Trichoplusia ni (the cabbage looper) is an economic pest in a wide variety of crops in North America, including the Fraser Valley of British Columbia, Canada, where it is a pest of field cole crops (i.e. Brussels sprouts, cauliflower, broccoli and cabbage). In commercial vegetable greenhouses in British Columbia, it was the primary lepidopteran pest for some years. During this time, cabbage loopers were controlled by the application of the microbial insecticide, *Bacillus thuringiensis kurstaki* (Btk) although more recently the chemical insecticides tebufenozide (Confirm®), chlorantraniliprole (Coragen®), chlorfenapyr (Pylon®) and the Naturalyte (biologically derived) insect control product, spinosad (Success®, Entrust®; a fermented by-product from the soil bacterium *Saccharpolyspora spinosa*) have been used. However, the usefulness of such control products may be short lived if strategies are not employed to deter the evolution of resistance.

5.2 Studies of Resistance in a Greenhouse Setting

The greenhouse environment provides ideal conditions for the rapid evolution of resistance to Btk and other insecticides.

* Corresponding author. E-mail address: myers@zoology.ubc.ca

Insect populations in greenhouses are relatively well contained, and grow rapidly as a result of being protected from external elements, high food plant quality and warm temperatures. These populations can also be exposed to intense selection pressures. Growing conditions in greenhouses have been implicated in at least two studies as contributing to the development of Btk resistance in the diamondback moth, *Plutella xylostella* (Hama *et al.*, 1992; Shelton *et al.*, 1993). In cabbage loopers, growers began to report variable control following Btk applications in the Fraser Valley of British Columbia in 2000. At this time, growers relied almost exclusively on Btk for cabbage looper control. In response to inadequate Btk efficacy, growers increased the frequency and rate of Btk applications and, eventually, they pursued emergency applications for the registration of chemical products to control *T. ni*. The reason for the failing efficacy of Btk was uncertain, and speculations arose about the quality of the product. However, reports of the poor effectiveness of an insecticidal spray are often the first signs of resistance development in a pest population (ffrench-Constant and Roush, 1990).

In 2000, a 3 year survey was initiated to determine whether genetic resistance to Btk was present in greenhouse *T. ni* populations. Populations of cabbage loopers residing in commercial vegetable greenhouses and in local broccoli fields were sampled for resistance and compared with a laboratory population that had not been exposed to Btk. Populations were surveyed periodically throughout the growing season each year for 3 years. The results of the study, together with additional genetic studies, demonstrated that populations of *T. ni* in commercial vegetable greenhouses had developed genetic resistance to Btk (Janmaat and Myers, 2003). Furthermore, resistance developed repeatedly within just 1 year in *T. ni* greenhouse populations in response to grower spray programmes, and the levels of the developed resistance were sufficient to disrupt *T. ni* control. When Btk was first registered for use in greenhouses, growers reported that they generally applied it according to the rates specified on the label, although self-reports by greenhouse growers suggest that applications largely in excess of levels recommended on Btk product labels were common in 2000–2002. The application of high Btk doses was then likely to have been a symptom of the evolution of Btk resistance and not its original cause.

What contributed to the initial evolution of Btk resistance in greenhouse *T. ni* populations? It is likely that the greenhouse environment played a significant role. Greenhouses simulate a tropical environment in temperate areas, in which insect populations are protected from extreme environmental conditions, grow rapidly, undergo multiple generations a year and avoid mortality over the winter. *T. ni* is considered to be a subtropical species that migrates from overwintering grounds in California to seasonally available habitats as far north as Canada (Mitchell and Chalfant, 1984). Cabbage loopers are unable to overwinter when the temperatures are below 10°C for a significant period, and consequently they are not likely to overwinter in British Columbia (Cervantes *et al.*, 2011). Commercial vegetable greenhouses in British Columbia generally operate on a yearly production cycle. At the end of the growing season, typically in December, all vegetation is removed and the greenhouse structure is cleaned to limit the carry-over of pests. Greenhouses generally remain unheated during this period and have an average temperature of 10°C (Caron and Myers, 2008). In spite of this, cabbage looper pupae have been shown to survive overwinter in greenhouses (Cervantes, 2005, Cervantes *et al.*, 2011), and pupal survivorship at 10°C is positively associated with Btk resistance (Caron and Myers, 2008). Genetic analysis using amplified fragment length polymorphism (AFLP) supports the conclusion that *T. ni* persists in some greenhouses through the winter, as larval collections from the same greenhouse before and after the winter clean-up in 2006 and 2007 showed genetic similarity and clustered together (Franklin *et al.*, 2010a). Inadequate clean-up of greenhouses may, therefore, allow resistant *T. ni* populations to persist through the winter and become problematic in the following growing season.

5.3 Genetic Structure of Greenhouse and Field Populations

The potential for Btk resistance to evolve is additionally exacerbated by the impact of greenhouses on the genetic structure of *T. ni*, which is a seasonal migrant in western North America. Migratory species typically display genetic homogeneity across large distances owing to the long-range movement of individuals (Peterson and Denno, 1998). Greenhouses are likely to limit individual movement and reduce the interactions between greenhouse insect populations and migrating field populations of *T. ni*. The genetic connectivity of *T. ni* populations collected from field sites across its migratory range was investigated using AFLP and sequence variation in mtDNA (Franklin *et al.*, 2010b). Limited genetic differentiation was detected among populations collected from California to British Columbia, which is consistent with the migratory movement of *T. ni* populations. Moreover, the isolation by distance (IBD) relationship showed no pattern across the migration route from California to British Columbia, thus further supporting the observation that *T. ni* moths disperse widely by travelling on wind currents (Franklin, 2009).

Another study compared the genetic structure of *T. ni* populations in British Columbian greenhouses and adjacent field populations (Franklin *et al.*, 2010a). Significant spatial genetic structure of populations occurred on a local scale, such that *T. ni* populations from the primary commercial greenhouse vegetable growing region in British Columbia separated into three genetically distinct groups. A significant isolation by distance relationship was observed among local *T. ni* greenhouse populations, which also demonstrates that movement among greenhouse populations is limited. Populations migrating from the south were shown to be a likely source of

migrants for some greenhouses, while other greenhouse populations of *T. ni* remained genetically distinct from the migrants. These results demonstrate that greenhouse populations often have little genetic connectivity with outside field populations and can rapidly diverge due to selection or random inbreeding events. This divergence occurs despite the presence of roof vents in greenhouse structures, which provide a potential entry point for *T. ni* immigrants from field populations. Furthermore, there was evidence for weak differentiation among field populations collected in British Columbia, suggesting that greenhouse *T. ni* populations may have influenced the genetic structure of the surrounding field populations. The levels of heterozygosity increased in greenhouse populations from spring to later in the growing season, possibly as a result of the influx of migrants from southern populations (Franklin *et al.*, 2010a). Despite the potential influx of moths into greenhouses, the frequency of immigration into greenhouses was not sufficient to offset the genetic differentiation of several of the greenhouse populations. If *T. ni* persists through the winter in a greenhouse, the greenhouse population may be of considerable size before the migrants arrive, in which case a small influx of migrants would have little effect on the overall population.

Genetic differentiation was positively correlated with the Btk resistance of the greenhouse population, thus demonstrating that differences in Btk use within greenhouses lead to further genetic differentiation among *T. ni* populations (Franklin *et al.*, 2010b). A positive relationship is expected if Btk use causes *T. ni* populations to go through severe population bottlenecks, resulting in intensified genetic drift. Across the years surveyed, on average, greenhouse populations were more resistant to Btk than field populations (Janmaat and Myers, 2003; Franklin and Myers, 2008), and levels of resistance varied significantly among greenhouse populations but not field populations (Franklin and Myers, 2008). This corresponds to the different selection pressures experienced by the greenhouse populations. Interestingly, field populations treated with Btk were less resistant than an untreated field population (Franklin and Myers, 2008), whereas Btk resistance levels in greenhouse populations were positively correlated with Btk use (Janmaat and Myers, 2003). This difference suggests that there is significant mixing of susceptible *T. ni* moths with treated field populations, but a limited mixing with greenhouse populations.

Although *T. ni* populations remain relatively isolated within greenhouse structures, the dispersal of moths between adjacent greenhouses was suggested. Increased levels of Btk resistance were observed in two untreated greenhouses that were less than 5 km from two treated greenhouses (Franklin and Myers, 2008). This movement between greenhouses was particularly apparent early in the growing season, before migrant moths had colonized surrounding fields. Moths generally occurred earlier in greenhouses than in fields, which is consistent with the finding that *T. ni* often survives the winter period within greenhouses. These persistent populations were likely to have been the source of resistant moths for surrounding greenhouses, and the associated problems with Btk resistance. One greenhouse that was likely colonized by resistant moths from a nearby greenhouse early in the growing season experienced significant problems with *T. ni* control, and the population was then subsequently exposed to nine additional Btk applications (Franklin and Myers, 2008). The findings demonstrate that greenhouse *T. ni* populations may be locally isolated from transient migratory populations, yet cannot be considered discrete entities.

5.4 Current Levels of Resistance

Current levels of Btk resistance have declined considerably since the original survey, as shown by a comparison between the resistance levels observed in 2000–2002 (Fig. 5.1) with those in 2005–2006 (Fig. 5.2). In fact, the levels of resistance observed in 2006, and those of the majority of populations surveyed in 2005, were similar to the levels of resistance of untreated greenhouse populations surveyed in 2001–2002. Resistance levels across greenhouses were highest and the least variable in 2000, after which there was a noticeable increase in variation in resistance and in the amount of Btk used among greenhouses. This suggests that growers responded to the finding of Btk resistance by altering their management methods.

5.5 Control Options

Prior to the original survey, growers had limited control options for *T. ni* and relied primarily on Btk. Presently, there are five insecticide products registered for use in greenhouses for the control of cabbage loopers and growers report that the options available are adequate for control (D. Woodske, British Columbia, 2014, personal communication). The decline in levels of Btk resistance is probably associated with a change in *T. ni* management, as tebufenozide was registered for *T. ni* control shortly after the original survey. Growers may have switched to alternative control products, although many still appear to be on an evolution of resistance treadmill. There are reports that many of the new control products suffer from reduced efficacy after several years of use, after which growers switch to other available products (A. Davenport, British Columbia, 2014, personal communication). Consequently, we question whether the message of Btk resistance been heard? What is required to move growers off the resistance treadmill and on to a more sustainable use of control products such as Btk?

An additional potential microbial control product for cabbage loopers is *Autograph californica* nucleopolyhedrovirus (AcMNPV). If this can be developed for use in greenhouses, it could provide another important tool. Btk-resistant cabbage loopers were found to be twice as susceptible to AcMNPV as loopers that had not been selected for resistance and so these two microbial products could work well together (Sarfraz *et al.*, 2010). Cabbage loopers consume much

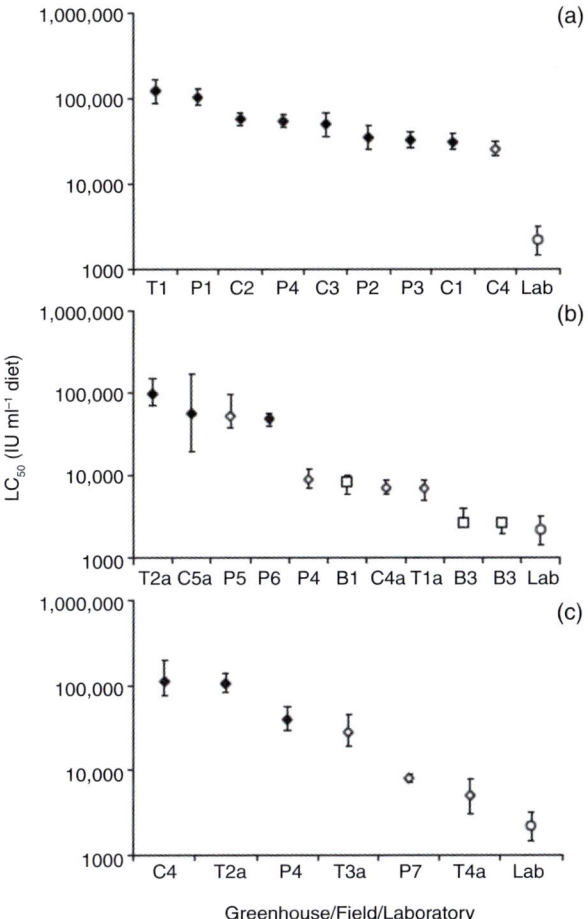

Fig. 5.1. Levels of resistance to Btk (*Bacillus thuringiensis kurstaki*) as shown by the Btk LC_{50} and 95% confidence intervals for each *Trichoplusia ni* population collected from greenhouses (C, P, T) or fields (B) in British Columbia, Canada, in (a) 2000, (b) 2001 and (c) 2002. Greenhouses in which Btk was applied are shown as ◆ and untreated greenhouses as ◇. Field collections are represented as □, and the reference susceptible laboratory colony is shown as O. C indicates cucumber crops, P, peppers, T, tomatoes and B, broccoli. This figure was originally published as Figure 1 in Janmaat and Myers (2003).

greater amounts of cucumber leaves than they do tomato and green pepper leaves, and this might make AcMNPV more effective in greenhouses growing cucumbers (Sarfraz *et al.*, 2011).

Another approach to reducing the resistance of cabbage loopers is to modify the toxins used in the sprays. Franklin *et al.* (2009) (see also Soberón *et al.*, Chapter 14) found that a hybrid strain of Btk and the genetically modified Bt toxins Cry1AbMod and Cry1AcMod that lacked helix α1 reduced

the resistance of a laboratory strain of cabbage loopers by more than 100-fold compared with the native toxins. This would require the development of new Bt products to become a resistance management tool.

5.6 Conclusion

The study that has been described here on the evolution of resistance to Btk in *T. ni* populations highlights the importance of

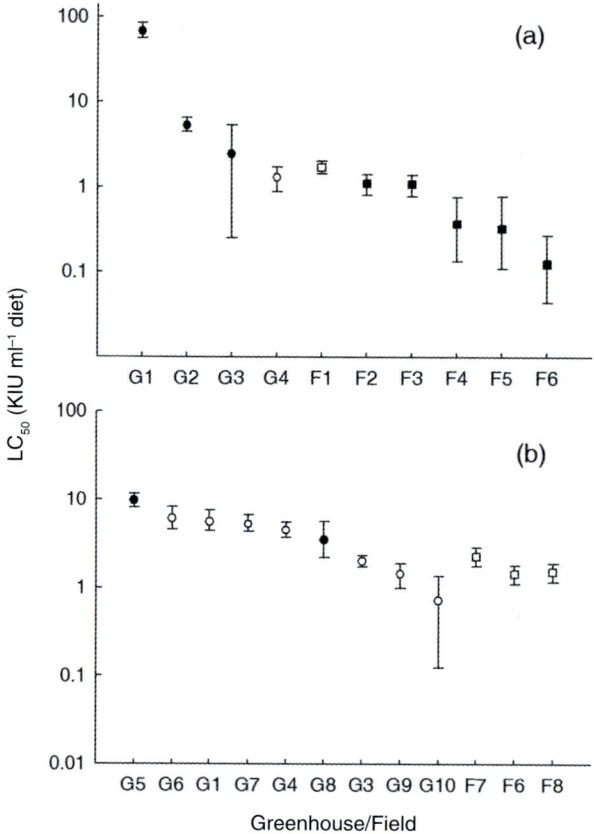

Fig. 5.2. Levels of resistance to Btk (*Bacillus thuringiensis kurstaki*) as shown by the Btk LC$_{50}$ and 95% confidence intervals for each *Ttichoplusia ni* population collected from greenhouses (G) and the field (F) in British Columbia, Canada, in (a) 2005 and (b) 2006. Greenhouse populations: ●, treated with Btk; ○, untreated. Field populations: ■, treated with Btk; □, untreated. Greenhouse crops were tomatoes or peppers (2005) and cucumbers or peppers (2006); field crops were broccoli or mixed crucifers (2005) and broccoli, mixed crucifers or rutabaga (2006). This figure was originally published as Figure 2 in Franklin and Myers (2008).

considering the genetic structure of pest populations when developing resistance management plans. Greenhouses were shown to greatly alter the genetic structure of *T. ni* populations, which contributed to the evolution of Btk resistance. Fortunately, the present example occurred at the northern limit of the range of a migratory species where the pest population is transient. The risk posed by the emigration of resistant moths out of greenhouses into field populations in British Columbia is fairly low as there is a limited period during which resistant moths can breed with ephemeral *T. ni* populations. The risk is much greater when field populations are resident and the emigration of resistant moths out of greenhouses can lead to a gradual increase of resistance allele frequencies in local populations. Yet, even in the northerly British Columbian climate, the resistant moths spread resistance between greenhouses. Therefore, to slow the evolution of resistance to Btk or other valuable control products, growers must look beyond their own facilities and implement coordinated regional resistance plans.

References

Caron, V. and Myers, J.H. (2008) Positive association between resistance to *Bacillus thuringiensis* and overwintering survival of cabbage loopers, *Trichoplusia ni* (Lepidoptera: Noctuidae). *Bulletin of Entomological Research* 98, 317–322.

Cervantes, V.M. (2005) Population ecology of *Trichoplusia ni* in greenhouses and the potential of *Autographa californica* nucleopolyhedrovirus for their control. MSc thesis, University of British Columbia, Vancouver, Canada.

Cervantes, V.M., Sarfraz, R.M. and Myers, J.H. (2011) Survival of cabbage looper, *Trichoplusia ni* (Lepidoptera: Noctuidae), through winter cleanups of commercial vegetable greenhouses: implications for resistance management. *Crop Protection* 30, 1091–1096.

Franklin, M.T. (2009) Influence of local and long-distance dispersal on patterns of Bt resistance in cabbage loopers, *Trichoplusia ni*. PhD thesis, University of British Columbia, Vancouver, Canada.

Franklin, M.T. and Myers, J.H. (2008) Refuges in reverse: the spread of *Bacillus thuringiensis* resistance to unselected greenhouse populations of cabbage loopers *Trichoplusia ni*. *Agricultural and Forest Entomology* 10, 119–127.

Franklin, M.T., Nieman, C.L., Janmaat, A.F., Soberón, M., Bravo, A. *et al.* (2009) Modified *Bacillus thuringiensis* toxins and a hybrid *B. thuringiensis* strain counter greenhouse-selected resistance in *Trichoplusia ni*. *Applied and Environmental Microbiology* 75, 5739–5741.

Franklin, M.T., Ritland, C.E. and Myers, J.H. (2010a) Spatial and temporal changes in genetic structure of greenhouse and field populations of cabbage looper, *Trichoplusia ni*. *Molecular Ecology* 19, 1122–1133.

Franklin, M.T., Ritland, C.E. and Myers, J.H. (2010b) Genetic analysis of cabbage loopers, *Trichoplusia ni* (Lepidoptera: Noctuidae), a seasonal migrant in western North America. *Evolutionary Applications* 4, 89–99.

ffrench-Constant, R.H. and Roush, R.T. (1990) Resistance detection and documentation: the relative role of pesticidal and biochemical assays. In: Roush, R.T. and Tabashnik, B.E. (eds) *Pesticide Resistance in Arthropods*. Chapman and Hall (imprint of Routledge), New York and London, pp. 4–38.

Hama H., Suzuki K. and Tanaka H. (1992) Inheritance and stability of resistance to *Bacillus thuringiensis* formulations of the diamondback moth, *Plutella xylostella* (Linnaeus) (Lepidoptera, Yponomeutidae). *Applied Entomology and Zoology* 27, 355–362.

Janmaat, A.F. and Myers, J.H. (2003) Rapid evolution and the cost of resistance to *Bacillus thuringiensis* in greenhouse populations of cabbage loopers, *Trichoplusia ni*. *Proceedings of the Royal Society B: Biological Sciences* 270, 2263–2270.

Mitchell, E.R. and Chalfant, R.B. (1984) Biology, behaviour and dispersal of adults. In: Lingren, P.D. and Green, G.L. (eds) *Suppression and Management of Cabbage Looper Populations*. Technical Bulletin No. 1684, US Department of Agriculture Agricultural Research Service, Washington, DC, pp. 14–18.

Peterson, M.A. and Denno, R.F. (1998) The influence of dispersal and diet breadth on patterns of genetic isolation by distance in phytophagous insects. *American Naturalist* 152, 428–446.

Sarfraz, R.M., Cervantes, V.M. and Myers, J.H. (2010) Resistance to *Bacillus thuringiensis* in cabbage looper (*Trichoplusia ni*) increases susceptibility to a nucleopolyhedrovirus. *Journal of Invertebrate Pathology* 105, 204–206.

Sarfraz, R.M., Cervantes, V.M. and Myers, J.H. (2011) The effect of host plant species on performance and movement behaviour of the cabbage looper *Trichoplusia ni* and their potential influences on infection by *Autographa californica* multiple nucleopolyhedrovirus. *Agricultural and Forest Entomology* 13, 157–164.

Shelton, A.M., Robertson, J.L., Tang, J.D., Perez, C., Eigenbrode, S.D. *et al.* (1993) Resistance of diamondback moth (Lepidoptera: Plutellidae) to *Bacillus thuringiensis* subspecies in the field. *Journal of Economic Entomology* 86, 697–705.

6

Different Models of the Mode of Action of Bt 3d-Cry Toxins

Alejandra Bravo,* Isabel Gómez, Gretel Mendoza, Meztlli Gaytán and Mario Soberón

Instituto de Biotecnología, Universidad Nacional Autónoma de México, Cuernavaca, Morelos, Mexico

Summary

Bacillus thuringiensis (Bt) produces insecticidal proteins that are active against different insect orders. In particular, the family of the three domain-Cry toxins (3d-Cry) have important applications in controlling insect pests as spray products or in transgenic Bt plants. At least three different models account for the mode of action of these proteins. Two of them are based on the pore formation activity of the toxin and propose that larvae are killed by osmotic shock to their midgut cells. Differences in these models are related to the processes of receptor interactions and membrane insertion. The third model proposes that midgut cells are killed through a signal transduction mechanism that is triggered after the binding of 3d-Cry toxins to a cadherin receptor. In this chapter, we will review these models in light of the different resistance mechanisms to Cry toxins that have been documented so far in several insect species.

6.1 Cry Toxins and their Use in Biological Control

Bacillus thuringiensis (Bt) is an important bacterial species due to its production of insecticidal proteins as inclusion bodies during the sporulation phase of growth (de Maagd *et al.*, 2003). A Cry toxin is 'a *Bacillus* *thuringiensis* parasporal inclusion protein that exhibits pesticide activity or some experimentally verifiable toxic effect to a target organism' (Crickmore *et al.*, 2014). Proteins with significant sequence similarity to other Cry toxins are also included in this nomenclature (Crickmore *et al.*, 2014). Until now, five different phylogenetic groups of insecticidal Bt proteins have been described (de Maagd *et al.*, 2003; Crickmore *et al.*, 2014). These proteins include the three-domain Cry (3d-Cry), the binary-like Cry (Bin-like Cry), the mosquitocidal-like Cry (Mtx-like Cry), the vegetative insecticidal proteins (Vip) and the cytolytic toxins (Cyt). These proteins show insecticidal activity against important crop pests, including coleopteran (3d-Cry, Bin-like Cry, Mtx-like Cry, Vip1-2 and Cyt), lepidopteran (3d-Cry, Vip3) and hemipteran (Mtx-like Cry and Vip1-2) insects. In addition, some Cry toxins display nematicidal activity (3d-Cry) and others are active against dipteran insect vectors of important human diseases (3d-Cry, Mtx-like Cry and Cyt). Finally, some 3d-Cry and Mtx-like Cry proteins, named parasporins, are toxic to some cancer cell lines (Mizuki *et al.*, 1999).

Among the Bt insecticidal proteins, the 3d-Cry toxins represent the biggest and most studied family of toxins. The structure of nine 3d-Cry proteins has been solved, and they show a similar conformation in being composed of three domains: domain I, which is formed of seven α-helices, and domains II

---

* Corresponding author. E-mail address: bravo@ibt.unam.mx

and III, which are composed of β-strands (de Maagd *et al.*, 2003) (Fig. 6.1). Different 3d-Cry toxins have been expressed in plants (Bt plants) and are now commercialized worldwide. Bt plants are resistant to the attack of some insect pests and their use has resulted in an important reduction in the use of chemical insecticides (James, 2013). Other Bt proteins, such as the Vip3, Bin-like Cry34/Cry35 and the Mtx-like Cry51 toxins, have been expressed in maize or cotton, and some of these are also commercially available (Jucovic *et al.*, 2008; Burkeness *et al.*, 2010; Baum *et al.*, 2012; Prasifka *et al.*, 2013).

6.2 Models of the Mechanism of Action of 3d-Cry Toxins

Here we describe and discuss three different models that have been proposed to explain the mechanism of action of 3d-Cry toxins. The three models start with same initial steps. The 3d-Cry protoxins accumulate in inclusion bodies during the sporulation phase of the bacteria. When ingested by susceptible insects, the protoxins are solubilized in the insect gut lumen. The solubilized protoxins are then cleaved by midgut proteases, leading to the production of a final protease-resistant core of 60 kDa that is composed of three domains (Fig. 6.1) (de Maagd *et al.*, 2003; Pardo-López *et al.*, 2013).

6.2.1 Pore formation model involving a pre-pore formation after sequential interaction with different receptors

In this model, it is proposed that 3d-Cry toxins form pores in the midgut cells of the insect, thereby destroying the gut tissue and killing the larvae. The process of insertion of the toxin into the membrane follows se-quential interactions with different proteins

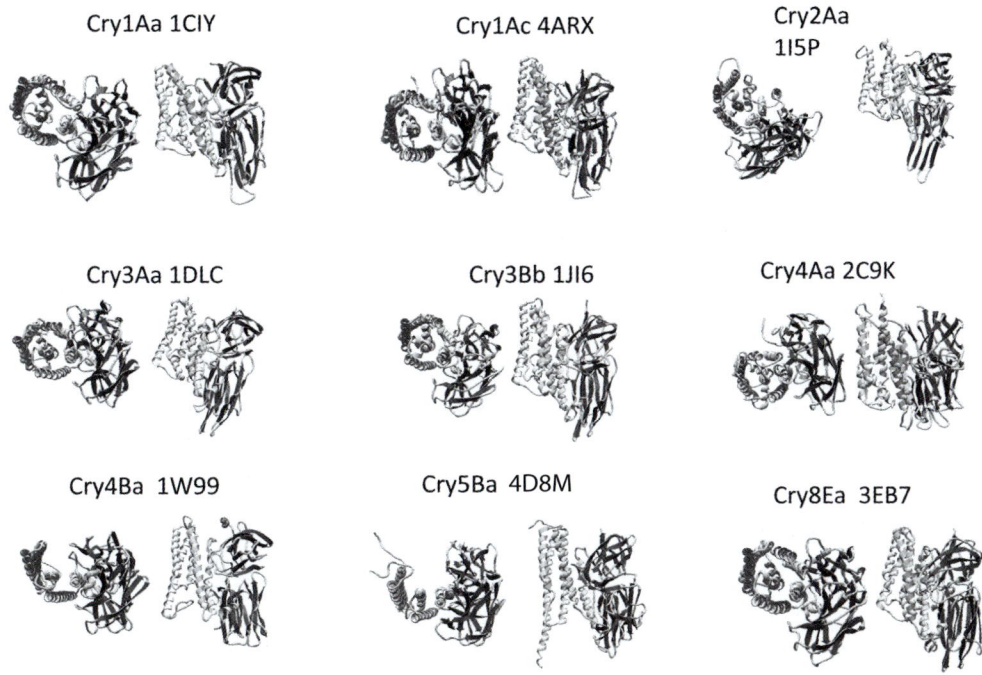

Fig. 6.1. Three-dimensional structure of nine 3d-Cry proteins. Two views of the three-dimensional structure of each of the crystallized proteins are shown. The Protein Data Bank (PDB) accession numbers (1CIY, 4ARX, 1I5P, 1DLC, 1JI6, 2C9K, 1W99, 4D8M and 3EB7; see http://www.pdb.org) are given after the name of each toxin.

that are present in the apical membrane of insect's midgut cells (Fig. 6.2a) (Pardo-López *et al.*, 2013). In the case of Cry1A toxins, it was proposed that loop-3 of domain II interacts with aminopeptidase N (APN) and that β-16 of domain III binds to alkaline phosphatase (ALP) (Pacheco *et al.*, 2009; Arenas *et al.*, 2010) in a low-affinity interaction. Both ALP and APN are anchored to the membrane by a glyco-sylphosphatidylinositol (GPI) anchor and are highly abundant on the cell surface. In the case of APN, this receptor has at least a 300-fold higher concentration than that of cadherin in *Trichoplusia ni* (Zhang *et al.*, 2012b). This binding step is important for localizing the toxin in close proximity to the membrane. Mutant Cry1A toxins in which this binding step is affected show an altered toxicity, indicating the importance of ALP and APN binding (Pacheco *et al.*, 2009;

Arenas *et al.*, 2010). The second binding interaction of the toxin is with a transmembrane cadherin-like protein. This binding step involves loops 2, 3 and α-8 from domain II of the Cry1A toxin in a high-affinity interaction with three different epitopes of the cadherin receptor. Toxins with mutations in these regions also resulted in non-toxic proteins (Gómez *et al.*, 2003, 2006). The binding with cadherin induces an extra cleavage in the toxin that results in the elimination of the N-terminal end, including helix α-1 of domain I (Gómez *et al.*, 2002). This cleavage exposes buried hydrophobic regions of the toxin and triggers oligomerization of the protein, forming a pre-pore oligomer outside the membrane. The pre-pore gains a 100–200 fold higher affinity to APN and ALP receptors (Pacheco *et al.*, 2009; Arenas *et al.*, 2010). Finally, the pre-pore is inserted into the

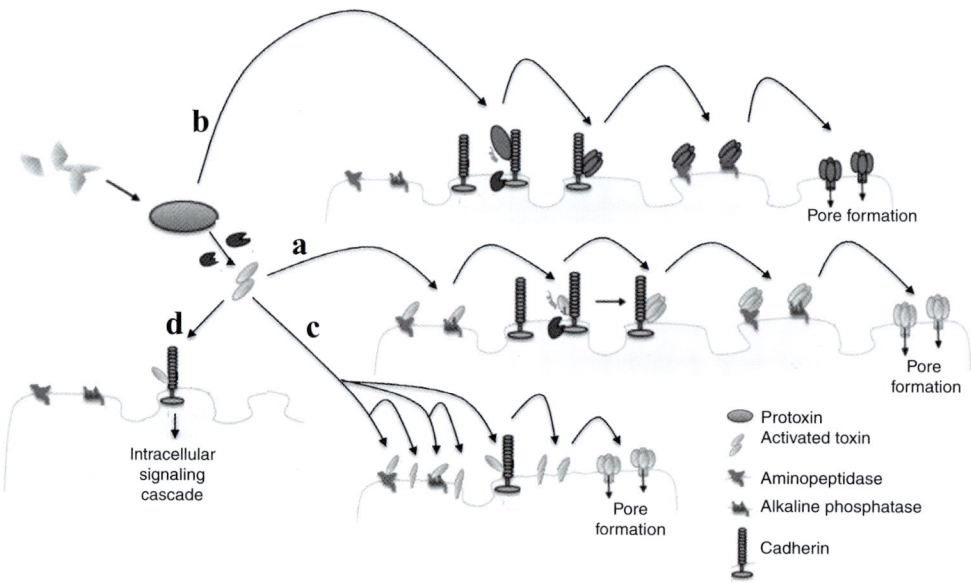

Fig. 6.2. Different models of the possible mechanisms of action of 3d-Cry proteins (A–D). In the sequential binding and pore formation model of activated Cry toxin (a), activated 3-d Cry1A toxin binds to alkaline phosphatase/aminopeptidase N (ALP/APN), then to cadherin, facilitating oligomerization and binding to ALP/APN before membrane insertion. In the sequential binding and pore formation model of Cry protoxin (b), Cry1A protoxin binding to cadherin facilitates oligomer pre-pore formation, binding to ALP/APN receptors and membrane insertion. In the pore formation model of monomeric activated toxin (c), activated 3-d Cry1A toxin binds to ALP, APN or cadherin, inserts into the membrane and oligomerizes in the membrane plane. In the signal transduction model (d), the interaction of Cry activated toxin with cadherin induces an intracellular signal transduction death mechanism.

membrane to form a pore that affects the permeability of the cells and causes death of the larvae (Bravo *et al.*, 2004; Gómez *et al.*, 2014). Mutant toxins that are unable to form transmembrane pores as a result of defects in oligomerization or membrane insertion are non-toxic to the larvae, even though they are still able to bind receptors (Vachon *et al.*, 2002; Jiménez-Juárez *et al.*, 2007; Girard *et al.*, 2008; Rodríguez-Almazán *et al.*, 2009). Figure 6.2A shows a diagram of this model.

Recently, an update of this model showed that both the 3d-Cry protoxin and the trypsin-activated toxin bind cadherin with high affinity (Figs 6.2a and 6.2b) (Gómez *et al.*, 2014). Two different pre-pores are produced after interaction of these molecules with cadherin. The two pre-pores are assembled before toxin insertion into the membrane (Gómez *et al.*, 2014). It was

shown that the characteristics of the resulting oligomeric structures are different. The oligomer that is formed after interaction of the protoxin with cadherin in the presence of proteases is SDS (sodium dodecyl sulfate) resistant and partially resistant to heat, and it inserts efficiently into synthetic liposomes, inducing ionic channel currents with stable conductance and high open probability (Figs 6.2b and 6.3a). In contrast, the SDS-resistant oligomer that is formed after interaction of the trypsin-activated toxin with cadherin is highly sensitive to heat and is less efficient at inserting into synthetic liposomes; further, the ionic channels that are induced show multiple subconducting states and low open probability (Figs 6.2a and 6.3b). Nevertheless, it was shown that the oligomeric structure formed from trypsin-activated toxin is able to insert efficiently into insect

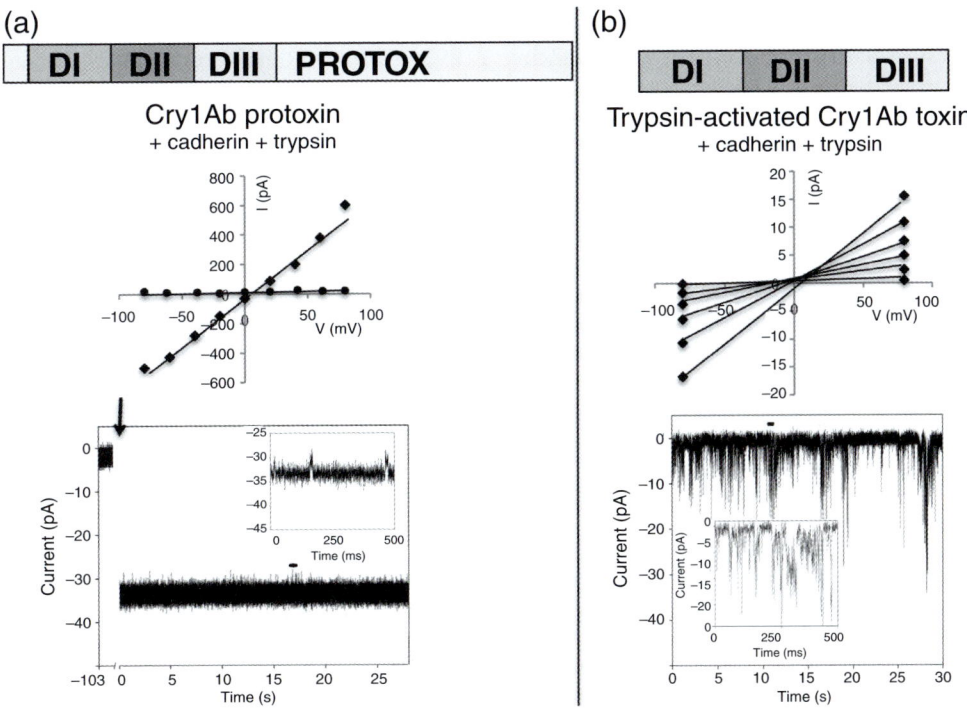

Fig. 6.3. Pore formation activity of oligomeric structures produced by Cry1Ab protoxin (Protox) (a) or activated toxin (b). DI, DII and DIII are the three Cry1Ab domains. (Reproduced with permission, from authors Gómez *et al.* (2014), *Biochemical Journal* 459, 383-396. © The Biochemical Society.)

membranes that contain other receptors, such as ALP and APN, thus supporting the model involving sequential interaction with different receptors (Gómez *et al.*, 2014). In addition, it was shown that both oligomeric structures participate in the toxicity of 3d-Cry proteins *in vivo,* because mutant toxins that are affected in the oligomerization of the activated toxin, but are still able to form oligomers from protoxin, showed reduced toxicity (Gómez *et al.*, 2014). These data indicated that different insects might have different receptors that would affect the insecticidal activity of 3d-Cry proteins, and might also have different proteases in their gut lumen. The rate of protoxin to toxin activation would depend on the type and amount of proteases that are present in each insect species which, in turn, would induce the formation of the two different oligomeric structures. This model suggests that the protoxin region may have a functional role that was selected during evolution as the protoxin triggers the formation of an alternative oligomeric structure that is important for the toxicity of 3d-Cry toxins in some insect pests.

6.2.2 Pore formation model involving insertion of monomeric Cry toxin into the membrane after interaction with a single receptor

In this model, it is proposed that the trypsin-activated Cry toxin inserts into the insect gut membrane after binding to any of the different Cry toxin-receptors previously described (APN, ALP or cadherin) resulting in irreversible binding of the activated toxin (Fig. 6.2c). The oligomerization of the toxin takes place in the membrane plane after the interaction of several monomeric-molecules, forming a pore that kills the larvae (Vachon *et al.*, 2012) (Fig. 6.2c). It has been argued that irreversible binding to brush border membrane vesicles (BBMVs) indicates that monomeric Cry toxins are able to insert into the BBMVs before oligomerization (Vachon *et al.* 2012). However, Gómez *et al.* (2014) showed that

the interaction of monomeric toxin with cadherin induces its oligomerization and that this oligomer gains affinity to APN and ALP and inserts efficiently into the membrane. The oligomeric structure formed from the activated toxin is highly sensitive to heat and it is completely disassembled into monomeric toxins if samples are heated above 60°C. For this reason, the oligomeric structure could only be observed if samples were not boiled before SDS-PAGE and Western blot analysis. Binding analysis of trypsin-activated toxin with BBMVs at 50°C showed that only the oligomeric structure is associated with the BBMVs (Fig. 6.4). These data indicate that irreversible interaction of the activated toxin with BBMVs is correlated with the insertion of a heat-sensitive oligomer into the membrane (Fig. 6.4) (Ihara and Himeno, 2008; Gómez *et al.*, 2014).

Additional data supporting this model is that activated Cry1A toxins induce pore formation activity into in planar lipid bilayer in the absence of receptors (Schwartz *et al.*, 1993, 1997a,b; Rausell *et al.*, 2004; Vachon *et al.*, 2012), although it is important to mention that the insertion of activated-toxin into synthetic membranes is highly in-efficient and, in some cases, pores failed to appear at all (Smedley *et al.*, 1997). In other cases, a special procedure to improve the in-corporation of the toxin into the bilayer, such as disruption and repainting of the bilayer, or microinjection of the toxin directly into the hole of the bilayer chamber followed by chloroform evaporation, have been reported (Schwartz *et al.*, 1993; Potvin *et al.*, 1998; Masson *et al.*, 1999; Putheeranurak *et al.*, 2004). The pores induced by trypsin-activated toxins displayed several sub-conducting states (Schwartz *et al.*, 1993, 1997a,b; Smedley *et al.*, 1997; Peyronnet *et al.*, 2001; Putheeranurak *et al.*, 2004), and different kinetic behaviours such as complex activity patterns, rapid flickering or slow gating and long-lasting subconducting states (Schwartz *et al.*, 1993; Peyronnet *et al.*, 2001; Putheeranurak *et al.*, 2004).

Other data that support this model include the fact that the pore formation

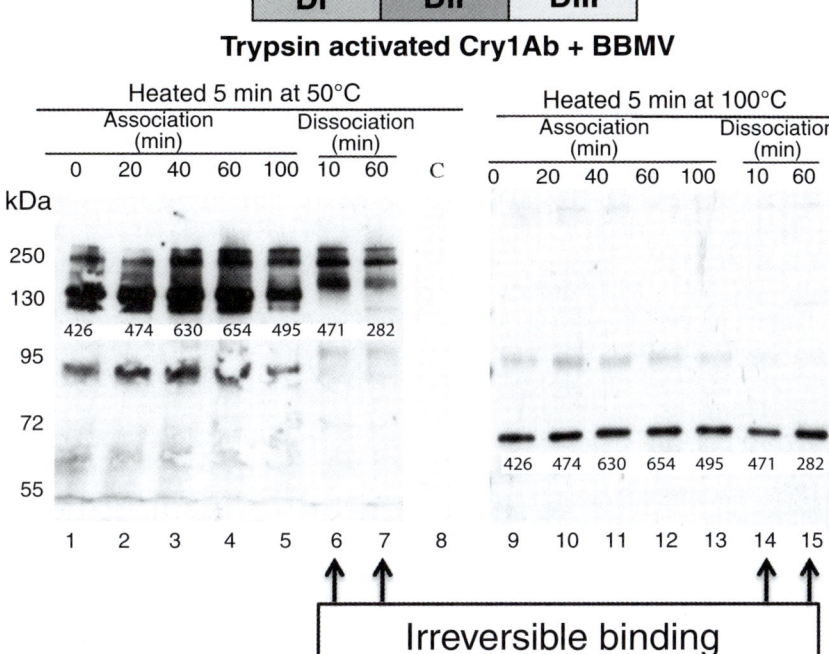

Fig. 6.4. Analysis of the irreversible binding of monomeric Cry1Ab activated toxin to the brush border membrane vesicles (BBMVs) after heat treatment at 50°C, or at 100°C. DI, DII and DIII are the three Cry1Ab domains. (Reproduced with permission, from authors Gómez *et al.* (2014) *Biochemical Journal* 459, 383–396. © The Biochemical Society.)

activity of activated Cry toxin is increased significantly in the presence of receptors. For example, the pore activity of Cry1A toxins measured in a planar lipid bilayer increased up to 250-fold in the presence of APN (Schwartz *et al.*, 1997a). Similarly, reconstituted lipid vesicles containing APN and ALP increased the release of $^{86}Rb^+$ from the vesicles 1000-fold after incubation with Cry1A toxin (Sangadala *et al.*, 1994). Finally, the addition of BBMVs isolated from different target insects containing toxin receptors also improves the pore formation activity of the monomeric toxin, demonstrating a 2–40 fold higher conductance (Lorence *et al.*, 1995; Martin and Wolfersberger 1995; Peyronnet *et al.*, 2001).

6.2.3 Signal transduction model after interaction of monomeric Cry toxin with cadherin receptor

This model was based on the analysis of Cry1Ab toxicity to a cell line originating from *T. ni* (High Five or TnH5) transformed to express the *Manduca sexta* cadherin protein. The expression of cadherin in this TnH5 cell line resulted in susceptibility to the Cry1Ab toxin (Zhang *et al.*, 2006). It was proposed that the interaction of 3d-Cry toxin with the cadherin receptor induces activation of an intracellular death-signal transduction pathway. In this process, a protein G is activated after the toxin binds to cadherin. The protein G, in turn, activates adenylate cyclase, resulting in increased levels of cyclic AMP that activate a protein kinase A (PKA), which induces cell necrosis (Zhang *et al.*, 2006) (Fig. 6.2d).

6.3 Mechanisms of Resistance to Cry Toxins

Insects resistant to 3d-Cry toxins have been selected in laboratory or field conditions. Alterations in insect midgut proteases that affect toxin activation or induce its degradation result in resistance (Keller *et al.*, 1996; Oppert *et al.*, 1997), as do glycolipid moieties or esterases that can sequestrate the toxin, reducing its interaction with receptors and its toxicity (Gunning *et al.*, 2005; Ma *et al.*, 2011). In addition, the induction of an elevated immune response results in insects that are tolerant to the effects of 3d-Cry toxins (Rahman *et al.*, 2004; Hernández-Martínez *et al.*, 2010). However, the most common mechanism of insect resistance to 3d-Cry toxins is linked to alterations in toxin-binding proteins (Ferré and Van Rie, 2002).

Mutations in the cadherin receptor are linked to resistance to Cry1A toxins in different Lepidoptera, such as *Heliothis virescens*, *Pectinophora gossypiella* and *Helicoverpa armigera* (Gahan *et al.*, 2001; Morin *et al.*, 2003; Xu *et al.*, 2005; Fabrick *et al.*, 2011). Lower expression of cadherin in *Diatraea saccharalis* was associated with resistance to Cry1Ab toxin; cadherin silencing with RNA interference (RNAi) was also associated with tolerance to this toxin in *D. saccharalis* and *M. sexta* (Soberón *et al.*, 2007; Yang *et al.*, 2011).

Mutations affecting the expression of the GPI-anchored receptors APN1 and ALP, as well as the silencing APN1 by RNAi, was correlated with resistance or tolerance to 3d-Cry in *Spodoptera exigua*, *H. armigera*, *H. virescens*, *S. frugiperda* and *S. litura* (Rajagopal *et al.*, 2002; Herrero *et al.*, 2005; Zhang *et al.*, 2009; Jurat-Fuentes *et al.*, 2011; Flores-Escobar *et al.*, 2013). Mutations in the aminopeptidase P of *Ostrinia nubilalis* were also associated with resistance to 3d-Cry toxins (Khajuria *et al.*, 2011).

Finally, mutations in the ABCC2 transporter in different lepidopteran species, such as *H. virescens*, *Plutella xylostella*, *Bombyx mori* and *T. ni* are also linked to a high level of resistance to Cry1A toxins (Gahan *et al.*, 2010; Baxter *et al.*, 2011; Atsumi *et al.*, 2012). ABC transporters participate in the transport of various substrates, such as xenobiotics and heavy metals. Their role in the mechanism of action of Cry toxins is not clearly understood, but it has been proposed that it could work as a functional binding receptor (Tanaka *et al.*, 2013).

6.4 Correlation among Mechanisms of Resistance to Cry Toxins and Mode of Action of the Toxin

The model of mode of action of 3d-Cry toxin that best explains the different resistance mechanisms to 3d-Cry toxin action is pore formation involving sequential interaction with different receptors (Figs 6.2a and 6.2b) (Section 6.2.1). In this case, the reduced expression or loss of only one receptor molecule in insect populations would result in Cry toxin resistance. In support of this, the silencing of a single receptor molecule (cadherin, APN or ALP) by RNAi in different insects resulted in high tolerance to Cry toxin (Gahan *et al.*, 2001; Rajagopal *et al.*, 2002; Morin *et al.*, 2003; Herrero *et al.*, 2005; Xu *et al.*, 2005; Soberón *et al.*, 2007; Zhang *et al.*, 2009; Fabrick *et al.*, 2011; Jurat-Fuentes *et al.*, 2011; Yang *et al.*, 2011; Flores-Escobar *et al.*, 2013). These data suggest that multiple receptor molecules participate in the toxicity of Cry proteins *in vivo* and that all of them are important because the loss or reduced expression of a single receptor results in resistance or tolerance to the action of Cry toxin.

Two distinct Cry1A pre-pore structures could be formed in solution after cadherin binding, depending on whether the trypsin-activated toxin or the protoxin bind this receptor (Gómez *et al.*, 2014) (Figs 6.2a and 6.2b). *In vivo*, the activation of protoxin to toxin would depend on the type and amount of proteases present in each insect (de Maagd et al., 2003). These different mechanisms of action involving protoxin or activated toxin could explain why some resistant populations, such as H. armigera and H. zea, showed a high level of resistance to activated toxins, but significantly reduced levels of resistance to protoxins (Xu et al., 2005; Anilkumar et al., 2008; Tabashnik et al., 2011; Caccia et al., 2012). These resistant populations are examples of insect species where the two pre-pores could have differential roles.

In the pore formation model that involves membrane-insertion of Cry monomer (Section 6.2.2), it is proposed that binding to a single receptor would be enough to induce insertion into the membrane and toxicity (Fig. 6.2c). This model is not supported by the finding that populations affected by the loss of a single receptor become resistant to the action of 3d-Cry toxin despite the fact that other receptors are still present in their midgut cells, although it is supported by the finding that the expression of single receptors in different cell lines induces susceptibility to Cry toxins. None the less, the heterologous expression of a single receptor such as cadherin, APN or ALP in different cell lines has also shown contradictory results. For instance, the expression of APN from different Lepidoptera in SF9 (from *Spodoptera frugiperda*) or DmS2 (from *Drosophila melanogaster*) insect cell lines did not induce susceptibility to Cry1Ac toxin, even though the toxin was able to bind to the transfected cells (Garner et al., 1999; Banks et al., 2003). In contrast, the expression of the ALPm isoform from *Aedes aegypti* in the SF9 cell line induced the binding of Cry4Ba toxin and provided susceptibility to this toxin, though at a high dose (Dechklar et al., 2011). Similarly, the expression of lepidopteran cadherins in the TnH5 and SF9 insect cell lines resulted in high sensitivity to Cry1Aa or Cry1Ab toxins, as observed by cell swelling and lysis (Nagamatsu et al., 1999; Flannagan et al., 2005; Zhang et al., 2006). It is important to note that these cell lines are naturally sensitive to other 3d-Cry toxins,

such as Cry1C, which is toxic to SF9 cells, and Cry1Ac, which is toxic to TnH5 cells (Kwa et al., 1998; Liu et al., 2004). These data imply that other toxin-binding molecules involved in the toxicity of Cry1C or Cry1Ac to these cell lines might be participating, along with the transfected cadherin receptor, to induce susceptibility to Cry1Ab or Cry1Aa toxins. Thus, the susceptibility conferred by expression of a single receptor does not exclude the involvement in Cry toxicity of other Cry-binding molecules. The expression of lepidopteran cadherin in other cell lines, such as the animal and human cell lines COS-7 and HEK293, induced moderate cytotoxicity to Cry1Ab when the toxin was used at high doses (Dorsch et al., 2002; Aimanova et al., 2006), and also in DmS2 cells (Hua et al., 2004; Jurat-Fuentes and Adang, 2006). Under these conditions, Cry1Ab causes morphological changes not related to pore formation or cell lysis. In the cadherin-transfected DmS2 cells, only 12–14% of the transfected cells were susceptible to the action of Cry1A toxin (Hua et al., 2004; Jurat-Fuentes and Adang, 2006). These data all suggest that the expression of the cadherin receptor in these cells may be triggering a different mechanism of action that is not related to pore formation activity, as has been proposed in the signalling transduction model (Fig. 6.2d).

In the signal transduction model (Section 6.2.3), it is proposed that an intracellular mechanism is triggered after the interaction of the Cry1A toxin with cadherin that results in necrotic cell death (Zhang et al., 2006) (Fig. 6.2d). In support of this model, the resistance of a population of H. armigera to Cry1Ac cotton in China was shown to be linked to a deletion in the cytoplasmic domain of the cadherin receptor (Zhang et al., 2012a). This suggests that signal transduction mechanisms could be involved in Cry1Ac toxicity in this insect. However, several non-toxic Cry1A mutants in which either toxin oligomerization or pore formation are affected are still able to bind the cadherin receptor with a similar affinity to that of the wild type toxin, suggesting that the binding of monomeric toxin to cadherin is not sufficient to induce toxicity

(Vachon *et al.*, 2002; Jiménez-Juárez *et al.*, 2007; Girard *et al.*, 2008; Rodríguez-Almazán *et al.*, 2009). Furthermore, the construction of Cry1AMod toxins without the amino-terminal end, including helix α-1, demonstrated that these mutant toxins were able to skip cadherin interaction and kill Cry1A-resistant insects linked to cadherin gene mutations or silenced in cadherin expression by RNAi (Soberón *et al.*, 2007). Nevertheless, the toxicity of Cry1AMod toxins against resistant populations was not fully recovered, as it showed higher LC_{50} values than those observed with the native Cry1A toxins against susceptible insects (Tabashnik *et al.*, 2011). This discrepancy could be related to the loss of a signal transduction component, because cadherin is not present in the resistant strains. Cry1AMod toxins also lost potency against susceptible cell lines. In addition, it was recently found that the oligomerization of the trypsin-activated Cry1AbMod toxin is affected (Gómez *et al.*, 2014). A possible explanation for these results is then that the loss of potency of Cry1AMod toxins against susceptible insects is related to the loss of a signal transduction component, but also to the loss of oligomer formation from the trypsin-activated molecule.

As mentioned above, mutations in an ABCC2 transporter were linked to resistance in several insect species (Baxter *et al.*, 2011). In the case of *B. mori*, a single tyrosine insertion on a predicted exposed loop region of this molecule was shown to be linked to resistance to Cry1Ab toxin, suggesting that this region in the ABBC2 transporter could be involved in Cry1Ab binding (Atsumi *et al.*, 2012). Heterologous expression of the *B. mori* ABCC2 transporter in SF9 cells conferred binding capacity to Cry1A toxins and toxicity, suggesting that ABCC2 is a functional receptor of Cry1A toxins (Tanaka *et al.*, 2013). Although not proven, it has been proposed that ABCC2 transporters are involved in facilitating oligomer membrane insertion, as has been proposed for ALP and APN in the sequential binding model (Heckel, 2012). Interestingly, Cry1AMod toxins were shown to counter resistance linked to ABCC2 mutations in resistant

insect populations (Franklin *et al.*, 2009; Tabashnik *et al.*, 2011). Thus, it is possible that ABCC2 may be involved in oligomer formation, as has been proposed for the cadherin receptor (Gómez *et al.*, 2002; Soberón *et al.*, 2007). In any case, both potential roles of ABBC2 support the sequential-binding mode of action of Cry1A toxins.

6.5 Conclusion

Cry toxins are versatile proteins with multiple modes of action. The characterization of different insect populations affected by Cry1A midgut-binding proteins have revealed that different insect gut proteins are involved in Cry toxicity. These include cadherins, APNs, ALPs and ABCC2 transporters. The resistant alleles characterized so far support the sequential binding model as the mechanism of action of Cry proteins. However, the signal transduction model is also supported by the role of the *H. armigera* allele in the cadherin cytoplasmic region. It is possible that the two mechanisms of action could have a differential role in different insect species. Cry toxin mutants in which toxin oligomerization is affected could be important tools in distinguishing between the roles of the two mechanisms among different insect species and different Cry toxins. Understanding the mode of action of Cry toxins will provide tools to counter resistance, thereby preserving the use of these important toxins in pest management for insect control.

References

Aimanova, K.G., Zhuang, M. and Gill, S.S. (2006) Expression of Cry1Ac cadherin receptor in insect midgut and cell lines. *Journal of Invertebrate Pathology* 92, 178–187.

Anilkumar, K.J., Rodrigo-Simón, A., Ferré, J., Pusztai-Carey M., Sivasupramaniam, S. *et al.* (2008) Production and characterization of *Bacillus thuringiensis* Cry1Ac-resistant cotton bollworm *Helicoverpa zea* (Boddie). *Applied and Environmental Microbiology* 74, 462–469.

Arenas, I., Bravo, A., Soberón, M. and Gómez, I. (2010) Role of alkaline phosphatase from *Manduca sexta* in the mechanism of action of *Bacillus thuringiensis* Cry1Ab toxin. *The Journal of Biological Chemistry* 285, 12497–12503.

Atsumi, S., Miyamato, K., Yamamoto, K., Narukawa, J., Kawai, S. *et al.* (2012) Single amino acid mutation in an ATP-binding cassette transporter causes resistance to Bt toxin Cry1Ab in the silkworm, *Bombyx mori. Proceedings of the National Academy of Sciences of the United States of America* 109, 1591–1598.

Banks, D.J., Hua, G. and Adang, M.J. (2003) Cloning of a *Heliothis virescens* 110 kDa aminopeptidase N and expression in *Drosophila* S2 cells. *Insect Biochemistry and Molecular Biology* 33, 499–508.

Baum, J.A., Suruku, U.R., Penn, S.R., Meyer, S.E., Subbarao, S. *et al.* (2012) Cotton plants expressing a hemipteran-active *Bacillus thuringiensis* crystal protein impact the development and survival of *Lygus hesperus* (Hemiptera: Miridae) nymphs. *Journal of Economic Entomology* 105, 616–624.

Baxter, S.W., Badenes-Pérez, F.R., Morrison, A., Vogel, H., Crickmore, N. *et al.* (2011) Parallel evolution of Bt toxin resistance in Lepidoptera. *Genetics* 189, 675–679.

Bravo, A., Gómez, I., Conde, J., Muñoz-Garay, C., Sánchez, J. *et al.* (2004) Oligomerization triggers binding of a *Bacillus thuringiensis* Cry1Ab pore-forming toxin to aminopeptidase N receptor leading to insertion into membrane microdomains. *Biochimica et Biophysica Acta* 1667, 38–46.

Burkeness, E.C., Dively, G., Patton, T., Morey, A.C. and Hutchison, W.D. (2010) Novel Vip3A *Bacillus thuringiensis* (Bt) maize approaches high-dose efficacy against *Helicoverpa zea* (Lepidoptera: Noctuidae) under field conditions: implications for resistance management. *GM Crops and Food* 1, 337–343.

Caccia, S., Moar, W.J., Chandrashekhar, J., Oppert, C., Anilkumar, K.J. *et al.* (2012) Association of Cry1Ac toxin resistance in *Helicoverpa zea* (Boddie) with increased alkaline phosphatase levels in midgut lumen. *Applied and Environmental Microbiology* 78, 5609–5698.

Crickmore, N., Baum, J., Bravo, A., Lereclus, D., Narva, K. *et al.* (2014) *Bacillus thuringiensis* toxin nomenclature. Available at: http://www.btnomenclature.info/ (accessed 15 October 2014).

de Maagd, R.A., Bravo, A., Berry, C., Crickmore, N. and Schnepf, H.E. (2003) Structure, diversity and evolution of protein toxins from spore-forming entomopathogenic bacteria. *Annual Review of Genetics* 37, 409–433.

Dechklar, M., Tiewsiri, K., Angsuthanasombat, C. and Pootanakit, K. (2011) Functional expression in insect cells of glycosylphosphatidylinositol-linked alkaline phosphatase from *Aedes aegypti* larval midgut: a *Bacillus thuringiensis* toxin receptor. *Insect Biochemistry and Molecular Biology* 41, 159–166.

Dorsch, J.A., Candas, M., Griko, N.B., Maaty, W.S.A., Midboe, E.G. *et al.* (2002) Cry1A toxins of *Bacillus thuringiensis* bind specifically to a region adjacent to the membrane proximal extracellular domain of Bt-R1 in *Manduca sexta*: involvement of a cadherin in the entomopathogenicity of *Bacillus thuringiensis. Insect Biochemistry and Molecular Biology* 32, 1025–1036.

Fabrick, J.A., Mathew, L.G., Tabashnik, B.E. and Li, X. (2011) Insertion of an intact CR1 retro-transposon in a cadherin gene liked with Bt resistance in the pink bollworm, *Pectinophora gossypiella. Insect Molecular Biology* 20, 651–665.

Ferré, J. and Van Rie, J. (2002) Biochemistry and genetics of insect resistance to *Bacillus thuringiensis. Annual Review of Entomology* 47, 501–533.

Flannagan, R.D., Yu, C.-G., Mathis, J.P., Meyer, T.E., Shi, X. *et al.* (2005) Identification cloning and expression of a Cry1Ab cadherin receptor from European corn borer, *Ostrinia nubilalis* (Hübner) (Lepidoptera: Crambidae). *Insect Biochemistry and Molecular Biology* 35, 33–40.

Flores-Escobar, B., Rodríguez-Magadan, H., Bravo, A., Soberón, M. and Gómez, I. (2013) Differential role of *Manduca sexta* aminopeptidase-N and alkaline phosphatase in the mode of action of Cry1Aa, Cry1Ab, and Cry1Ac toxins from *Bacillus thuringiensis. Applied and Environmental Microbiology* 79, 4543–4550.

Franklin, M.T., Nieman, C.L., Janmaat, A.F., Soberón, M., Bravo, A. *et al.* (2009) Modified *Bacillus thuringiensis* toxins and a hybrid *B. thuringiensis* strain counter greenhouse-selected resistance in *Trichoplusia ni. Applied and Environmental Microbiology* 75, 5739–5741.

Gahan, L.J., Gould, F. and Heckel, D.G. (2001) Identification of a gene associated with Bt resistance in *Heliothis virescens. Science* 293, 857–860.

Gahan, L.J., Pauchet, Y., Vogel, H. and Heckel, D.G. (2010) An ABC transporter mutation is correlated with insect resistance to *Bacillus thuringiensis* Cry1Ac toxin. *PLoS Genetics* 6(12): e1001248.

Garner, K.J., Hiremath, S., Lehtoma, K. and Valaitis, A.P. (1999) Cloning and complete sequence characterization of two gypsy moth aminopeptidase-N cDNAs, including the receptor for *Bacillus thuringiensis* Cry1Ac toxin. *Insect Biochemistry and Molecular Biology* 29, 527–535.

Girard. F., Vachon, V., Préfontaine, G., Marceau, L., Su, Y. *et al.* (2008) Cysteine scanning mutagenesis of α4, a putative pore forming helix of the *Bacillus thuringiensis* insecticidal toxin Cry1Aa. *Applied and Environmental Microbiology* 74, 2565–2572.

Gómez, I., Sánchez, J., Miranda, R., Bravo, A. and Soberón, M. (2002) Cadherin-like receptor binding facilitates proteolytic cleavage of helix α-1 in domain I and oligomer pre-pore formation of *Bacillus thuringiensis* Cry1Ab toxin. *FEBS Letters* 513, 242–246.

Gómez, I., Dean, D.H., Bravo, A. and Soberón, M. (2003) Molecular basis for *Bacillus thuringiensis* Cry1Ab toxin specificity: two structural determinants in the *Manduca sexta* Bt-R1 receptor interact with loops α-8 and 2 in domain II of Cy1Ab toxin. *Biochemistry* 42, 10482–10489.

Gómez, I., Arenas, I., Benitez, I., Miranda-Ríos, J., Becerril, B. *et al.* (2006) Specific epitopes of domains II and III of *Bacillus thuringiensis* Cry1Ab toxin involved in the sequential interaction with cadherin and aminopeptidase-N receptors in *Manduca sexta*. *The Journal of Biological Chemistry* 281, 34032–34039.

Gómez, I., Sanchez, J., Muñoz-Garay, C., Matus, V., Gill, S.S. *et al.* (2014) *Bacillus thuringiensis* Cry1A toxins are versatile-proteins with multiple modes of action: two distinct pre-pores are involved in toxicity. *Biochemical Journal* 459, 383–396.

Gunning, R.V., Dang, H.T., Kemp, F.C., Nicholson, I.C. and Moores, G.D. (2005) New resistance mechanism in *Helicoverpa armigera* threatens transgenic crops expressing *Bacillus thuringiensis* Cry1Ac toxin. *Applied and Environmental Microbiology* 71, 2558–2563.

Heckel, D.G. (2012) Learning the ABCs of Bt: ABC transporters and insect resistance to *Bacillus thuringiensis* provide clues to a crucial step in toxin mode of action. *Pesticide Biochemistry and Physiology* 104, 103–110.

Hernández-Martínez, P., Navarro-Cerrillo, G., Caccia, S., de Maagd, R.A., Moar, W.J. *et al.* (2010) Constitutive activation of the midgut response to *Bacillus thuringiensis* in Bt-resistant *Spodoptera exigua*. *PLoS ONE* 5(9): e12795.

Herrero, S., Gechev, T., Bakker, P.L., Moar, W.J. and de Maagd, R.A. (2005) *Bacillus thuringiensis* Cry1Ca-resistant *Spodoptera exigua* lacks expression of one of four aminopeptidase N genes. *BMC Genomics* 6:96.

Hua, G., Jurat-Fuentes, J.L. and Adang, M.J. (2004)

Fluorescent based assay establish *Manduca sexta* Bt-R1 cadherin as receptor for multiple *Bacillus thuringiensis* Cry1A toxins in *Drosophila* S2 cells. *Insect Biochemistry and Molecular Biology* 34, 193–202.

Ihara, H. and Himeno, M. (2008) Study of the irreversible binding of *Bacillus thuringiensis* Cry1Aa to brush border membrane vesicles from *Bombyx mori* midgut. *Journal of Invertebrate Pathology* 98, 177–183.

James, C. (2013) *Global Status of Commercialized Biotech/GM Crops: 2013*. ISAAA Brief No. 46, International Service for the Acquisition of Agri-Biotech Applications, Ithaca, New York.

Jiménez-Juárez, N., Muñoz-Garay, C., Gómez, I., Saab-Rincon, G., Damian-Almazo, J.Y. *et al.* (2007) *Bacillus thuringiensis* Cry1Ab mutants affecting oligomer formation are non toxic to *Manduca sexta* larvae. *The Journal of Biological Chemistry* 282, 21222–21229.

Jucovic, M., Walters, F.S., Warren, G.W., Palekar, N.V. and Chen, J.S. (2008) From enzyme to zymogen: engineering Vip2, an ADP-ribosyltransferase from *Bacillus cereus*, for conditional toxicity. *Protein Engineering, Design and Selection* 21, 631–638.

Jurat-Fuentes, J.L. and Adang, M.J. (2006) The *Heliothis virescens* cadherin protein expressed in *Drosophila* S2 cells functions as a receptor for *Bacillus thuringiensis* Cry1A but not Cry1Fa toxins *Biochemistry* 45, 9688–9695.

Jurat-Fuentes, J.L., Karumbaiah, L., Jakka, S.R.K., Ning, C., Liu, C. *et al.* (2011) Reduced levels of membrane-bound alkaline phosphatase are common to lepidopteran strains resistant to Cry toxins from *Bacillus thuringiensis*. *PLoS ONE* 6(3): e17606.

Keller, M., Sneh, B., Strizhov, N., Prudovsky, E., Regev, A. *et al.* (1996) Digestion of delta-endotoxin by gut proteases may explain reduced sensitivity of advanced instar larvae of *Spodoptera littoralis* to Cry1C. *Insect Biochemistry and Molecular Biology* 26, 365–373.

Khajuria, C., Buschman, L.L., Chen, M.S., Siegfried, B.D. and Zhu, K.Y. (2011) Identification of a novel aminopeptidase P-like gene (*OnAPP*) possibly involved in Bt toxicity and resistance in a major corn pest (*Ostrinia nubilalis*). *PLoS ONE* 6(8): e23983.

Kwa, M.S.G., de Maagd, R.A., Stiekema, W.J., Vlak, J.M. and Bosch, D. (1998) Toxicity and binding properties of the *Bacillus thuringiensis* delta-endotoxin Cry1C to cultured insect cells. *Journal of Invertebrate Pathology* 71, 121–127.

Liu, K., Zheng, B., Hong, H., Jiang, C., Peng, R. *et al.* (2004) Characterization of cultured insect cells selected by *Bacillus thuringiensis* crystal toxins. *In Vitro Cellular and Developmental Biology – Animal* 40, 312–317.

Lorence, A., Darszon, A., Díaz, C., Liévano, A., Quintero, R. *et al.* (1995) Delta-endotoxins induce cation channels in *Spodoptera frugiperda* brush border membrane in suspension and in planar lipid bilayers. *FEBS Letters* 360, 353–356.

Ma, G., Rahman, M.M., Grant, W., Schmidt, O. and Asgari, S. (2011) Insect tolerance to the crystal toxins Cry1Ac and Cry2Ab is mediated by binding of monomeric toxin to lipophorin glycolipids causing oligomerization and sequestration reactions. *Developmental and Comparative Immunology* 37, 184–192.

Martin, F.G. and Wolfersberger, M.G. (1995) *Bacillus thuringiensis* δ-endotoxin and larval *Manduca sexta* midgut brush border membrane vesicles act synergistically to cause very large increases in the conductance of planar lipid bilayers. *Journal of Experimental Biology* 198, 91–96.

Masson, L., Tabashnik, B.E., Liu, Y.-B., Brousseau, R. and Schwartz, J.L. (1999) Helix 4 of the *Bacillus thuringiensis* Cry1Aa toxin lines the lumen of the ion cannel. *The Journal of Biological Chemistry* 274, 31996–32000.

Mizuki, E., Ohba, M, Akao, T., Yamashita, S., Saitoh, H. *et al.* (1999) Unique activity associated with non-insecticidal *Bacillus thuringiensis* parasporal inclusions: *in vitro* cell-killing action on human cancer cells. *Journal of Applied Microbiology* 86, 477–486.

Morin, S., Biggs, R.W., Shriver, L., Ellers-Kirk, C., Higginson, D. *et al.* (2003) Three cadherin alleles associated with resistance to *Bacillus thuringiensis* in pink bollworm. *Proceedings of the National Academy of Sciences of the United States of America* 100, 5004–5009.

Nagamatsu, Y., Koiki, T., Sasaki, K., Yoshimoto, A. and Furukawa, Y. (1999) The cadherin like protein is essential to specificity determination and cytotoxic action of the *Bacillus thuringiensis* insecticidal Cry1Aa toxin. *FEBS Letters* 460, 385–390.

Oppert, B., Kramer, K.J., Beeman, R.W., Johnson, D. and McGaughey, W.H. (1997) Proteinase-mediated insect resistance to *Bacillus thuringiensis* toxins. *The Journal of Biological Chemistry* 272, 23473–23476.

Pacheco, S., Gómez, I., Arenas, I., Saab-Rincon, G., Rodríguez-Almazán, C. *et al.* (2009) Domain II loop 3 of *Bacillus thuringiensis* Cry1Ab toxin is involved in a "ping pong" binding mechanism with *Manduca sexta* aminopeptidase-N and cadherin receptors. *The Journal of Biological Chemistry* 284, 32750–32757.

Pardo-López, L., Soberón, M. and Bravo, A. (2013) *Bacillus thuringiensis* insecticidal 3-domain Cry toxins: mode of action, insect resistance and consequences for crop protection. *FEMS Microbiology Reviews* 37, 3–22.

Peyronnet, O., Vachon, V., Schwartz, J.L. and Laprade, R. (2001) Ion channels in planar lipid bilayers by the *Bacillus thuringiensis* toxin Cry1Aa in the presence of gypsy moth (*Lymantria dispar*) brush border membrane. *The Journal of Membrane Biology* 184, 45–54.

Potvin, L., Laprade, R. and Schwartz, J.L. (1998) Cry1Ac, a *Bacillus thuringiensis* toxin, triggers extracellular Ca^{2+} influx and Ca^{2+} release from intracellular stores in Cf1 cells (*Choristoneura fumiferana*, Lepidoptera). *Journal of Experimental Biology* 201, 1851–1858.

Prasifka, P.L., Rule, D.M., Storer, N.P., Nolting, S.P. and Hendrix, W.H. (2013) Evaluation of corn hybrids expressing Cry34Ab1/Cry35Ab1 and Cry3BbL against the western corn rootworm (Coleoptera: Chrysomelidae). *Journal of Economic Entomology* 106, 823–829.

Putheeranurak, T., Uawithya, P., Potvin, L., Angsuthanasombat, C. and Schwartz, J.L. (2004) Ion channels in planar lipid bilayers by the dipteran specific Cry4B *Bacillus thuringiensis* toxins and its α1–α5 fragment. *Molecular Membrane Biology* 21, 67–74.

Rahman, M.M., Roberts, H.L., Sarjan, M., Asgari, S. and Schmidt, O. (2004) Induction and transmission of *Bacillus thuringiensis* tolerance in the flour moth *Ephestia kuehniella*. *Proceedings of the National Academy of Sciences of the United States of America* 101, 2696–2699.

Rajagopal, R., Sivakumar, S., Agrawai, N., Malhotra, P. and Bhatnagar, R.K. (2002) Silencing of midgut aminopeptidase N of *Spodoptera litura* by double-stranded RNA establishes its role as *Bacillus thuringiensis* toxin receptor. *The Journal of Biological Chemistry* 277, 46849–46851.

Rausell, C., Muñoz-Garay, C., Miranda-CassoLuengo, R., Gómez, I., Rudiño-Piñera, E. *et al.* (2004) Tryptophan spectroscopy studies and black lipid bilayer analysis indicate that the oligomeric structure of Cry1Ab toxin from *Bacillus thuringiensis* is the membrane-insertion intermediate. *Biochemistry* 43, 166–174.

Rodríguez-Almazán, C., Zavala, L.E., Muñoz-Garay, C., Jiménez-Juárez, N., Pacheco, S. *et al.* (2009) Dominant negative mutants of *Bacillus thuringiensis* Cry1Ab toxin function as anti-toxins: demonstration of the role of oligomerization in toxicity. *PLoS ONE* 4(5): e5545.

Sangadala, S., Walters, F.S., English, L.H. and Adang, M.J. (1994) A mixture of *Manduca sexta* aminopeptidase and phosphatase enhances *Bacillus thuringiensis* insecticidal Cry1Ac toxin binding and [86]Rb[+]-K[+] efflux *in vitro. The Journal of Biological Chemistry* 269, 10088–10092.

Schwartz, J.L., Garneau, L., Savaria, D., Masson, L., Brousseau, R. *et al.* (1993) Lepidopteran specific crystal toxins from *Bacillus thuringiensis* form cation and anion selective channels in planar lipid bilayers. *The Journal of Membrane Biology* 132, 53–62.

Schwartz, J.L., Lu, Y.J., Söhnlein, P., Brousseau, R., Laprade, R. *et al.* (1997a) Ion channels formed in planar lipid bilayers by *Bacillus thuringiensis* toxins in the presence of *Manduca sexta* midgut receptors. *FEBS Letters* 412, 270–276.

Schwartz, J.L., Potvin, L., Chen, X.J., Brousseau, R., Laprade, R. *et al.* (1997b) Single-site mutations in the conserved alternative arginine region affect ionic channels formed by Cry1Aa a *Bacillus thuringiensis* toxin. *Applied and Environmental Microbiology* 63, 3978–3984.

Smedley, D.P., Armstrong, G. and Ellar, D.J. (1997) Channel activity caused by *Bacillus thuringiensis* delta-endotoxin preparation depends on the method of activation. *Molecular Membrane Biology* 14, 13–18.

Soberón, M., Pardo-López, L., López, I., Gómez, I., Tabashnik, B. *et al.* (2007) Engineering modified Bt toxins to counter insect resistance. *Science* 318, 1640–1642.

Tabashnik, B.E., Huang, F., Ghimire, M.N., Leonard, B.R., Siegfried, B.D. *et al.* (2011) Efficacy of genetically modified Bt toxins against insects with different mechanisms of resistance. *Nature Biotechnology* 29, 1128–1131.

Tanaka, S., Miyamoto K., Noda, H., Jurat-Fuentes, J.L., Yoshizawa, Y. *et al.* (2013) The ATP-binding cassette transporter subfamily C member 2 in *Bombyx mori* larvae is a functional receptor for Cry toxins from *Bacillus thuringiensis. The FEBS Journal* 280, 1782–1794.

Vachon, V., Prefontaine, G., Coux, F., Rang, C., Marceau, L. *et al.* (2002) Role of helix 3 in pore formation by *Bacillus thuringiensis* insecticidal toxin Cry1Aa. *Biochemistry* 41, 6178–6184.

Vachon. V., Laprade, R. and Schwartz, J.L. (2012) Current models of the mode of action of *Bacillus thuringiensis* insecticidal crystal proteins: a critical review. *Journal of Invertebrate Pathology* 111, 1–12.

Xu, X., Yu, L. and Wu, Y. (2005) Disruption of a cadherin gene associated with resistance to Cry1Ac δ-endotoxin of *Bacillus thuringiensis* in *Helicoverpa armigera. Applied and Environmental Microbiology* 71, 948–954.

Yang, Y., Zhu, Y.C., Ottea, J., Husseneder, C., Leonard, B.R. *et al.* (2011) Down regulation of a gene for cadherin, but not alkaline phosphatase, associated with Cry1Ab resistance in sugarcane borer *Diatraea saccharalis. PLoS ONE* 6: e25783.

Zhang, H., Wu, S., Yang, Y., Tabashnik, B.E. and Wu, Y. (2012a) Non-recessive Bt toxin resistance conferred by an intracellular cadherin mutation in field-selected populations of cotton bollworm. *PLoS ONE* 7(12): e53418.

Zhang, S., Cheng, H., Gao, Y., Wang, G., Liang, G. *et al.* (2009) Mutation of an aminopeptidase N gene is associated with *Helicoverpa armigera* resistance to *Bacillus thuringiensis* Cry1Ac toxin. *Insect Biochemistry and Molecular Biology* 39, 421–429.

Zhang, X., Candas, M., Grinko, N.B., Taussig, R. and Bulla, L.A. (2006) A mechanism of cell death involving an adenylyl cyclase/PKA signaling pathway is induced by the Cry1Ab toxin of *Bacillus thuringiensis. Proceedings of the National Academy of Sciences of the United States of America* 103, 9897–9902.

Zhang, X., Tiewsiri, K., Kain, W., Huang, L. and Wang, P. (2012b) Resistance of *Trichoplusia ni* to *Bacillus thuringiensis* toxin Cry1Ac is independent of alteration of the cadherin-like receptor for Cry toxins. *PLoS ONE* 7(5): e35991.

7

Roles of Insect Midgut Cadherin in Bt Intoxication and Resistance

Jeffrey A. Fabrick[1]* and Yidong Wu[2]

[1]USDA Agricultural Research Service, US Arid Land Agricultural Research Center, Maricopa, Arizona, USA; [2]Department of Entomology, Nanjing Agricultural University, Nanjing, People's Republic of China

Summary

Genetically engineered crops producing *Bacillus thuringiensis* (Bt) proteins for insect control target major insect pests. Bt crops have improved yields and their use reduces the risks associated with the application of conventional insecticides. However, the evolution of resistance to Bt toxins by target pests threatens the long-term success of such transgenic crops. Insects resistant to Bt Cry toxins have been selected in the laboratory and field-evolved resistance has been reported for economically important insects in several regions of the world. Although the mechanisms of resistance have not been reported for all cases, the most common mechanism involves changes in larval midgut target sites that probably reduce binding to Bt toxins. The binding of Cry toxins to midgut cadherin represents an important step in Bt intoxication for many insects and mutations in the cadherin gene can result in resistance to Bt toxins. Here, we highlight the roles that insect midgut cadherins play in Bt Cry intoxication and review cases where changes in cadherin are involved with resistance to Cry toxins. Furthermore, we emphasize the importance of understanding the underlying molecular basis of Bt intoxication and resistance to Bt, and the implications of fundamental knowledge for resistance management strategies.

7.1 Introduction

The cadherins constitute a large family of cell surface transmembrane proteins, conserved among metazoan organisms, that play fundamental roles in development, morphogenesis, cell sorting and migration, cell signalling and the maintenance of structural integrity (Hulpiau and van Roy, 2011). Specific functions associated with cadherins include embryonic cell layer separation and the formation of tissue boundaries, synapse formation, neuron growth and connectivity, the establishment of cell polarity, mechanotransduction, cell adhesion, cell signalling and physical homeostasis (Halbleib and Nelson, 2006; Brasch *et al.*, 2012).

Given their breadth of function, it follows that cadherins are produced in various shapes and sizes. Here, we provide an overview of vertebrate and invertebrate cadherins. We specifically highlight the roles that midgut cadherin receptors for *Bacillus thuringiensis* (Bt) toxins, i.e. BtR cadherins (or BtRs), play in Bt Cry intoxication. Furthermore, we review cases where changes in BtR cadherins are associated with resistance to Cry toxins and how a comprehensive understanding of Bt intoxication and resistance has implications for designing resistance management strategies to Bt crops and biopesticides.

* Corresponding author. E-mail address: jeff.fabrick@ars.usda.gov

7.2 The Cadherin Superfamily

Cadherins are defined by the presence of an extracellular region consisting of cadherin repeat (CR) domains, a transmembrane domain and an intracellular cytoplasmic (IC) domain. The extracellular region includes a variable number of CRs, which contain the conserved motifs that include alanine (Ala), arginine (Arg), asparagine (Asn), aspartate (Asp), glutamate (Glu), phenylalanine (Phe) and proline (Pro), as well as variable amino acids (X): Asp-Arg-Glu; Asp-X-Asn-Asp-Asn-Ala-Pro-X-Phe; and Asp-X-Asp (Takeichi, 1990). Each of these CRs consists of about 110 amino acids and forms a unique immunoglobulin-like β-sandwich fold (Fig. 7.1a). The interface between these CR domains harbours calcium-binding sites that are important for the adhesive properties of cadherin, whose name arose from the contraction of 'calcium-dependent adherent protein'. The calcium-binding sites rigidify the ectodomain structure, which is important for dimerization and protection from proteolysis (Takeichi, 1991) (Fig. 7.1b).

The cadherin superfamily includes hundreds of members across species ranging from unicellular organisms to vertebrates. Vertebrate cadherins are classified into several families, including the classical cadherins, desmosomal cadherins, truncated cadherins, protein kinase cadherins, protocadherins, FAT-like cadherins, the seven-pass transmembrane/Flamingo cadherins and the calsyntenins (Takeichi, 2007) (Fig. 7.2a).

Classical cadherins are single-span (pass) transmembrane proteins composed of five extracellular CR domains and a conserved cytoplasmic domain. They confer calcium-dependent cell–cell adhesion through adhesive dimerization both with cadherins within the same plasma membrane (*cis* dimers) and with those from opposing cells (*trans* dimers) (Fig. 7.1b) (Brasch *et al.*, 2012). These extracellular interactions mediate homophilic cell adhesion through intracellular interactions of the cadherin cytoplasmic domain with proteins and cytoskeletal components, such as actin filaments (Fig. 7.1c) (Brasch *et al.*, 2012).

While vertebrate classical cadherin function is well studied, members from other cadherin families perform numerous essential functions, including cell adhesion (desmosomal cadherins, truncated cadherins, protocadherins, FAT-like cadherins), tissue morphogenesis (protocadherins, FAT-like cadherins), cell signalling (protocadherins, protein kinase (PK) cadherins and calsyntenins), cell polarization (FAT-like cadherins, seven-pass transmembrane cadherins) and others (Morishita and Yagi, 2007). Hence, cadherins are likely to have originated to mediate mechanical cell–cell adhesion in simple organisms and their activities subsequently diversified, with parallel increases in their numbers and structural variations, to morphogenetic processes required in more complex organisms (Morishita and Yagi, 2007).

Despite their importance for multicellularity, cadherins are lacking from non-metazoan multicellular organisms (e.g. fungi and plants) (Abedin and King, 2008). Cadherins are found in the unicellular choanoflagellates, which are the earliest predecessors of metazoans (Abedin and King, 2008). Whereas mammalian genomes have over 100 genes belonging to the cadherin superfamily, basal metazoan organisms generally have fewer cadherin genes. For example, a lancelet, a sea anemone and a placozoan have 30, 16 and eight cadherin genes, respectively (Hulpiau and van Roy, 2011). The *Caenorhabditis elegans* genome encodes 13 cadherins, including representatives of the major cadherin families that are conserved between insects and vertebrates: the classic, FAT-like, Flamingo and calsyntenin families (Pettitt, 2005). Within insects, a total of 17 cadherin genes are found in *Drosophila melanogaster* (Hill *et al.*, 2001), 38 genes in *Anopheles gambiae* (Moita *et al.*, 2005), 19 in *Tribolium castaneum* (B.S. Oppert, Kansas, 2014, personal communication) and 13 in *Bombyx mori* (Duan *et al.*, 2010).

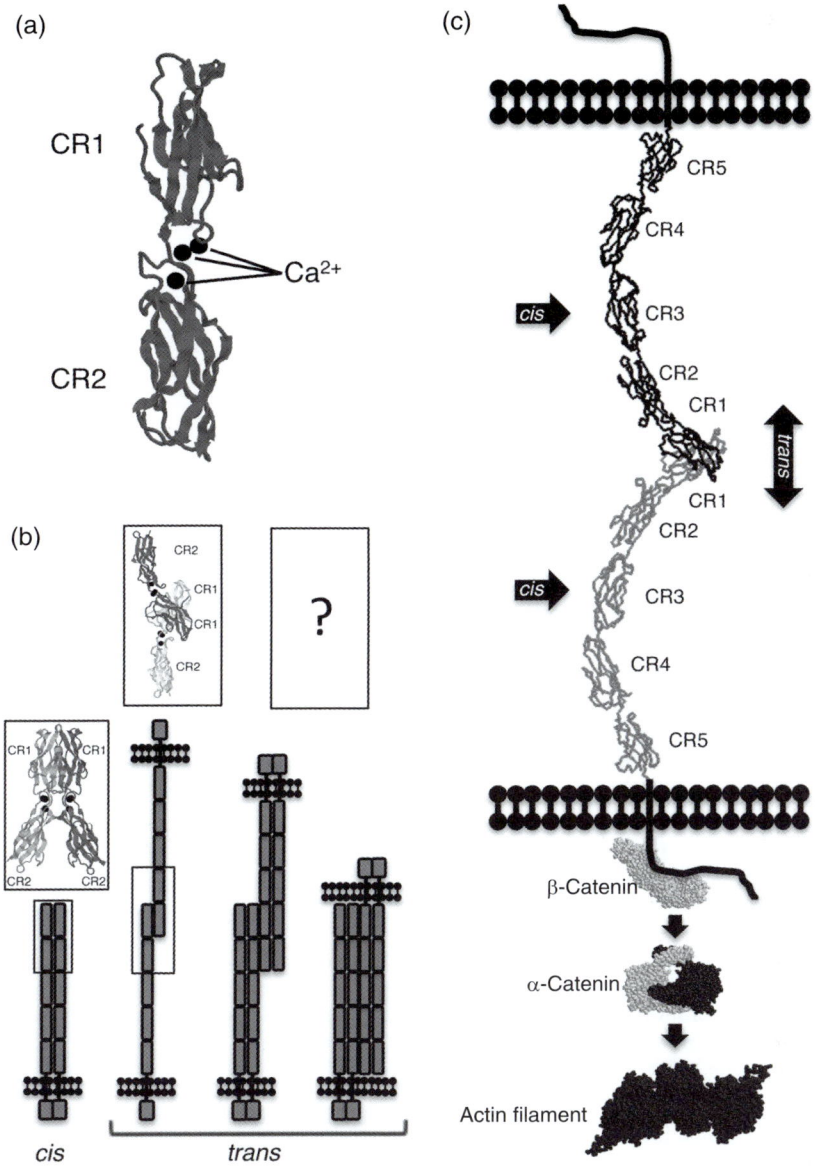

Fig. 7.1. Structure of the cadherin repeat domain and the architecture of classical cadherins forming adhesive dimers. (a) Cadherin repeats (CRs) assume an immunoglobulin-like β-sandwich fold; these occur in tandem and are separated by a linker region that harbours three calcium-binding sites (*Mus musculus* E-cadherin; Protein Data Bank (PDB) accession number (ID), 2QVF). (b) Vertebrate classical cadherins form *cis* and *trans* dimers. A *cis* dimer consists of two cadherin molecules laterally associated within the same plasma membrane (example shown is *M. musculus* E-cadherin; PDB ID, 1EDH), whereas cadherins that associate from opposing cells are *trans* dimers (example shown is *Gallus gallus* VE-cadherin; PDB ID, 3PPE). Both *cis* and *trans* interactions mediate homophilic cell adhesion. Although structures are not yet available (indicated by question mark), the extent of lateral overlap between the extracellular regions may increase the magnitude of the intercellular contact and simultaneously decrease the intercellular distance. (c) Cadherins bridge the intermembrane space between cells and the cytoplasmic domain of classical cadherins interacts with the cytoskeleton through specific adapter proteins, such as the catenins (example shown for *M. musculus* N-cadherin; PDB ID, 3Q2W).

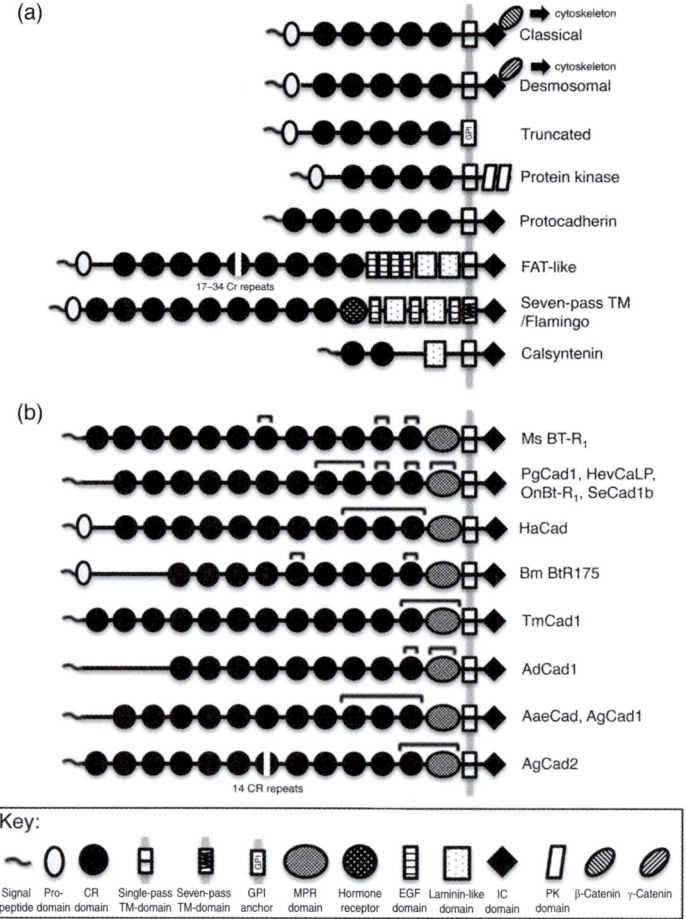

Fig. 7.2. Schematic domain organization of vertebrate cadherin families and BtR cadherins (BtRs) that function as *Bacillus thuringiensis* (Bt) Cry toxin receptors. (a) Cadherins are characterized by the presence of two or more extracellular cadherin repeat (CR) domains. Classical and desmosomal cadherins have five CR domains, but have distinct intracellular cytoplasmic (IC) domains that link through intermediate proteins to the intracellular cytoskeleton. Truncated cadherins resemble classical cadherins, but lack a transmembrane domain and attach to cell membrane via a glycosylphosphatidylinositol (GPI) anchor. Protein kinase cadherins have four CR domains and an intracellular protein kinase (PK) domain involved in signal transduction. Protocadherins are the largest subgroup of cadherins and can have 5–27 CRs. FAT-like cadherins have a large extracellular region composed of 17–34 tandem CRs, epidermal growth factor (EGF) motifs, and laminin domains. Seven-pass transmembrane (TM) cadherins are anchored in the membrane by a seven-pass transmembrane domain and have nine CRs, EGF, laminin-like and Flamingo hormone receptor-like domains. Calsyntenins have two CRs, a single-pass TM domain and a cytoplasmic domain. (b) BtRs are receptors of Cry Bt toxins from Lepidoptera (*Manduca sexta* BT-R₁, Ms BT-R₁; *Pectinophora gossypiella* cadherin 1, PgCad1; *Heliothis virescens* cadherin, HevCaLP; *Ostrinia nubilalis* Bt-R₁, OnBt-R₁; *Spodoptera exigua* cadherin 1, SeCad1b; *Helicoverpa armigera* cadherin, HaCad; *Bombyx mori* BtR175, Bm BtR175), Coleoptera (*Tenebrio molitor* cadherin 1, TmCad1; *Alphitobius diaperinus* cadherin 1, AdCad1) and Diptera (*Aedes aegypti* cadherin, AaeCad; *Anopheles gambiae* cadherin 1, AgCad1 or BT-R₃; *An. gambiae* cadherin 2, AgCad2). BtRs all have similar predicted domain organization, including 9–14 CRs, a membrane proximal region (MPR), a single TM domain and a cytoplasmic domain. One or more BtRs are listed together because they share predicted domain structures. Brackets indicate experimentally determined toxin binding regions (TBRs) and are shown only for the first named BtR (even though experimental evidence for indicated TBRs may not be available or in agreement between all co-listed BtRs). The wide vertical grey lines represent the cell membrane.

7.3 Insect BtR Cadherins

The Lepidoptera, Coleoptera and Diptera possess phylogenetically unique cadherins, otherwise known as 'cadherin-like protein', 'Cad', 'CADR' or '12-cadherin domain' that were co-opted by the bacterium *B. thuringiensis* as a receptor of Bt endotoxins (see below). As already noted, here we refer to these insect midgut cadherins as 'BtRs'. BtRs do not align well with established vertebrate cadherin families. Although the *Manduca sexta* BT-R$_1$ was previously classified as an atypical cadherin from the protocadherin group (Midboe *et al.*, 2003), the sequence conservation between BtRs and protocadherins is extremely low. Furthermore, protocadherins represent the largest family of vertebrate cadherins (Morishita and Yagi, 2007). Consequently, the initial characterization by Vadlamudi *et al.* (1995) and Candas *et al.* (2002) of Bt-R$_1$ as a 'new type of insect cadherin' remains accurate, with the BtRs representing a novel family of cadherins that evolved independently from other known cadherins.

7.3.1 Inherent BtR function

While the roles of BtRs as functional Bt receptors have been extensively studied (see below), the fundamental roles of these proteins in insect physiology are not known. Because of their shared sequence similarities with metazoan cadherins with described functions, it has been inferred that the BtRs have similar three-dimensional structures and therefore share roles in cell signalling, cell adhesion and the maintenance of cell integrity (Dorsch *et al.*, 2002; Midboe *et al.*, 2003). The *M. sexta* BT-R$_1$ has two cell-adhesion recognition sequences (HAV) and two cell-attachment sequences (RGD and LDV), from which it was inferred that it is a heterophilic surface adhesion protein with the capacity to bind other cadherins and/or other cell-adhesion protein partners (Dorsch *et al.*, 2002). However, no direct evidence for a role in cell adhesion has been demonstrated. Williams *et al.* (2011) provide indirect evidence of BtR involvement in the maintenance of cell–cell integrity. They showed that pink bollworm (*Pectinophora gossypiella*) larvae harbouring mutations in the midgut cadherin PgCad1 exhibited greater permeation by the phytochemical gossypol than larvae lacking such mutations. Thus, the altered cadherin protein may facilitate increased uptake of the plant defence compound, perhaps through compromised cell integrity.

Expression of BtRs in *B. mori*, *M. sexta*, *An. gambiae* and *Aedes aegypti* larvae is primarily within the midgut epithelial microvilli, and correlates with Bt Cry toxin binding sites. The *B. mori* cadherin BtR175 is most abundant on columnar cell microvilli in the posterior midgut, but not in the lateral membranes where cell–cell contacts are prevalent (Hara *et al.*, 2003). BT-R$_1$ is localized at the base of microvilli throughout the entire midgut and at the apex of microvilli in the middle and posterior regions of the midgut (Chen *et al.*, 2005). Both *An. gambiae* cadherin 1 (AgCad1) and *Ae. aegypti* cadherin (AaeCad) are located primarily within the microvilli apices in the posterior midgut as well as in the *Ae. aegypti* apical gastric caecae, which co-localizes with Cry toxin binding (Hua *et al.*, 2008; Chen *et al.*, 2009). The expression of BtRs at the apical ends of microvilli suggests that these proteins may play a role beyond cell–cell contact, and resembles that of the *Drosophila* Cad99C, which specifically localizes to the apices of the microvilli of ovarian follicle cells (D'Alterio *et al.*, 2005). In this case, whereas the overexpression of Cad99C leads to a dramatic increase of microvillus length and the overproduction of microvilli bundles, the loss of Cad99C results in short, abnormal microvilli (D'Alterio *et al.*, 2005). Even though a number of insect species harbour cadherin mutations, and RNA interference (RNAi) has been used to reduce the production of cadherin protein experiments have not demonstrated that changes in cadherins correlate with shortened/abnormal microvilli or altered junctions between cells.

Although BtRs are produced within midgut epithelium, only limited expression profiling and functional analyses have been performed. Hence, BtRs may play unknown, pleiotropic roles throughout development and in different tissues. For example, the pink bollworm cadherin PgCad1 may have an additional cryptic function in reproduction, as mRNA is present in testes, and mutations in PgCad1 affect apyrene sperm transfer (Carrière et al., 2009). What is more, Yang et al. (2009) showed that knock-down of Plutella xylostella cadherin by RNAi affects fecundity, egg hatch, pupal weight and adult eclosion, indicating cryptic functional roles of BtRs. Comprehensive expression profiling and functional analysis are needed to determine the inherent biochemical and physiological roles played by BtRs in insects.

7.3.2 Molecular and biochemical characteristics of BtRs

The genetic structures of lepidopteran BtR genes are highly conserved. Bel and Escriche (2006) showed that the genomic structure for midgut cadherins from Ostrinia nubilalis, Helicoverpa armigera and B. mori consists of 35 exons joined by 34 introns. The pink bollworm cadherin gene (PgCad1) shares a similar exon/intron structural pattern with 34 exons and 33 introns (Fabrick et al., 2011), though the 5′ untranslated region has not been assessed for the putative intron 1 identified in other lepidopterans (Bel and Escriche, 2006). These genes are large, encompassing 19.6, 20.0 and 41.8 kb for cadherins from O. nubilalis, H. armigera and B. mori, respectively (Bel and Escriche, 2006). PgCad1 and the M. sexta cadherin (BT-R$_1$) are similarly large, as ascertained by Southern blot analysis (Franklin et al., 1997; Fabrick et al., 2011). While there is thought to be a single functional copy of the midgut cadherin gene in Lepidoptera, Franklin et al. (1997) implicated the presence of a second and related pseudogene in M. sexta. Furthermore, two An. gambiae cadherin genes, AgCad1 (also named BT-R$_3$) and AgCad2, encode different BtR cadherin proteins that function as receptors for Cry4Ba and Cry11Ba, respectively (Hua et al., 2008, 2013; Ibrahim et al., 2013).

Transcriptional analysis of lepidopteran cadherin indicates that expression is primarily localized within the larval midgut epithelium (Midboe et al., 2003; Carrière et al., 2009). Cadherin transcripts and/or proteins are most abundant in the posterior end of the midgut, where columnar epithelial cells are abundant and have well-developed microvilli that extend into the gut lumen (Hara et al., 2003; Midboe et al., 2003; Chen et al., 2005, 2009; Aimanova et al., 2006; Hua et al., 2008).

BtRs are transmembrane proteins of 175–250 kDa composed of four domains: (i) an extracellular domain consisting of repetitive CRs, (ii) a membrane proximal region (MPR), (iii) a single transmembrane domain, and (iv) an IC domain (Fig. 7.2b). Several BtRs have been cloned and characterized, but all are predicted to have similar domain structures and vary primarily in the number of extracellular CR domains (Fig. 7.2b). M. sexta BT-R$_1$ was the first BtR to be cloned and characterized (Vadlamudi et al., 1995; Dorsch et al., 2002; Hua et al., 2004b). BtRs that function as Bt Cry receptors were subsequently identified from three insect orders, including: HevCaLP from Heliothis virescens (Gahan et al., 2001; Xie et al., 2005); OnBt-R$_1$ from O. nubilalis (Flannagan et al., 2005); HaCad from H. armigera (Wang et al., 2005; Zhang et al., 2012b); BtR175 from B. mori (Nagamatsu et al., 1998a; Atsumi et al., 2008); PgCad1 from P. gossypiella (Morin et al., 2003; Fabrick and Tabashnik, 2007); SeCad1b from Spodoptera exigua (Park and Kim, 2013; Ren et al., 2013; Chen et al., 2014); TmCad1 from Tenebrio molitor (Fabrick et al., 2009); AdCad1 from Alphitobius diaperinus (Hua et al., 2014); AgCad1 (also known as BT-R$_3$) and AgCad2 from An. gambiae (Hua et al., 2013; Ibrahim et al., 2013); and AaeCad from Ae. aegypti (Chen et al., 2009).

Cadherin binding sites for Cry toxins map primarily to the CR domains adjacent to the membrane-proximal regions of the protein (Nagamatsu et al., 1999; Gómez et al., 2001, 2002a, 2003; Dorsch et al., 2002; Hua et al., 2004a, 2008; Wang et al., 2005; Xie et al., 2005; Fabrick and Tabashnik, 2007; Fabrick et al., 2009; Ibrahim et al.,

2013) (Fig. 7.2b). Cry1A toxin binding to BT-R$_1$ is reported in three specific toxin binding regions (TBRs) or epitopes, including TBR1 in CR7, TBR2 in CR11 and TBR3 in CR12 (Dorsch et al., 2002; Gómez et al., 2002a; Hua et al., 2004a) (Fig. 7.2b). At least two TBRs that bind Cry1Aa are found in BtR175, including one in CR5 (that has a homologous sequence to TBR1 of BT-R$_1$) and one in CR9 (Nagamatsu et al., 1999; Gómez et al., 2001). Recombinant peptides from PgCad1 corresponding to CR8-CR9, CR10, CR11 and MPR bound to both activated Cry1Ac and Cry1Ac protoxin, indicating multiple TBRs in this BtR cadherin (Fabrick and Tabashnik, 2007). Among the coleopteran BtR cadherins, TBRs are found within the final CR domain and the MPR (Fabrick et al., 2009; Hua et al., 2014). AgCad, AaeCad and HaCad all have Cry toxin binding regions localized within their final three CR domains (CR9-CR11) (Wang et al., 2005; Chen et al., 2009; Ibrahim et al., 2013). Hence, the critical TBR required for the binding and toxicity of Cry toxins is predominantly found in the final CR domain (Fig. 7.2b). The proximity to the membrane surface suggests that BtRs play a role in concentrating toxin at the membrane surface, which promotes oligomerization and pore formation (see Section 7.3.3).

Chen et al. (2007) discovered that a recombinant cadherin fragment corresponding to the CR12-MPR from BT-R$_1$ increased Cry1 toxicity when fed to several species of lepidopterous larvae. The TBR3 in CR12 was shown to be critical for toxin binding and toxin synergy (Chen et al., 2007). Because the peptide alone was inactive against larvae and because it must be mixed with the Cry1 toxin before delivery, they hypothesized that the observed synergy resulted from an interaction of the toxin with the BtR fragment that either increased interaction with receptors or accelerated oligomerization and insertion into the membrane (Chen et al., 2007). It has since been demonstrated that the peptide fragment

increases toxin oligomerization of Cry toxins, which may increase the formation of pre-pore oligomers and enhance cytotoxicity through pore formation (Gómez et al., 2002b; Fabrick et al., 2009; Liu et al., 2009; Pacheco et al., 2009b; Peng et al., 2010).

Cry toxin potentiation by BtR peptides has been observed in lepidopterans, dipterans and coleopterans (Chen et al., 2007; Hua et al., 2008, 2013, 2014; Liu et al., 2009; Pacheco et al., 2009b; Park et al., 2009; Peng et al., 2010; Gao et al., 2011). Some cadherin fragments can synergize Cry toxins in species other than those from which the BtR was isolated (Chen et al., 2007; Park et al., 2009; Gao et al., 2011). In fact, the fragment corresponding to CR12-MPR from the T. molitor cadherin 1 (TmCad1) enhances mortality and/or reduces the time to kill larvae from several unrelated beetle species (Gao et al., 2011) and in a lepidopteran (Oppert et al., 2008/2013). These results suggest similarities in the Cry toxin mode of action are shared among different insect orders.

Post-translational processing of BtRs is probably important for cadherin protein folding and function. BtRs have membrane signal peptides that are important for membrane localization on the cell surface (Nagamatsu et al., 1998b; Wang et al., 2005), though proteolytic removal of the signal peptide may not occur in all cases (Pigott and Ellar, 2007). Some BtRs may be produced as proproteins that require proteolytic processing for maturation (Pigott and Ellar, 2007). Candas et al. (2002) showed that calcium directly influences the structural integrity of BT-R$_1$ as bound calcium protects the ectodomain from proteolytic cleavage. This suggests that calcium-dependent protection and proteolytic processing may regulate the functional properties of the cadherin ectodomain. Furthermore, the glycosylation of BtRs may be important for their inherent function, and most likely contributes to minor discrepancies between calculated and observed molecular weights (Vadlamudi et al., 1993; Nagamatsu et al., 1998a).

7.3.3 Role of BtRs in Cry Bt toxin mode of action

Insecticidal crystalline (Cry) proteins from *B. thuringiensis* are widely used as bio-pesticides or produced in genetically engineered crops to control some major insect pests. Commercial Cry toxin-based products have been available for >50 years, despite limited knowledge of their modes of action. Recently, progress in characterizing Cry intoxication has implicated at least two mechanisms by which insect midgut cells are specifically killed. These include the sequential/pore formation model and the signal transduction model. Both models require the action of BtRs (see also Chapter 6, Bravo *et al.*).

The pore formation model

Although Cry toxins are used to target lepidopteran, dipteran and coleopteran pests, the pore formation mode of action is best characterized for Cry1 toxin in lepidopterous larvae (Gómez *et al.*, 2014) (Fig. 7.3). For Cry1 intoxication, the protein must be ingested either as an insoluble Cry protoxin contained as protein inclusion body within the Bt bacterium or as a soluble recombinant protein produced by a trans-genic Bt plant. As the protoxin passes through the insect alimentary canal, it is solubilized within the alkaline environment and proteolytically activated by endogenous endopeptidases present in the midgut (Choma *et al.*, 1990). Activated monomeric Cry toxin binds with relatively low affinity to glycosylphosphatidylinositol (GPI)-anchored aminopeptidase N (APN) or alkaline phos-phatase (ALP) proteins in the midgut epithelium (Gómez *et al.*, 2006; Pacheco *et al.*, 2009a). These binding interactions con-centrate the activated toxin on the membrane where it binds with high affinity to the BtR. This interaction with cadherin facilitates proteolytic removal of helix α-1 of domain I from the toxin and promotes the formation of toxin oligomeric structures, known as pre-pore oligomers (Gómez *et al.*, 2002b; Atsumi *et al.*, 2008). Pre-pore oligomers bind with high affinity to the GPI-anchored receptors (APN and ALP), which are thought to aid the insertion of the pre-pore structure into the membrane, thereby causing pore formation and cell lysis (Pardo-López *et al.*, 2006). A recent extension to the pore formation model proposes the ABC transporter protein ABCC2 aids in irreversible insertion of the pre-pore oligomer into the membrane through co-ordinated opening and closing of the ABC transporter channel (Gahan *et al.*, 2010).

The signal transduction model

In the signal transduction model, the toxicity of Cry proteins is caused by activation of an intracellular pathway that leads to cell death (Zhang *et al.*, 2006, 2008; Ibrahim *et al.*, 2013) (Fig. 7.3). In this model, the steps leading up to cadherin binding are identical to those of the pore formation model, except that the Cry1A toxin monomer binds to the cadherin receptor and activates an intracellular Mg^{2+}-dependent signal transduction pathway (Fig. 7.3). The univalent toxin binding to cadherin stimulates heterotrimeric G protein and adenylyl cyclase (also known as adenylate cyclase) to increase intracellular cAMP. cAMP activates a protein kinase A that is thought to propagate a secondary messenger system that results in changes to the cytoskeleton and ion fluxing, ultimately leading to cell death (Zhang *et al.*, 2005, 2006).

Due to the technical limitations of work-ing with intact insect midgut epithelium, neither model has been definitively validated within actual insect midgut tissue. The signal transduction pathway is based exclusively on experiments performed in non-midgut insect cell cultures and the induction of cell death signalling by Cry toxins and has not been verified *in vivo*. Further, while more empirical data has been amassed in support of the pore formation model, no direct evidence have been provided for intact living cells or insects. So it is possible that neither model occurs *in vivo* and that variations or new undiscovered modes of action exist. Alternatively, both models may function

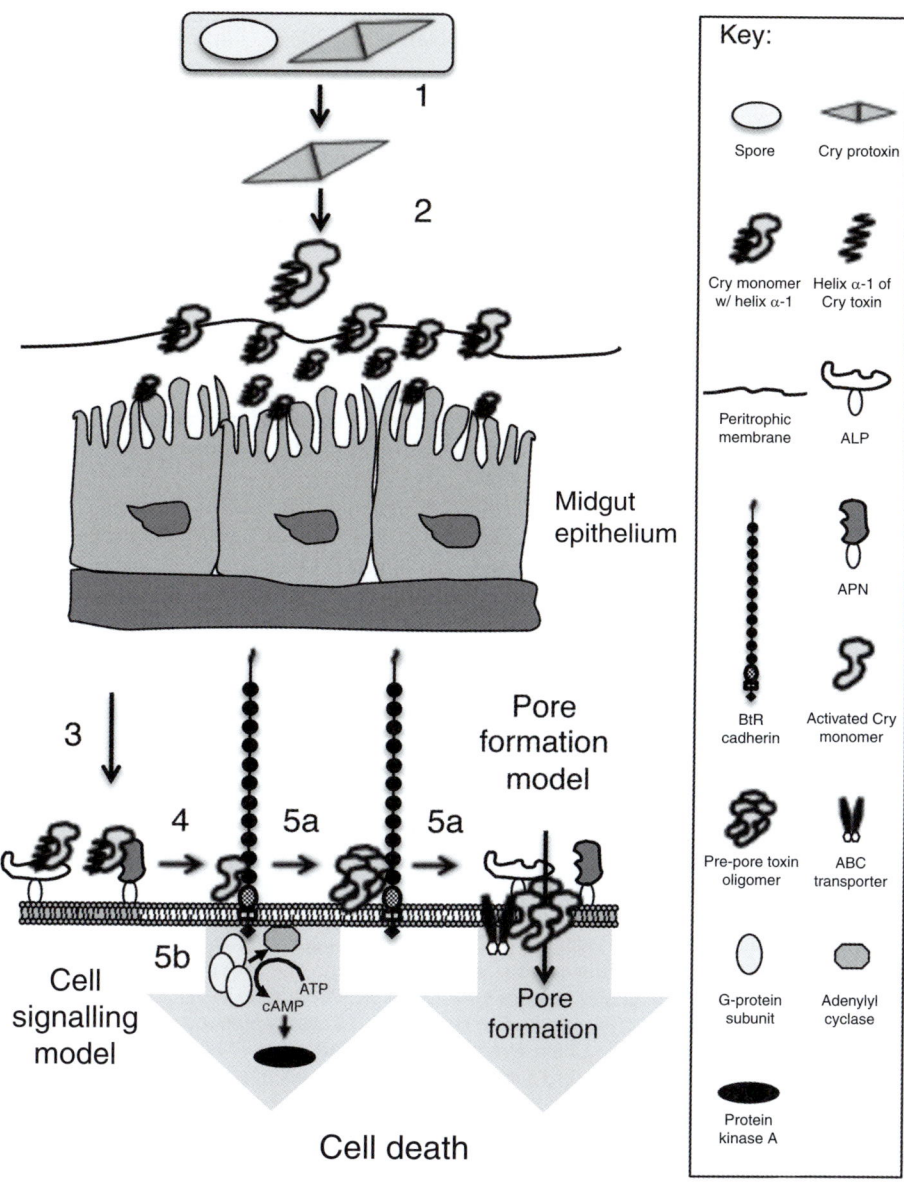

Fig. 7.3. The role of lepidopteran BtRs (*Bacillus thuringiensis* Cry toxin cadherin receptors) in the mode of action of Cry1 toxins. The five steps for Bt intoxication (1–5) include the following: (1) Ingestion of Bt Cry protoxin either solubilized from the Bt bacterium or from a transgenic Bt plant; (2) proteolytic activation of the protoxin by midgut endopeptidases; (3) binding of monomeric Cry1 toxin to GPI (glycosylphosphatidylinositol)-anchored aminopeptidase N (APN) and/or alkaline phosphatase (ALP); (4) binding of monomeric Cry toxin to BtR and proteolytic removal of the domain I α-helix; and (5), either (5a) oligomerization of Cry monomers and binding of the oligomer to GPI-anchored APN or ALP, which leads to insertion of the pre-pore toxin oligomer via ABC transporter into the cell membrane to form pores and ultimately leads to cell death (sequential/pore formation model), or (5b) binding of monomeric Cry toxin to cadherin, which activates an intracellular signal transduction pathway that leads to cell death (cell signalling model) (5b).

simultaneously and some form of both mechanisms may be responsible for Cry intoxication. Hence, deciphering the mechanisms by which Bt toxins function is of utmost importance in order to implement better 'biorationale' (biologically rational) pest control tools, as well as to delay the evolution of resistance to the currently available Bt technologies.

7.4 BtRs and Resistance to Bt Toxins

7.4.1 BtR mutations

Although several mechanisms of resistance to Bt toxins are known, the most common type involves mutations that reduce the binding of Bt toxins to larval midgut proteins (Caccia *et al.*, 2010; Jurat-Fuentes *et al.*, 2011). Cadherin mutations that confer resistance to Bt toxins have been isolated from *H. virescens* (Gahan *et al.* 2001), *P. gossypiella* (Morin *et al.*, 2003; Fabrick and Tabashnik, 2012; Fabrick *et al.*, 2014) and *H. armigera* (Xu *et al.*, 2005; Yang *et al.*, 2007;

Zhao *et al.*, 2010; Zhang *et al.*, 2011, 2012a,b; Nair *et al.*, 2013) (Fig. 7.4).

Genetic mapping indicated that a BtR mutation from the YHD2 strain of *H. virescens* was genetically linked with Bt resistance (Gahan *et al.*, 2001). A tight linkage was observed between resistance to Cry1Ac in the *H. virescens* YHD2 strain and the resistance (*r*) allele *r1* of *HevCaLP* (Gahan *et al.*, 2001). The insertion of an LTR (long terminal repeat)-type retrotransposon disrupts *HevCaLP* and causes the introduction of a premature stop codon in the cadherin coding sequence. The truncated HevCaLP *r1* protein probably lacks regions important for toxin binding and membrane localization (Fig. 7.4).

Seventeen cadherin resistance alleles (*r1–r17*) are known from laboratory and field-collected *H. armigera* in China and India (Fig. 7.4). The *r1* mutation of *HaCad* (synonymous with *Ha_BtR*) was identified from the laboratory-selected GYBT strain (Xu *et al.*, 2005), and later from three resistant strains isolated from the field-selected Anyang population in Henan Province of northern

Fig. 7.4. Predicted BtR (*Bacillus thuringiensis* toxin cadherin receptor) proteins from *Heliothis virescens*, *Helicoverpa armigera* and *Pectinophora gossypiella* cadherin alleles. A single resistance (*r1*) allele from *H. virescens* in the USA was the first allele producing a mutant cadherin protein (HevCaLP) that was mapped and genetically linked with resistance to Cry1Ac Bt toxin (Gahan *et al.*, 2001). A total of 17 resistance alleles encoding mutated/truncated *H. armigera* cadherin (HaCad) proteins have been isolated from *H. armigera* in northern China and western India (Xu *et al.*, 2005; Yang *et al.*, 2007; Zhao *et al.*, 2010; Zhang *et al.*, 2012a,b; Nair *et al.*, 2013). The HaCad *r1–r15* alleles produce cadherin protein isoforms obtained from *H. armigera* in China. Products from HaCad *r10* to *r14* (indicated by *) differ from wild type HaCad protein only by amino acid substitutions, as the alleles have no premature stop codons, deletions or insertions. HaCad *r16* and *r17*, previously named by Nair *et al.* (2013), are here renamed from *r9* and *r10*, respectively. The HaCad *s* and *r16* protein sequences differ by three amino acid substitutions (R1296T, arginine to threonine at position 1296; R1308K, arginine to lysine at 1308; A1313S, alanine to serine at 1313) and *r16* has an insertion of a single codon encoding N1341 (asparagine at 1341, indicated by ★). Twelve cadherin resistance alleles (*r1–r12*) encoding mutant *P. gossypiella* cadherin (PgCad) proteins were isolated in the USA and western India (Morin *et al.*, 2003; Tabashnik *et al.*, 2004; 2005a; Carrière *et al.*, 2006; Fabrick and Tabashnik, 2012; Fabrick *et al.*, 2014). Four of these cadherin alleles (*r1–r4*) are genetically linked with resistance to Cry1Ac in laboratory-selected strains from Arizona and another eight (*r5–r12*) are associated with resistance to Cry1Ac Bt cotton in fields from western India. Nineteen different isoforms (*r5A, r5B,* etc.) of mutant *PgCad1* alleles *r5–r12* are shown. Predicted proteins are shown for cDNA of the susceptible (*s*) allele and resistant (*r*) mutant alleles. The amino-terminal membrane signal sequence, putative prodomain, cadherin repeats domain (CR), membrane proximal region (MPR), transmembrane (TM) region, and cytoplasmic domain (IC) are shown. Truncated structures indicate proteins predicted from cDNA with premature stop codons. Grey indicates missing regions of proteins caused by deletions. The wide vertical grey lines represent the cell membrane.

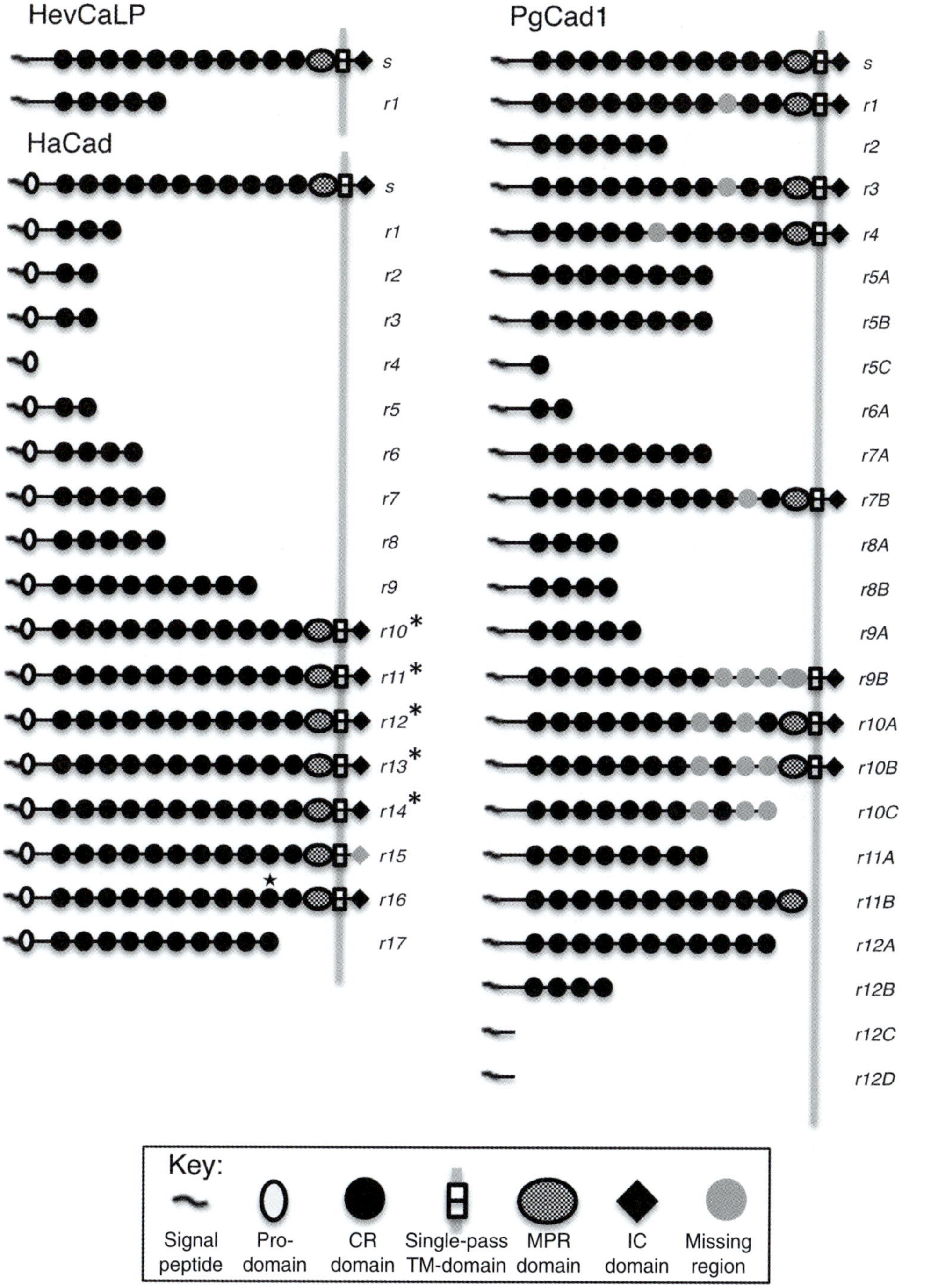

HevCaLP
s
r1
HaCad
s
r1
r2
r3
r4
r5
r6
r7
r8
r9
r10*
r11*
r12*
r13*
r14*
r15
r16
r17
PgCad1
s
r1
r2
r3
r4
r5A
r5B
r5C
r6A
r7A
r7B
r8A
r8B
r9A
r9B
r10A
r10B
r10C
r11A
r11B
r12A
r12B
r12C
r12D
Key:
Signal peptide
Pro-domain
CR domain
Single-pass TM-domain
MPR domain
IC domain
Missing region

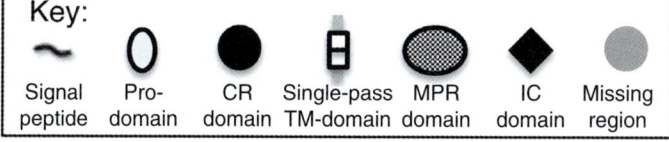

China (Zhang *et al.*, 2012a). The *r1* allele features a deletion in the *HaCad* gene corresponding to exons 8–24, which causes a frame shift and the introduction of a premature stop codon in the coding sequence (Xu *et al.*, 2005; Yang *et al.*, 2006). The coding sequences of alleles *r2*, *r3* and *r5–r8* are disrupted by the insertion of putative transposable DNA elements and all encode a truncated HaCad protein (Yang *et al.*, 2007; Zhao *et al.*, 2010) (Fig. 7.4). Whereas *HaCad r9* is produced by incorrect mRNA splicing between exons 24 and 25 (Zhang et al., 2012a), alleles *r4* and *r10–r14* have mutations causing amino acid substitutions (Zhao *et al.*, 2010, Zhang *et al.*, 2012a). Nair *et al.* (2013) identified two *r* alleles from *H. armigera* collected from western India. To avoid repetitive nomenclature, here we rename alleles *r9* and *r10* of Nair *et al.* (2013) to *r16* and *r17*, respectively. The coding sequence of the *r16* allele differs from that of the wild type susceptible (*s*) allele by the insertion of a single codon producing an additional amino acid residue at position 1340 in CR10 (Nair *et al.*, 2013). The *r17* mRNA encodes a truncated HaCad protein missing 413 amino acid residues beyond CR10 (Nair *et al.*, 2013). The *r15* allele isolated from field-selected *H. armigera* is unique because it is a non-recessive mutation resulting from the loss of 55 amino acids in the intracellular domain of HaCad (Zhang *et al.*, 2012b). All other cadherin mutations from *H. armigera* are recessive and affect extracellular regions. The *r15* mutation indicates that the cytoplasmic domain is important for Cry intoxication and therefore supports contributions by both the pore formation and signal transduction mechanisms in the efficacy of Bt toxins (Jurat-Fuentes and Adang, 2006; Zhang *et al.*, 2012b).

A total of 12 *PgCad1* resistance alleles were isolated in *P. gossypiella* from the USA and India (Morin *et al.*, 2003; Fabrick and Tabashnik, 2012; Fabrick *et al.*, 2014) (Fig. 7.4). The four US mutations (*r1–r4*) are genetically linked with resistance to Bt cotton producing Cry1Ac in laboratory-selected strains from Arizona (Morin *et al.*,

2003; Tabashnik *et al.*, 2004, 2005a; Carrière *et al.*, 2006; Fabrick and Tabashnik, 2012). While the *r2* allele has a deletion of 202 bp that introduces a premature stop codon, the others (*r1*, *r3* and *r4*) have only a single deletion of 24, 126 and 15 bp, respectively (Morin *et al.*, 2003; Fabrick and Tabashnik, 2012). The *r2* deletion results in a protein truncated following the CR6 domain; the protein products from the *r1*, *r3* and *r4* alleles lack 8, 42, and five amino acids, respectively. The 126 bp deletion from the *r3* allele results from the insertion of a non-LTR chicken-repeat retrotransposon (*CR1-1_Pg*) that causes splicing out of exon 21 from mRNA (Fabrick *et al.*, 2011). Recently, eight *PgCad1* resistance alleles (*r5–r12*) were isolated from *P. gossypiella* collected from Bt cotton fields in India (Fabrick *et al.*, 2014). From these eight alleles, a total of 19 transcript isoforms each containing a premature stop codon, a deletion of at least 99 bp, or both, were identified. Seven of the eight disrupted alleles involved alternative splicing of mRNA, which represents a novel genetic mechanism by which pests generate genetic diversity and accelerate the evolution of resistance to Bt crops (Fabrick *et al.*, 2014).

7.4.2 Reduced transcription/expression

Cadherin-based resistance to Cry1A toxin is associated with altered gene expression (due to spontaneous mutations and those introduced by the transposition of mobile DNA) and changes to mRNA (caused by alternative splicing and downregulation of transcription). Downregulation of BtRs is implicated in at least two cases of resistance. First, Bonin *et al.* (2009) found polymorphisms in *AaeCad* of *Ae. aegypti* that are consistent with positive selection for Bt resistance. Furthermore, cadherin gene expression was lower in a resistant strain of *Ae. aegypti* than in a susceptible strain (Bonin *et al.*, 2009). In the sugarcane borer, *Diatraea saccharalis*, Yang *et al.* (2011) showed that cadherin transcript levels were lower in a Cry1Ab-resistant strain than in a Cry1Ab-susceptible strain. Experiments using RNAi to knock down BtR expression have demonstrated a corresponding decrease in susceptibility to Cry intoxication in several insects (Soberón *et al.*, 2007; Fabrick *et al.*, 2009; Yang *et al.*, 2011; Hua *et al.*, 2013; Park and Kim, 2013; Ren *et al.*, 2013; Chen *et al.*, 2014). These results support the role of BtRs as functional receptors of Cry toxins and implicate reduced expression as a mechanism of resistance.

7.4.3 DNA screening of BtR mutations

The monitoring of pest populations for field-evolved resistance to Bt crops is an important component of management strategies to delay the onset of resistance. Laboratory bioassays of target pests collected from Bt crops and surrounding non-Bt hosts are the primary tools used to detect field-evolved resistance (Tabashnik *et al.*, 2009).

A complementary approach to laboratory bioassays is DNA-based molecular monitoring for known resistance marker genes. PCR-based DNA screening for BtR resistance alleles was implemented for three major insect pests in Bt cotton, including *H. virescens* (Gahan *et al.*, 2007) and *P. gossypiella* (Morin *et al.*, 2004) in the USA and *H. armigera* in China (Zhang *et al.*, 2013). The screening of more than 7000 field-collected *H. virescens* failed to detect a single *r1* HevCaLP allele (Gahan *et al.*, 2007). Similarly, in the southwest USA, beginning in 2001, the screening of approximately 10,000 *P. gossypiella* for *r1*, *r2* or *r3* cadherin alleles indicated no resistance alleles in the field (Tabashnik *et al.*, 2005b, 2006, 2010; B.E. Tabashnik, Arizona, 2014, personal communication). With *P. gossypiella*, PCR screening results were consistent with a low resistance allele frequency in the field as determined from diet bioassays, indicating that resistance to Bt cotton did not increase (Tabashnik *et al.*, 2005b, 2006, 2010). In *H. armigera* from northern China, DNA-based screening to detect the *r15* cadherin allele identified three field-collected specimens (out of 876) that were heterozygous at the *HaCad* locus (Zhang *et al.*, 2013). However,

the recent discovery of diverse cadherin mutations associated with resistance to Cry1Ac in field-selected populations of *P. gossypiella* in India (Fabrick *et al.*, 2014) and of *H. armigera* in China (Xu *et al.*, 2005; Yang *et al.*, 2006, 2007; Zhao *et al.*, 2010; Zhang *et al.*, 2012a) implies that the monitoring of resistance in some populations by screening cadherin DNA for specific resistance alleles may be ineffective.

The molecular and biochemical mechanisms of resistance to Bt toxins and the crops that produce them can be highly variable. Comprehensive examination of these mechanisms requires pest-specific investigation of biology, genetics, ecology, biochemistry and physiology from specific isolation sites. This requires enormous investment of time and capital resources, which is often impractical or unattainable, although technological and economic advances in functional genomics and next-generation sequencing offers promise for genome- or transcriptome-wide genetic screening of samples and more comprehensive detection and evaluation of pesticide resistance traits. Nevertheless, the significance of such technical advances in the screening of pest resistance to Bt crops remains uncertain.

7.5 Conclusion

Bt crops are planted on millions of hectares throughout the world. As a result, their benefits are realized by an ever-increasing populace. However, field-evolved resistance to Bt crops continues to threaten these long-term benefits (Tabashnik *et al.*, 2013). Thus, as reliance on Bt crops increases, the need to preserve existing technologies and/or find new complementary approaches is also intensified. In order to improve current strategies to delay resistance, we must better understand how Bt crops work and how resistance evolves. We are only now making strides towards understanding the intricacies of the mode of action of Bt and of Bt resistance at the molecular and biochemical level. For example, BtRs appear to play a central role in Bt's mode of action and Bt resistance in several insects. Without such fundamental knowledge, advances such as the development of modified or 'Mod' toxins (reviewed in Chapter 14, Soberón et al.) and the identification of cadherin resistance alleles would not have been possible.

Fundamental questions remain on the mode of action of Bt, the effective implementation and preservation of Bt technologies and the evolution of pest resistance. Some questions lie within the apparent differences in mechanisms between insect species and can only be answered by in-depth knowledge of each specific case. These include the following: What roles do cadherin, ABCC, APN, and ALP play and do they play more prevalent roles in some insects than others? Will cadherin-based resistance or changes to steps in the Bt mode of action become prevalent in Bt fields? Can we implement strategies to delay the evolution of resistance? How will our understanding of the basis of resistance ultimately have an impact on efforts to design resistance management strategies?

In developed countries, the technology and infrastructure exist to design, build and implement transgenic strategies for pest control, but a major challenge is to find tools with unique modes of action that complement or replace the current generation of Bt crops. There are certainly untapped resources of novel plant protectant proteins (such as Bt toxins) yet to be found, but society also benefits from action to extend the life of safe and effective pest management tools, such as Bt crops, and to delay the evolution of resistance.

Acknowledgements

We thank Dale Spurgeon for helpful comments. Mention of trade names or commercial products in this article is solely for the purpose of providing specific information and does not imply recommendation or endorsement by the US Department of Agriculture (USDA). The USDA is an equal opportunity provider and employer.

References

Abedin, M. and King, N. (2008) The premetazoan ancestry of cadherins. *Science* 319, 946–948.

Aimanova, K.G., Zhuang, M. and Gill, S.S. (2006) Expression of Cry1Ac cadherin receptors in insect midgut and cell lines. *Journal of Invertebrate Pathology* 92, 178–187.

Atsumi, S., Inoue, Y., Ishizaka, T., Mizuno, E., Yoshizawa, Y. et al. (2008) Location of the *Bombyx mori* 175kDa cadherin-like protein-binding site on *Bacillus thuringiensis* Cry1Aa toxin. *The FEBS Journal* 275, 4913–4926.

Bel, Y. and Escriche, B. (2006) Common genomic structure for the Lepidoptera cadherin-like genes. *Gene* 381, 71–80.

Bonin, A., Paris, M., Tetreau, G., David, J.P. and Despres, L. (2009) Candidate genes revealed by a genome scan for mosquito resistance to a bacterial insecticide: sequence and gene expression variations. *BMC Genomics* 10:551.

Brasch, J., Harrison, O.J., Honig, B. and Shapiro, L. (2012) Thinking outside the cell: how cadherins drive adhesion. *Trends in Cell Biology* 22, 299–310.

Caccia, S., Hernández-Rodríguez, C.S., Mahon, R.J., Downes, S. et al. (2010) Binding site alteration is responsible for field-isolated resistance to *Bacillus thuringiensis* Cry2A insecticidal proteins in two *Helicoverpa* species. *PLoS ONE* 5(4): e9975.

Candas, M., Francis, B.R., Griko, N.B., Midboe, E.G. and Bulla, L.A. (2002) Proteolytic cleavage of the developmentally important cadherin BT-R1 in the midgut epithelium of *Manduca sexta*. *Biochemistry* 41, 13717–13724.

Carrière, Y., Ellers-Kirk, C., Biggs, R.W., Nyboer, M.E., Unnithan, G.C. et al. (2006) Cadherin-based resistance to *Bacillus thuringiensis* cotton in hybrid strains of pink bollworm: fitness costs and incomplete resistance. *Journal of Economic Entomology* 99, 1925–1935.

Carrière, Y., Showalter, A.M., Fabrick, J.A., Sollome, J., Ellers-Kirk, C. et al. (2009) Cadherin gene expression and effects of Bt resistance on sperm transfer in pink bollworm. *Journal of Insect Physiology* 55, 1058–1064.

Chen, J., Brown, M.R., Hua, G. and Adang, M.J. (2005) Comparison of the localization of *Bacillus thuringiensis* Cry1A δ-endotoxins and their binding proteins in larval midgut of tobacco hornworm, *Manduca sexta*. *Cell and Tissue Research* 321, 123–129.

Chen, J., Hua, G., Jurat-Fuentes, J.L., Abdullah, M.A. and Adang, M.J. (2007) Synergism of *Bacillus thuringiensis* toxins by a fragment of a toxin-binding cadherin. *Proceedings of the*

National Academy of Sciences of the United States of America 104, 13901–13906.

Chen, J., Aimanova, K.G., Fernandez, L.E., Bravo, A., Soberón, M. et al. (2009) Aedes aegypti cadherin serves as a putative receptor of the Cry11Aa toxin from Bacillus thuringiensis subsp. israelensis. Biochemical Journal 424, 191–200.

Chen, R.-R., Ren, X.-L., Han, Z.-J., Mu, L.-L., Li, G.-Q. et al. (2014) A cadherin-like protein from the beet armyworm Spodoptera exigua (Lepidoptera: Noctuidae) is a putative Cry1Ac receptor. Archives of Insect Biochemistry and Physiology 86, 58–71.

Choma, C.T., Surewicz, W.K., Carey, P.R., Pozsgay, M., Raynor, T. et al. (1990) Unusual proteolysis of the protoxin and toxin from Bacillus thuringiensis. Structural implications. European Journal of Biochemistry 189, 523–527.

D'Alterio, C., Tran, D.D.D., Au Yeung, M.W.Y., Hwang, M.S.H., Li, M.A. et al. (2005) Drosophila melanogaster Cad99C, the orthologue of human Usher cadherin PCDH15, regulates the length of microvilli. Journal of Cell Biology 171, 549–558.

Dorsch, J.A., Candas, M., Griko, N.B., Maaty, W.S.A., Midboe, E.G. et al. (2002) Cry1A toxins of Bacillus thuringiensis bind specifically to a region adjacent to the membrane-proximal extracellular domain of BT-R1 in Manduca sexta: involvement of a cadherin in the entomopathogenicity of Bacillus thuringiensis. Insect Biochemistry and Molecular Biology 32, 1025–1036.

Duan, J., Li, R., Cheng, D., Fan, W., Zha, X. et al. (2010) SilkDB v2.0: a platform for silkworm (Bombyx mori) genome biology. Nucleic Acids Research 38, D453–D456.

Fabrick, J.A. and Tabashnik, B.E. (2007) Binding of Bacillus thuringiensis toxin Cry1Ac to multiple sites of cadherin in pink bollworm. Insect Biochemistry and Molecular Biology 37, 97–106.

Fabrick, J.A. and Tabashnik, B.E. (2012) Similar genetic basis of resistance to Bt toxin Cry1Ac in boll-selected and diet-selected strains of pink bollworm. PLoS ONE 7(4): e35658.

Fabrick, J., Oppert, C., Lorenzen, M.D., Morris, K., Oppert, B. et al. (2009) A novel Tenebrio molitor cadherin is a functional receptor for Bacillus thuringiensis Cry3Aa toxin. The Journal of Biological Chemistry 284, 18401–18410.

Fabrick, J.A., Mathew, L.G., Tabashnik, B.E. and Li, X. (2011) Insertion of an intact CR1 retrotransposon in a cadherin gene linked with Bt resistance in the pink bollworm, Pectinophora gossypiella. Insect Molecular Biology 20, 651–665.

Fabrick, J.A., Ponnuraj, J., Singh, A., Tanwar, R.K., Unnithan, G.C. et al. (2014) Alternative splicing and highly variable cadherin transcripts associated with field-evolved resistance of pink bollworm to Bt cotton in India. PLoS One 9(5): e97900.

Flannagan, R.D., Yu, C.G., Mathis, J.P., Meyer, T.E., Shi, X. et al. (2005) Identification, cloning and expression of a Cry1Ab cadherin receptor from European corn borer, Ostrinia nubilalis (Hübner) (Lepidoptera: Crambidae). Insect Biochemistry and Molecular Biology 35, 33–40.

Franklin, S.E., Young, L., Watson, D., Cigan, A., Meyer, T. et al. (1997) Southern analysis of BT-R$_1$, the Manduca sexta gene encoding the receptor for the Cry1Ab toxin of Bacillus thuringiensis. Molecular and General Genetics 256, 517–524.

Gahan, L.J., Gould, F. and Heckel, D.G. (2001) Identification of a gene associated with Bt resistance in Heliothis virescens. Science 293, 857–860.

Gahan, L.J., Gould, F., López, J.D., Micinski, S. and Heckel, D.G. (2007) A polymerase chain reaction screen of field populations of Heliothis virescens for a retrotransposon insertion conferring resistance to Bacillus thuringiensis toxin. Journal of Economic Entomology 100, 187–194.

Gahan, L.J., Pauchet, Y., Vogel, H. and Heckel, D.G. (2010) An ABC transporter mutation is correlated with insect resistance to Bacillus thuringiensis Cry1Ac toxin. PLoS Genetics 6(12): e1001248.

Gao, Y., Jurat-Fuentes, J.L., Oppert, B., Fabrick, J.A., Liu, C. et al. (2011) Increased toxicity of Bacillus thuringiensis Cry3Aa against Crioceris quatuordecimpunctata, Phaedon brassicae and Colaphellus bowringi by a Tenebrio molitor cadherin fragment. Pest Management Science 67, 1076–1081.

Gómez, I., Oltean, D.I., Gill, S.S., Bravo, A. and Soberón, M. (2001) Mapping the epitope in cadherin-like receptors involved in Bacillus thuringiensis Cry1A toxin interaction using phage display. The Journal of Biological Chemistry 276, 28906–28912.

Gómez, I., Miranda-Ríos, J., Rudiño-Piñera, E., Oltean, D.I., Gill, S.S. et al. (2002a) Hydropathic complementarity determines interaction of epitope [869]HITDTNNK[876] in Manduca sexta Bt-R$_1$ receptor with loop 2 of domain II of Bacillus thuringiensis Cry1A toxins. The Journal of Biological Chemistry 277, 30137–3013743.

Gómez, I., Sánchez, J., Miranda, R., Bravo, A. and Soberón, M. (2002b) Cadherin-like receptor binding facilitates proteolytic cleavage of helix α-1 in domain I and oligomer pre-pore formation of *Bacillus thuringiensis* Cry1Ab toxin. *FEBS Letters* 513, 242–246.

Gómez, I., Dean, D.H., Bravo, A. and Soberón, M. (2003) Molecular basis for *Bacillus thuringiensis* Cry1Ab toxin specificity; two structural determinants in the *Manduca sexta* Bt-R1 receptor interact with loops α-8 and 2 in domain II of Cry1Ab toxin. *Biochemistry* 42, 10482–10489.

Gómez, I., Arenas, I., Benitez, I., Miranda-Ríos, J., Becerril, B. *et al.* (2006) Specific epitopes of domains II and III of *Bacillus thuringiensis* Cry1Ab toxin involved in the sequential interaction with cadherin and aminopeptidase-N receptors in *Manduca sexta*. *The Journal of Biological Chemistry* 281, 34032–34039.

Gómez, I., Sánchez, J., Muñoz-Garay, C., Matus, V., Gill, S.S. *et al.* (2014) *Bacillus thuringiensis* Cry1A toxins are versatile-proteins with multiple modes of action: two distinct pre-pores are involved in toxicity. *Biochemical Journal* 459, 383–396.

Halbleib, J.M. and Nelson, W.J. (2006) Cadherins in development: cell adhesion, sorting, and tissue morphogenesis. *Genes and Development* 20, 3199–3214.

Hara, H., Atsumi, S., Yaoi, K., Nakanishi, K., Higurashi, S. *et al.* (2003) A cadherin-like protein functions as a receptor for *Bacillus thuringiensis* Cry1Aa and Cry1Ac toxins on midgut epithelial cells of *Bombyx mori* larvae. *FEBS Letters* 538, 29–34.

Hill, E., Broadbent, I.D., Chothia, C. and Pettitt, J. (2001) Cadherin superfamily proteins in *Caenorhabditis elegans* and *Drosophila melanogaster*. *Journal of Molecular Biology* 305, 1011–1024.

Hua, G., Jurat-Fuentes, J.L. and Adang, M.J. (2004a) Bt-R1a extracellular cadherin repeat 12 mediates *Bacillus thuringiensis* Cry1Ab binding and cytotoxicity. *The Journal of Biological Chemistry* 279, 28051–28056.

Hua, G., Jurat-Fuentes, J.L. and Adang, M.J. (2004b) Fluorescent-based assays establish *Manduca sexta* Bt-R$_1$a cadherin as a receptor for multiple *Bacillus thuringiensis* Cry1A toxins in *Drosophila* S2 cells. *Insect Biochemistry and Molecular Biology* 34, 193–202.

Hua, G., Zhang, R., Abdullah, M.A.F. and Adang, M.J. (2008) *Anopheles gambiae* cadherin AgCad1 binds the Cry4Ba toxin of *Bacillus thuringiensis israelensis* and a fragment of AgCad1 synergizes toxicity. *Biochemistry* 47, 5101–5110.

Hua, G., Zhang, Q., Zhang, R., Abdullah, A.M., Linser, P.J. *et al.* (2013) AgCad2 cadherin in *Anopheles gambiae* larvae is a putative receptor of Cry11Ba toxin of *Bacillus thuringiensis* subsp.

jegathesan. *Insect Biochemistry and Molecular Biology* 43, 153–161.

Hua, G., Park, Y. and Adang, M.J. (2014) Cadherin AdCad1 in *Alphitobius diaperinus* larvae is a receptor of Cry3Bb toxin from *Bacillus thuringiensis*. *Insect Biochemistry and Molecular Biology* 45, 11–17.

Hulpiau, P. and van Roy, F. (2011) New insights into the evolution of metazoan cadherins. *Molecular Biology and Evolution* 28, 647–657.

Ibrahim, M.A., Griko, N.B. and Bulla, L.A. (2013) Cytotoxicity of the *Bacillus thuringiensis* Cry4B toxin is mediated by the cadherin receptor BT-R3 of *Anopheles gambiae*. *Experimental Biology and Medicine* 238, 755–764.

Jurat-Fuentes, J.L. and Adang, M.J. (2006) Cry toxin mode of action in susceptible and resistant *Heliothis virescens* larvae. *Journal of Invertebrate Pathology* 92, 166–171.

Jurat-Fuentes, J.L., Karumbaiah, L., Jakka, S.R.K., Ning, C., Liu, C. *et al.* (2011) Reduced levels of membrane-bound alkaline phosphatase are common to lepidopteran strains resistant to Cry toxins from *Bacillus thuringiensis*. *PLoS ONE* 6(3): e17606.

Liu, C., Wu, K., Wu, Y., Gao, Y., Ning, C. *et al.* (2009) Reduction of *Bacillus thuringiensis* Cry1Ac toxicity against *Helicoverpa armigera* by a soluble toxin-binding cadherin fragment. *Journal of Insect Physiology* 55, 686–693.

Midboe, E.G., Candas, M. and Bulla, L.A. (2003) Expression of a midgut-specific cadherin BT-R$_1$ during the development of *Manduca sexta* larva. *Comparative Biochemistry and Physiology – B: Biochemistry and Molecular Biology* 135, 125–137.

Moita, C., Simoes, S., Moita, L.F., Jacinto, A. and Fernandes, P. (2005) The cadherin superfamily in *Anopheles gambiae*: a comparative study with *Drosophila melanogaster*. *Comparative and Functional Genomics* 6, 204–216.

Morin, S., Biggs, R.W., Sisterson, M.S., Shriver, L., Ellers-Kirk, C. *et al.* (2003) Three cadherin alleles associated with resistance to *Bacillus thuringiensis* in pink bollworm. *Proceedings of the National Academy of Sciences of the United States of America* 100, 5004–5009.

Morin, S., Henderson, S., Fabrick, J.A., Carrière, Y., Dennehy, T.J. *et al.* (2004) DNA-based detection of Bt resistance alleles in pink bollworm. *Insect Biochemistry and Molecular Biology* 34, 1225–1233.

Morishita, H. and Yagi, T. (2007) Protocadherin family: diversity, structure, and function. *Current Opinion in Cell Biology* 19, 584–592.

Nagamatsu, Y., Toda, S., Yamaguchi, F., Ogo, M., Kogure, M. *et al.* (1998a) Identification of

Bombyx mori midgut receptor for *Bacillus thuringiensis* insecticidal CryIA(a) toxin. *Bioscience, Biotechnology and Biochemistry* 62, 718–726.

Nagamatsu, Y., Toda, S., Koike, T., Miyoshi, Y., Shigematsu, S. *et al.* (1998b) Cloning, sequencing, and expression of the *Bombyx mori* receptor for *Bacillus thuringiensis* insecticidal CryIA(a) toxin. *Bioscience, Biotechnology and Biochemistry* 62, 727–734.

Nagamatsu, Y., Koike, T., Sasaki, K., Yoshimoto, A. and Furukawa, Y. (1999) The cadherin-like protein is essential to specificity determination and cytotoxic action of the *Bacillus thuringiensis* insecticidal CryIAa toxin. *FEBS Letters* 460, 385–390.

Nair, R., Kalia, V., Aggarwal, K.K. and Gujar, G.T. (2013) Variation in the cadherin gene sequence of Cry1Ac susceptible and resistant *Helicoverpa armigera* (Lepidoptera: Noctuidae) and the identification of mutant alleles in resistant strains. *Current Science* 104, 215–223.

Oppert, B.S., Jurat-Fuentes, J.L., Fabrick, J.A. and Oppert, C. (2008/2013) Novel cadherin receptor peptide for potentiating Bt biopesticides. *US Patent Provisional No. 60/988,919*. Patent No.: US 8,354,371 B2; Date of Patent: Jan. 15, 2013. US Patent and Trademark Office, Alexandria, Virginia.

Pacheco, S., Gómez, I., Arenas, I., Saab-Rincon, G., Rodríguez-Almazán, C. *et al.* (2009a) Domain II loop 3 of *Bacillus thuringiensis* Cry1Ab toxin is involved in a "ping pong" binding mechanism with *Manduca sexta* aminopeptidase-N and cadherin receptors. *The Journal of Biological Chemistry* 284, 32750–32757.

Pacheco, S., Gómez, I., Gill, S.S., Bravo, A. and Soberón, M. (2009b) Enhancement of insecticidal activity of *Bacillus thuringiensis* Cry1A toxins by fragments of a toxin-binding cadherin correlates with oligomer formation. *Peptides* 30, 583–588.

Pardo-López, L., Gómez, I., Muñoz-Garay, C., Jiménez-Juárez, N., Soberón, M. *et al.* (2006) Structural and functional analysis of the pre-pore and membrane-inserted pore of Cry1Ab toxin. *Journal of Invertebrate Pathology* 92, 172–177.

Park, Y. and Kim, Y. (2013) RNA interference of cadherin gene expression in *Spodoptera exigua* reveals its significance as a specific Bt target. *Journal of Invertebrate Pathology* 114, 285–291.

Park, Y., Abdullah, M.A., Taylor, M.D., Rahman, K. and Adang, M.J. (2009) Enhancement of *Bacillus thuringiensis* Cry3Aa and Cry3Bb toxicities to coleopteran larvae by a toxin-binding fragment of an insect cadherin. *Applied and Environmental Microbiology* 75, 3086–3092.

Peng, D., Xu, X., Ye, W., Yu, Z. and Sun, M. (2010) *Helicoverpa armigera* cadherin fragment enhances Cry1Ac insecticidal activity by facilitating toxin-oligomer formation. *Applied Microbiology and Biotechnology* 85, 1033–1040.

Pettitt, J. (2005) The cadherin superfamily. *WormBook*, 1–9.

Pigott, C.R. and Ellar, D.J. (2007) Role of receptors in *Bacillus thuringiensis* crystal toxin activity. *Microbiology and Molecular Biology Reviews* 71, 255–281.

Ren, X.-L., Chen, R.-R., Zhang, Y., Ma, Y., Cui, J.-J. *et al.* (2013) A *Spodoptera exigua* cadherin serves as a putative receptor for *Bacillus thuringiensis* Cry1Ca toxin and shows differential enhancement of Cry1Ca and Cry1Ac toxicity. *Applied and Environmental Microbiology* 79, 5576–5583.

Soberón, M., Pardo-López, L., López, I., Gómez, I., Tabashnik, B.E. *et al.* (2007) Engineering modified Bt toxins to counter insect resistance. *Science* 318, 1640–1642.

Tabashnik, B.E., Liu, Y.B., Unnithan, D.C., Carrière, Y., Dennehy, T.J. *et al.* (2004) Shared genetic basis of resistance to Bt toxin Cry1Ac in independent strains of pink bollworm. *Journal of Economic Entomology* 97, 721–726.

Tabashnik, B.E., Biggs, R.W., Higginson, D.M., Henderson, S., Unnithan, D.C. *et al.* (2005a) Association between resistance to Bt cotton and cadherin genotype in pink bollworm. *Journal of Economic Entomology* 98, 635–644.

Tabashnik, B.E., Dennehy, T.J. and Carrière, Y. (2005b) Delayed resistance to transgenic cotton in pink bollworm. *Proceedings of the National Academy of Sciences of the United States of America* 102, 15389–15393.

Tabashnik, B.E., Fabrick, J.A., Henderson, S., Biggs, R.W., Yafuso, C.M. *et al.* (2006) DNA screening reveals pink bollworm resistance to Bt cotton remains rare after a decade of exposure. *Journal of Economic Entomology* 99, 1525–1530.

Tabashnik, B.E., van Rensburg, J.B. and Carrière, Y. (2009) Field-evolved insect resistance to Bt crops: definition, theory, and data. *Journal of Economic Entomology* 102, 2011–2025.

Tabashnik, B.E., Sisterson, M.S., Ellsworth, P.C., Dennehy, T.J., Antilla, L. *et al.* (2010) Suppressing resistance to Bt cotton with sterile insect releases. *Nature Biotechnology* 28, 1304–1307.

Tabashnik, B.E., Brevault, T. and Carrière, Y. (2013) Insect resistance to Bt crops: lessons from the first billion acres. *Nature Biotechnology* 31, 510–521.

Takeichi, M. (1990) Cadherins: a molecular family important in selective cell–cell adhesion. *Annual Review of Biochemistry* 59, 237–252.

Takeichi, M. (1991) Cadherin cell adhesion receptors as a morphogenetic regulator. *Science* 251, 1451–1455.

Takeichi, M. (2007) The cadherin superfamily in neuronal connections and interactions. *Nature Reviews Neuroscience* 8, 11–20.

Vadlamudi, R.K., Ji, T.H. and Bulla, L.A. (1993) A specific binding protein from *Manduca sexta* for the insecticidal toxin of *Bacillus thuringiensis* subsp. *berliner*. *The Journal of Biological Chemistry* 268, 12334–12340.

Vadlamudi, R.K., Weber, E., Ji, I., Ji, T.H. and Bulla, L.A. (1995) Cloning and expression of a receptor for an insecticidal toxin of *Bacillus thuringiensis*. *The Journal of Biological Chemistry* 270, 5490–5494.

Wang, G., Wu, K., Liang, G. and Guo, Y. (2005) Gene cloning and expression of cadherin in midgut of *Helicoverpa armigera* and its Cry1A binding region. *Science in China Series C: Life Sciences* 48, 346–356.

Williams, J.L., Ellers-Kirk, C., Orth, R.G., Gassmann, A.J., Head, G. *et al.* (2011) Fitness cost of resistance to Bt cotton linked with increased gossypol content in pink bollworm larvae. *PLoS ONE* 6(6): e21863.

Xie, R., Zhuang, M., Oltean, D.I., Gill, S.S., Ross, L.S. *et al.* (2005) Single amino acid mutations in the cadherin receptor from *Heliothis virescens* affect its toxin binding ability to Cry1A toxins. *The Journal of Biological Chemistry* 280, 8416–8425.

Xu, X., Yu, L. and Wu, Y. (2005) Disruption of a cadherin gene associated with resistance to Cry1Ac delta-endotoxin of *Bacillus thuringiensis* in *Helicoverpa armigera*. *Applied and Environmental Microbiology* 71, 948–954.

Yang, Y., Chen, H., Wu, S., Yang, Y., Xu, X. *et al.* (2006) Identification and molecular detection of a deletion mutation responsible for a truncated cadherin of *Helicoverpa armigera*. *Insect Biochemistry and Molecular Biology* 36, 735–740.

Yang, Y., Chen, H., Wu, Y., Yang, Y. and Wu, S. (2007) Mutated cadherin alleles from a field population of *Helicoverpa armigera* confer resistance to *Bacillus thuringiensis* toxin Cry1Ac. *Applied and Environmental Microbiology* 73, 6939–6944.

Yang, Y., Zhu, Y.C., Ottea, J., Husseneder, C., Leonard, B.R. *et al.* (2011) Down regulation of a gene for cadherin, but not alkaline phosphatase, associated with Cry1Ab resistance in the sugarcane borer *Diatraea saccharalis*. *PLoS ONE* 6(10): e25783.

Yang, Z.X., Wen, L.Z., Wu, Q.J., Wang, S.L., Xu, B.Y. *et al.* (2009) Effects of injecting cadherin gene dsRNA on growth and development in diamondback moth *Plutella xylostella* (Lep.: Plutellidae). *Journal of Applied Entomology* 133, 75–81.

Zhang, H., Yin, W., Zhao, J., Jin, L., Yang, Y. *et al.* (2011) Early warning of cotton bollworm resistance associated with intensive planting of Bt cotton in China. *PLoS ONE* 6(8): e22874.

Zhang, H., Tian, W., Zhao, J., Jin, L., Yang, J. *et al.* (2012a) Diverse genetic basis of field-evolved resistance to Bt cotton in cotton bollworm from China. *Proceedings of the National Academy of Sciences of the United States of America* 109, 10275–10280.

Zhang, H., Wu, S., Yang, Y., Tabashnik, B.E. and Wu, Y. (2012b) Non-recessive Bt toxin resistance conferred by an intracellular cadherin mutation in field-selected populations of cotton bollworm. *PLoS ONE* 7(12): e53418.

Zhang, H., Tang, M., Yang, F., Yang, Y. and Wu, Y. (2013) DNA-based screening for an intracellular cadherin mutation conferring non-recessive Cry1Ac resistance in field populations of *Helicoverpa armigera*. *Pesticide Biochemistry and Physiology* 107, 148–152.

Zhang, X., Candas, M., Griko, N.B., Rose-Young, L. and Bulla, L.A. (2005) Cytotoxicity of *Bacillus thuringiensis* Cry1Ab toxin depends on specific binding of the toxin to the cadherin receptor BT-R$_1$ expressed in insect cells. *Cell Death and Differentiation* 12, 1407–1416.

Zhang, X., Candas, M., Griko, N., Taussig, R. and Bulla L.A. (2006) A mechanism of cell death involving an adenylyl cyclase/PKA signaling pathway is induced by the Cry1Ab toxin of *Bacillus thuringiensis*. *Proceedings of the National Academy of Sciences of the United States of America* 103, 9897–9902.

Zhang, X., Griko, N.B., Corona, S.K. and Bulla, L.A. (2008) Enhanced exocytosis of the receptor BT-R$_1$ induced by the Cry1Ab toxin of *Bacillus thuringiensis* directly correlates to the execution of cell death. *Comparative Biochemistry and Physiology – B: Biochemistry and Molecular Biology* 149, 581–588.

Zhao, J., Jin, L., Yang, Y. and Wu, Y. (2010) Diverse cadherin mutations conferring resistance to *Bacillus thuringiensis* toxin Cry1Ac in *Helicoverpa armigera*. *Insect Biochemistry and Molecular Biology* 40, 113–118.

8 Mechanism of Cry1Ac Resistance in Cabbage Loopers – A Resistance Mechanism Selected in Insect Populations in an Agricultural Environment

Ping Wang*

Department of Entomology, Cornell University, New York State Agricultural Experiment Station, Geneva, New York, USA

Summary

The development of resistance to *Bacillus thuringiensis* (Bt) in insect populations in agriculture not only depends on the level of resistance conferred by a selected resistance mechanism, but also on the fitness cost associated with the resistance mechanism under specific ecological and environmental conditions. Bt resistance in the cabbage looper (*Trichoplusia ni*), which was identified by Janmaat and Myers (2003), is a case of Bt resistance evolved in an agricultural system, and is used in this chapter to review and discuss the mechanism of Cry1Ac resistance that is selected in an agricultural environment.

8.1 Introduction

Resistance of insects to pesticide sprays in agriculture has been observed for a century (Melander, 1914). Under selection pressure by pesticide applications, thousands of cases of pesticide resistance in hundreds of arthropod species have been recorded (Mota-Sánchez *et al.*, 2008). Since the first report of insect resistance to *Bacillus thuringiensis* (Bt) in 1985 (McGaughey, 1985), the potential for the development of insect resistance to Bt has been well demonstrated by the laboratory selection of various insects with resistance to Bt toxins (Tabashnik, 1994; Ferré and Van Rie, 2002; Bravo and Soberón, 2008). Insect resistance to Bt toxins from both Bt sprays and transgenic Bt crops has now been reported in field populations of a number of species (Tabashnik *et al.*, 1990, 2009; Shelton *et al.*, 1993; Janmaat and Myers, 2003; van Rensburg, 2007; Downes *et al.*, 2010; Storer *et al.*, 2010; Zhang *et al.*, 2011; Wan *et al.*, 2012; Gassmann *et al.*, 2014). The occurrence of increasing numbers of cases of field-evolved resistance confirms the potential for the development of insect resistance to Bt toxins in the field and indicates the rising risk of its occurrence with the increasing application of Bt toxins for insect control, if adequate resistance management programmes are not in place.

Laboratory selections of Bt-resistant insect populations have greatly facilitated the study of Bt resistance in insects and enabled the building of the main body of the current understanding of the various mechanisms of Bt resistance (Oppert *et al.*, 1997; Gahan *et al.*, 2001; Griffitts and Aroian, 2005; Pardo-López *et al.*, 2013). Bt-resistant lepidopteran strains established

* Corresponding author. E-mail address: pingwang@cornell.edu

by laboratory selections showed resistance to different Bt toxins at different levels, and exhibited various cross-resistance patterns (Ferré and Van Rie, 2002; Tabashnik *et al.*, 2003). Biochemical and molecular studies have indicated that resistance to Bt in insects is complex and that the mechanisms of Bt resistance in different insects and strains can be diverse (Griffitts and Aroian, 2005; Heckel *et al.*, 2007; Pardo-López *et al.*, 2013). For resistance management in agriculture, it is important to understand the resistance mechanisms that may be selected in agricultural systems as a means of conferring resistance in the field. Current understanding of insect resistance to Bt toxins has indicated that laboratory-selected Bt resistance does not always confer resistance to Bt transgenic plants, and that the Bt resistance developed in field insect populations may involve a mechanism different from those found in laboratory-selected resistance (Tabashnik *et al.*, 2003; Baxter *et al.*, 2005; Zhang *et al.*, 2012b).

In the field, cases of insect resistance or increased frequency of resistant alleles to either Bt formulations or Bt crops have been reported in a number of lepidopteran pests, including *Plutella xylostella, Trichoplusia ni, Busseola fusca, Spodoptera frugiperda, Helicoverpa zea, H. armigera, H. punctigera* and *Pectinophora gossypiella* (Tabashnik *et al.*, 1990, 2009; Janmaat and Myers, 2003; van Rensburg, 2007; Matten *et al.*, 2008; Dhurua and Gujar, 2011; Zhang *et al.*, 2011, 2012a; Wan *et al.*, 2012). Field-evolved and laboratory-selected resistant insects may exhibit similar resistance characteristics. For example, high-level resistance to the Bt toxins Cry1Ab or Cry1Ac conferred by reduced toxin binding to the host midgut receptors has been found to be the major type of resistance in both laboratory-selected and field-selected resistant insect populations. However, the underlying molecular basis conferring the resistance can be distinctively different between the laboratory-selected and field-selected resistant insect populations (Morin *et al.*, 2003; Baxter *et al.*, 2005, 2011; Xu *et al.*, 2005; Yang *et al.*, 2006; Tiewsiri and Wang, 2011; Zhang *et al.*, 2012b). Hence, it is crucially important to understand the

molecular genetic basis of Bt resistance in insect populations evolved in agricultural situations in order to provide fundamental knowledge for insect resistance management in agriculture.

8.2 Resistance of the Cabbage Looper to the Bt toxin Cry1Ac

The cabbage looper (*T. ni*) is an important agricultural pest that is widely distributed in temperate regions in Africa, Asia, Europe and the Americas. Although *T. ni* is a major pest of cruciferous crops, its hosts include over 160 plants in 36 families, many of which are important crops (Lingren and Green, 1984). *T. ni* is considered to be a secondary pest on cotton in the USA but, if uncontrolled, it could cause severe yield loss as much as 92% (Schwartz, 1983). Bt resistance in *T. ni* populations has been found in commercial greenhouses in British Columbia, Canada, that exhibited various levels of resistance to a sprayable formulation of *B.t. kurstaki* (Btk), DiPel®, of up to 160-fold (Janmaat and Myers, 2003). *T. ni* is one of only two species that have evolved resistance to Bt under selective pressure from Bt sprays in agricultural practice (Tabashnik *et al.*, 1990; Shelton *et al.*, 1993; Janmaat and Myers, 2003). Thus, Bt-resistant populations of *T. ni* are a unique biological system for studying the mechanisms of field-evolved Bt resistance. The characterization of a Bt-resistant greenhouse population of *T. ni*, GLEN-DiPel, determined that the DiPel-resistance trait was polygenic and incompletely recessive (Janmaat *et al.*, 2004). The incompletely recessive inheritance of DiPel resistance in *T. ni* is similar to most cases of insect resistance to Bt. The polygenic inheritance so demonstrated is indicative of multiple resistance mechanisms to the multiple toxins in Bt sprays.

The DiPel-resistant *T. ni* populations were highly resistant to the toxin Cry1Ac, a major Cry toxin in Btk (Kain *et al.*, 2004). Resistance to Cry1Ac in *T. ni* exhibits typical 'Mode 1' type resistance (Kain *et al.*, 2004), i.e. a high level of resistance to one or more Cry1A toxins, recessive inheritance, reduced binding of one or more Cry1A toxins to the midgut brush border membranes and little or no cross-resistance to Cry1C toxin (Tabashnik *et al.*, 1998). Mode 1-type resistance is the most common type of Bt resistance and has been identified in both laboratory-selected and field-evolved resistant strains from numerous insect species (Tabashnik *et al.*, 1994, 1998; González-Cabrera *et al.*, 2003; Wang *et al.*, 2007). Nevertheless, the underlying genetic mechanisms conferring Mode 1-type resistance selected under different situations, e.g. field versus laboratory, can be different (Baxter *et al.*, 2005, 2011; Tiewsiri and Wang, 2011). Therefore, studying the mechanism of Bt resistance in *T. ni* will shed light on understanding the development of Bt resistance in field insect populations.

Cry1Ac resistance in *T. ni* is an autosomal monogenic trait (Kain *et al.*, 2004; Wang *et al.*, 2007). A backcross strain of *T. ni*, GLEN-Cry1Ac-BCS, generated by introgression of the Cry1Ac resistance trait into a susceptible inbred laboratory strain showed a high level of Cry1Ac resistance, similar to that of the original DiPel-resistant GLEN population, and could survive on transgenic Cry1Ac broccoli and Cry1Ac cotton plants. For analysis of the resistance mechanism using comparative biochemical and molecular approaches, it is desirable to have a resistant backcross strain near isogenic to a susceptible strain to facilitate identification of resistance-associated biochemical and molecular alterations. Introgression of the Cry1Ac resistance trait into a highly inbred susceptible laboratory strain has been proven effective in minimizing non-resistance-associated variations and thereby allowing comparative biochemical analysis to identify biochemical and molecular changes that are associated with Bt resistance in *T. ni* (Wang *et al.*, 2007; Tiewsiri and Wang, 2011).

8.3 Mechanism of Cry1Ac Resistance in the Cabbage Looper

The intoxication pathways of Bt toxins in insects involve a complex cascade of toxin-midgut protein interactions (Bravo *et al.*, 2004; Heckel, 2012; Pardo-López *et al.*, 2013). Alteration of any step in the pathway can potentially lead to Bt resistance. It has been reported that the toxicity of Bt toxins in the insect midgut can be affected by reduced solubilization of the Cry protein crystals (Schnepf *et al.*, 1998), insufficient proteolytic activation or excessive degradation of Bt toxins by midgut proteinases (Oppert *et al.*, 1997; Shao *et al.*, 1998; Li *et al.*, 2004; Karumbaiah *et al.*, 2007), reduced permeability of the midgut peritrophic membrane to the toxin (Hayakawa *et al.*, 2004), elevated immune response (Rahman *et al.*, 2004) and increased sequestering of toxin in the midgut (Gunning *et al.*, 2005). Nevertheless, numerous studies on Bt resistance have indicated that reduced binding of toxins to the midgut brush border membranes is a primary mechanism for high level Bt resistance (Heckel *et al.*, 2007; Pardo-López *et al.*, 2013). Currently identified midgut proteins that may serve as receptors for Cry toxins include the midgut cadherin, aminopeptidase Ns (APNs), the membrane-bound alkaline phosphatase (mALP), an ABC (ATP Binding Cassette) transporter and several other midgut proteins and glycolipids (Pigott and Ellar, 2007; Pardo-López *et al.*, 2013). The identification of Bt resistance in *T. ni* in commercial greenhouses provided an opportunity to investigate Bt resistance mechanisms that may be selected in an agricultural environment.

8.3.1 Midgut proteinases

Midgut proteases in lepidopteran larvae are primarily serine proteinases and the alteration of midgut proteinases could contribute to Bt resistance in insects (Oppert *et al.*, 1997; Li *et al.*, 2004). In the *T. ni* larval midgut, serine proteinases are highly active at an alkaline pH (pH 10) (Li *et al.*, 2009). By SDS-PAGE based proteinase zymographic analysis, midgut proteinase variations could be detected within the original Cry1Ac-resistant greenhouse *T. ni* strain, GLEN-Cry1Ac, and between the resistant and the susceptible strains; however, the observed variations of the midgut proteinase activity profiles were confirmed not to be associated with Bt resistance (Wang et al., 2007). In addition, when an examination was made of both the activation Cry1Ac protoxin and the degradation of activated Cry1Ac by larval midgut fluid from susceptible and resistant strains of T. ni, there was no significant difference between the resistant and susceptible strains in either toxin activation and degradation in the midgut (Wang et al., 2007). Therefore, alteration of proteinase activities is not the mechanism of Cry1Ac resistance selected in T. ni populations in greenhouses.

8.3.2 Midgut esterases

Upregulated production of midgut esterases to bind and sequester Cry1Ac toxin has been reported to be a mechanism of resistance to Cry1Ac in H. armigera (Gunning et al., 2005). This midgut esterase-mediated resistance mechanism has not been observed in T. ni. In the Cry1Ac-resistant T. ni strain, the larval midgut esterase activity and esterase isoenzyme composition do not differ from those in its near-isogenic susceptible strain (Wang et al., 2007).

8.3.3 Haemolymph melanization activity

Heightened immune response, as determined by in vitro haemolymph melanization activity and visualization of melanization in the midgut and the midgut peritrophic membrane, has been proposed to be a mechanism by which Bt resistance is conferred (Rahman et al., 2004; Ma et al., 2005). In T. ni, the in vitro melanization activity of haemolymph plasma from both the susceptible and the resistant T. ni larvae was determined to be low, and no activity difference was observed between the two strains (Wang et al., 2007). Melanization or darkening of the midgut or the peritrophic membrane does not occur in Cry1Ac-resistant T. ni larvae.

8.3.4 Binding of Cry1Ac to midgut brush border membranes

Binding of a Cry toxin to the midgut brush border membrane is a key process in the intoxication pathway of Cry toxins. The association of reduced binding of a Cry toxin to the insect midgut brush border membrane with resistance was first observed in a Bt-resistant strain of Plodia interpunctella (Van Rie et al., 1990). It has become well known that reduced binding of toxins to the midgut brush border membranes is a primary mechanism for high-level Bt resistance (Heckel et al., 2007; Pardo-López et al., 2013). In T. ni larvae, there are specific binding sites in the midgut brush border membranes for Cry1Ac and Cry1Ab (Estada and Ferré, 1994; Iracheta et al., 2000; Wang et al., 2007). A binding analysis of Cry1Ac and Cry1Ab toxins to the midgut brush border membrane vesicles (BBMVs) confirmed that the toxins bound to these specific binding sites in the BBMVs from the susceptible larvae, but neither Cry1Ab and Cry1Ac bound to the BBMVs from the larvae of the Cry1Ac-resistant strain GLEN-Cry1Ac-BCS (Wang et al., 2007). The GLEN-Cry1Ac-BCS larvae were highly resistant to Cry1Ac, but showed no significant cross-resistance to Cry1C (Wang et al., 2007). So the resistance to Cry1Ac in T. ni is a case of Mode 1-type Bt resistance.

Mode 1-type resistance is conferred by the alteration of the midgut binding sites, or receptors, for Cry1Ac. The midgut cadherin, APNs, mALP and an ABC transporter are the primary midgut proteins that have been proposed to serve as the receptors to interact with Cry toxins in the cascade of the intoxication pathways (Griffitts and Aroian, 2005; Heckel et al., 2007; Pardo-López et al., 2013). These putative receptor proteins play different physiological functions in the midgut, so alterations to them may result in different types or different levels of negative fitness consequences. Therefore, alterations of the different receptors may differentially respond to selections for Bt resistance in different situations. The Cry1Ac resistance evolved in T. ni represents a case of resistance-conferring alteration of midgut binding sites for Cry1Ac selected in an agricultural environment.

8.3.5 Midgut cadherin

The midgut cadherin is a known Bt toxin-binding protein with high-binding affinity for Cry toxins in the monomeric form (Gómez *et al.*, 2003) and serves as an important receptor for Cry toxins (Francis and Bulla, 1997; Nagamatsu *et al.*, 1999; Bravo *et al.*, 2004). Mutations of the cadherin gene have been identified as linked with resistance to Cry1Ab or Cry1Ac. In a laboratory-selected Bt-resistant *Heliothis virescens* strain, the resistance was found to be associated with disruption of the cadherin gene by insertion of a retrotransposon (Gahan *et al.*, 2001). Similar cadherin mutations have also been identified in Cry1Ac-resistant *P. gossypiella* and *H. armigera* (Morin *et al.*, 2003; Xu *et al.*, 2005). The *T. ni* midgut cadherin, a 194.7 kDa protein with 1733 amino acid residues, shares the same sequence characteristics as other known lepidopteran midgut cadherins, containing 11 cadherin repeats followed by a membrane-proximal domain in the extracellular region, a transmembrane region and a cytoplasmic tail at the C-terminus (Zhang *et al.*, 2012b). Sequence motifs identified as Cry toxin-binding regions from other lepidopterans are also present in the *T. ni* cadherin (Zhang *et al.*, 2013). The *T. ni* cadherin gene is highly polymorphic. Single nucleotide polymorphisms (SNPs), insertion mutations and deletion mutations have all been identified in the *T. ni* cadherin gene (Zhang *et al.*, 2013). In addition to gene sequence polymorphisms, differential splicing of the cadherin transcript also occurs in the expression of the cadherin gene in *T. ni* (Zhang *et al.*, 2013).

The high variability of the cadherin in *T. ni* could potentially be the genetic basis for the selection of cadherin-mediated Bt resistance (Zhang *et al.*, 2013). However, the Cry1Ac resistance developed in *T. ni* greenhouse populations has been identified as independent of the alteration of the midgut cadherin (Zhang *et al.*, 2012b).

Genetic linkage analysis of the cadherin alleles with Cry1Ac resistance in *T. ni* determined that the cadherin gene was not genetically associated with greenhouse-selected Cry1Ac resistance in *T. ni*. Analyses of cadherin expression in the *T. ni* midgut at both the mRNA and protein levels further confirmed that there is no quantitative difference of the cadherin between susceptible and Cry1Ac-resistant *T. ni* larvae. Moreover, Cry1Ac binds similarly to the cadherin from the Cry1Ac-susceptible and Cry1Ac-resistant *T. ni* larvae (Zhang *et al.*, 2012b). In addition, genetic mapping using amplified fragment length polymorphism (AFLP) markers confirmed that the gene controlling Cry1Ac resistance and the cadherin gene reside on two different chromosomes in *T. ni* (Baxter *et al.*, 2011). Thus, the resistance to Cry1Ac evolved in greenhouse populations of *T. ni* is not conferred by cadherin alteration.

It is noteworthy that among the cadherin alleles identified in *T. ni*, some are predicted to lack the membrane domain to localize in the midgut brush border membranes and so lose any function as a receptor for Cry toxins (Zhang *et al.*, 2013). Such alleles would be expected to confer cadherin-mediated resistance, but were found to be low in abundance and were not selected for Cry1Ac resistance in *T. ni*. Why these loss-of-function mutations were not selected for resistance to Cry1Ac has yet to be understood, but it is possible that they may be associated with a very strong fitness cost.

8.3.6 Alkaline phosphatase

The midgut mALP from *H. virescens* has been identified as a potential receptor for the Bt toxin Cry1Ac (Jurat-Fuentes and Adang, 2004). This mALP is a glycoprotein glycosylphosphatidylinositol (GPI)-anchored to the midgut brush border membranes, and the terminal GalNAc on mALP serves as the binding site for the toxin. It has been shown that a decreased level of mALP in the midgut directly correlated with resistance to the Bt toxin in *H. virescens*. Additionally, reduced mALP activity has also been found in Cry-resistant *H. armigera* and *S. frugiperda* (Jurat-Fuentes *et al.*, 2011). The mALP in *T. ni* has a predicted molecular weight 61.4 kDa with 564 amino acid residues (Baxter *et al.*, 2011). Analysis of mALP activity in the midgut BBMVs from Cry1Ac-susceptible and Cry1Ac-resistant *T. ni* larvae determined that there was no mALP activity change in Cry1Ac-resistant *T. ni* (Wang *et al.*, 2007). Similarly, a quantitative comparative proteomic analysis of the midgut BBMV proteins from Cry1Ac-susceptible and Cry1Ac-resistant *T. ni* larvae showed that there was no significant difference in mALP quantity between the two strains (Tiewsiri and Wang, 2011). Genetic mapping of the Cry1Ac resistance has also determined that the mALP gene is not on the same chromosome as the Cry1Ac resistance gene in *T. ni* (Baxter *et al.*, 2011). Thus, greenhouse-evolved Cry1Ac resistance in *T. ni* is not associated with the mALP.

8.3.7 Aminopeptidase N

Insect APNs are a multi-gene family of GPI-anchored membrane proteins (Adang, 2013). Midgut APNs are the first identified midgut receptors for Cry toxins (Knight et al., 1994; Sangadala *et al.*, 1994; Gill *et al.*, 1995). The role of an APN as a receptor for Cry1Ac has been shown by the transformation of *Drosophila*, which was not susceptible to Cry1Ac, with an APN gene from *Manduca sexta*. The resulting transgenic *Drosophila* with the *M. sexta* APN transgene became susceptible to Cry1Ac, indicating the functional role of the APN in Bt toxicity (Gill and Ellar, 2002). In addition, Cry1Ac-induced pore formation in the midgut brush border membranes from *T. ni* larvae was found to depend on the APN activity on the brush border membranes (Lorence *et al.*, 1997). In a Cry1C-resistant *Spodoptera exigua* strain, it was found that the expression of one APN was completely lacking (Herrero *et al.*, 2005). Therefore, the alteration of APNs could potentially be a mechanism for Bt resistance in insects.

In *T. ni*, six APNs have been identified in the larval midgut by the cloning of

complementary DNA (cDNA) and proteomic analysis (Wang *et al.*, 2005; Tiewsiri and Wang, 2011). A comparative analysis of proteins in the midgut BBMV proteins from the susceptible and the near-isogenic Cry1Ac-resistant larvae identified that the Cry1Ac-resistant *T. ni* strain lacked a 110 kDa protein from the BBMV proteins (Tiewsiri and Wang, 2011). Liquid chromatography-tandem mass spectrometry (LC-MS/MS) quantitative proteomic analysis and Western blot analysis with APN1-specific antibodies determined that the missing protein was the intact 110 kDa APN1 that Cry1Ac could bind to (Tiewsiri and Wang, 2011). Further LC-MS/MS analysis of midgut BBMV protein bands ranging from 33 to 250 kDa resolved by SDS-PAGE, identified another differentially expressed BBMV protein, APN6, in resistant *T. ni* larvae; APN6 was rare in BBMV proteins from the susceptible strain, but was detected in multiple protein bands with a relatively higher abundance in BBMV proteins from the resistant strain (Tiewsiri and Wang, 2011).

The midgut BBMV proteins from *T. ni* larvae have been globally analysed to identify proteins that are differentially present between Cry1Ac-susceptible and Cry1Ac-resistant *T. ni* larvae; the analysis used the non-gel-based quantitative proteomic technique, iTRAQ (isobaric tags for relative and absolute quantitation)-based 2D-LC-MS/MS analysis (Tiewsiri and Wang, 2011). Over 1400 proteins could be identified from the midgut BBMVs of *T. ni* larvae and their relative abundances were determined. Quantitative analysis of the BBMV proteins from *T. ni* larvae identified two proteins that were significantly different in quantity between Cry1Ac-susceptible and Cry1Ac-resistant *T. ni* – the amounts of APN1 and APN6 in the resistant strain were 0.11 times and 6.0 times, respectively, of those found in the susceptible strain (Tiewsiri and Wang, 2011).

The significant decrease in APN1 and increase in APN6 in the midgut of resistant *T. ni* larvae have been confirmed to be regulated at transcription level. The expression of APN1 and APN6 genes in the midgut of resistant larvae was downregulated to 2.6% and upregulated to 3900%, respectively, at mRNA level. The other four APNs, APN2–APN5, were found to be unchanged in the resistant *T. ni* larvae at both protein and mRNA levels (Tiewsiri and Wang, 2011). Importantly, Cry1Ac resistance in *T. ni* was determined to be associated with the differential expression of APN1 and APN6 by a linkage analysis (Tiewsiri and Wang, 2011). So the Mode 1-type resistance selected in greenhouse populations of *T. ni* by Bt sprays is associated with differential alteration of APN1 and APN6 in the midgut, which is distinctly different from the cadherin gene mutation-based mechanism previously identified in three laboratory-selected insects (Gahan *et al.*, 2001; Morin *et al.*, 2003; Xu *et al.*, 2005).

Although the greenhouse-selected Cry1Ac resistance in *T. ni* is associated with downregulation of APN1 and upregulation of APN6, genetic linkage analysis of the APN genes with resistance determined that all six APN genes were clustered in one linkage group and had no genetic linkage with resistance (Tiewsiri and Wang, 2011). An additional genetic mapping study of Cry1Ac resistance in *T. ni* further confirmed that the APN genes and the resistance gene are localized on different chromosomes (Baxter *et al.*, 2011). Therefore, resistance to Cry1Ac in *T. ni* is controlled by a *trans*-regulatory mechanism, leading to the absence of the full size (110 kDa) toxin-binding APN1 in the midgut brush border membranes and, as a result, the loss of binding sites for the toxin.

8.3.8 ABC transporter

ABC transporters are a large superfamily of transmembrane proteins. A mutation in an ABC transporter gene, *ABCC2*, has been identified to be genetically associated with Cry1Ac resistance in *H. virescens* (Gahan *et al.*, 2010). ABC transporter proteins have not been identified as Cry toxin-binding proteins by the biochemical analysis of midgut proteins from any insects, but their functional role as a Cry toxin receptor has been proposed and is supported by

experimental data from the functional expression of the *Bombyx mori ABCC2* gene in cell culture and the introduction of a susceptible allele of this gene into a resistant strain of *B. mori* to rescue its susceptibility to Cry1Ab (Atsumi *et al.*, 2012; Heckel, 2012; Tanaka *et al.*, 2013). The ABCC2 protein from *T. ni*, which is orthologous to the *H. virescens* ABCC2, is a protein of 150 kDa with similar domain architecture and sequence characteristics to the ABCC2 from other lepidopterans (Gahan *et al.*, 2010; Baxter *et al.*, 2011; Atsumi *et al.*, 2012). By genetic mapping, Cry1Ac resistance in *T. ni* was mapped to the *ABCC2* gene locus in a linkage group homologous to *B. mori* chromosome 15 (Baxter *et al.*, 2011). Notably, Cry1Ac resistance in *P. xylostella* selected by Bt sprays in open fields has also been mapped to the *ABCC2* locus, but is independent of the cadherin gene (Baxter *et al.*, 2005, 2011).

Although Cry1Ac resistance in *T. ni* has been mapped to the *ABCC2* locus region in *T. ni*, whether mutations in *ABCC2* or in another gene in the same region control the resistance and whether or how the altered expression of APN1 and APN6 is conferred by the mutation in an ABC transporter have yet to be understood.

8.4 Conclusion

The intoxication pathways of Bt toxins in insects are complex and the mechanisms of Bt resistance can be diverse. For the sustained application of Bt for insect pest control, it is important to understand the resistance mechanisms that have evolved in insect populations in agricultural situations to provide fundamental knowledge for the management of insect resistance in agriculture. Bt resistance in *T. ni* was selected in an agricultural situation and the resistant *T. ni* could not only survive Bt sprays on vegetable crops, but also on Bt broccoli and Bt cotton plants. Consequently, Bt resistance in *T. ni* is conferred by a mechanism that threatens the continuing success of Bt technology in agriculture.

Cry1Ac resistance in *T. ni* is a typical example of Mode 1-type Bt resistance.

However, the loss of midgut binding sites for Cry1Ac in *T. ni* is associated with downregulation of APN1 and upregulation of APN6, which is different from the cadherin mutation-associated Mode 1-type resistance identified in *H. virescens*, *H. armigera* and *P. gossypiella* (Gahan *et al.*, 2001; Morin *et al.*, 2003; Xu *et al.*, 2005). The midgut cadherin gene in *T. ni* populations is highly polymorphic and differential slicing of its transcripts also occurs. Even so, cadherin-mediated resistance was not selected for Cry1Ac resistance in greenhouse *T. ni* populations. The alteration of APN expression in Cry1Ac-resistant *T. ni* is regulated by a *trans*-regulatory mechanism yet to be known, and the resistance is localized to an ABC transporter gene locus region. Bt resistance in *T. ni* is a unique case for studying the molecular mechanism of Bt resistance that has evolved in agricultural systems. Cases of field-evolved Bt resistance have been increasingly reported, but the detailed molecular mechanisms of such resistance remain to be understood.

Acknowledgements

This work was supported in part by the Biotechnology Risk Assessment Grant Program competitive grant no. 2012-33522-19791 from the USDA National Institute of Food and Agriculture and Agricultural Research Service, and by the Cornell University Agricultural Experiment Station federal formula funds received from the USDA Cooperative State Research, Education, and Extension Service.

References

Adang, M.J. (2013) Insect aminopeptidase N. In: Rawlings, N.D. and Salvesen, G. (eds) *Handbook of Proteolytic Enzymes*. Academic Press, San Diego, California, pp. 405–409.

Atsumi, S., Miyamoto, K., Yamamoto, K., Narukawa, J., Kawai, S. *et al.* (2012) Single amino acid mutation in an ATP-binding cassette transporter gene causes resistance to Bt toxin Cry1Ab in the silkworm, *Bombyx mori*. *Proceedings of the National Academy of Sciences of the United States of America* 109, E1591–E1598.

Baxter, S.W., Zhao, J.Z., Gahan, L.J., Shelton, A.M., Tabashnik, B.E. *et al.* (2005) Novel genetic basis of field-evolved resistance to Bt toxins in *Plutella xylostella*. *Insect Molecular Biology* 14, 327–334.

Baxter, S.W., Badenes-Peréz, F.R., Morrison, A., Vogel, H., Crickmore, N. *et al.* (2011) Parallel evolution of *Bacillus thuringiensis* toxin resistance in Lepidoptera. *Genetics* 189, 675–679.

Bravo, A. and Soberón, M. (2008) How to cope with insect resistance to Bt toxins? *Trends in Biotechnology* 26, 573–579.

Bravo, A., Gómez, I., Conde, J., Muñoz-Garay, C., Sánchez, J. *et al.* (2004) Oligomerization triggers binding of a *Bacillus thuringiensis* Cry1Ab pore-forming toxin to aminopeptidase N receptor leading to insertion into membrane microdomains. *Biochimica et Biophysica Acta* 1667, 38–46.

Dhurua, S. and Gujar, G.T. (2011) Field-evolved resistance to Bt toxin Cry1Ac in the pink bollworm, *Pectinophora gossypiella* (Saunders) (Lepidoptera: Gelechiidae), from India. *Pest Management Science* 67, 898–903.

Downes, S., Parker, T. and Mahon, R. (2010) Incipient resistance of *Helicoverpa punctigera* to the Cry2Ab Bt toxin in Bollgard II cotton. *PloS One* 5(9): e12567.

Estada, U. and Ferré, J. (1994) Binding of insecticidal crystal proteins of *Bacillus thuringiensis* to the midgut brush border of the cabbage looper, *Trichoplusia ni* (Hubner) (Lepidoptera: Noctuidae), and selection for resistance to one of the crystal proteins. *Applied and Environmental Microbiology* 60, 3840–3846.

Ferré, J. and Van Rie, J. (2002) Biochemistry and genetics of insect resistance to *Bacillus thuringiensis*. *Annual Review of Entomology* 47, 501–533.

Francis, B.R. and Bulla, L.A. Jr (1997) Further characterization of BT-R$_1$, the cadherin-like receptor for Cry1Ab toxin in tobacco hornworm (*Manduca sexta*) midguts. *Insect Biochemistry and Molecular Biology* 27, 541–550.

Gahan, L.J., Gould, F. and Heckel, D.G. (2001) Identification of a gene associated with Bt resistance in *Heliothis virescens*. *Science* 293, 857–860.

Gahan, L.J., Pauchet, Y., Vogel, H. and Heckel, D.G. (2010) An ABC transporter mutation is correlated with insect resistance to *Bacillus thuringiensis* Cry1Ac toxin. *PLoS Genetics* 6(12): e1001248.

Gassmann, A.J., Petzold-Maxwell, J.L., Clifton,

E.H., Dunbar, M.W. *et al.* (2014) Field-evolved resistance by western corn rootworm to multiple *Bacillus thuringiensis* toxins in transgenic maize. *Proceedings of the National Academy of Sciences of the United States of America* 111, 5141–5146.

Gill, M. and Ellar, D. (2002) Transgenic *Drosophila* reveals a functional *in vivo* receptor for the *Bacillus thuringiensis* toxin Cry1Ac1. *Insect Molecular Biology* 11, 619–625.

Gill, S.S., Cowles, E.A. and Francis, V. (1995) Identification, isolation, and cloning of a *Bacillus thuringiensis* CryIAc toxin-binding protein from the midgut of the lepidopteran insect *Heliothis virescens*. *The Journal of Biological Chemistry* 270, 27277–27282.

Gómez, I., Dean, D.H., Bravo, A. and Soberón, M. (2003) Molecular basis for *Bacillus thuringiensis* Cry1Ab toxin specificity: two structural determinants in the *Manduca sexta* Bt-R$_1$ receptor interact with loops α-8 and 2 in domain II of Cy1Ab toxin. *Biochemistry* 42, 10482–10489.

González-Cabrera, J., Escriche, B., Tabashnik, B.E. and Ferré, J. (2003) Binding of *Bacillus thuringiensis* toxins in resistant and susceptible strains of pink bollworm (*Pectinophora gossypiella*). *Insect Biochemistry and Molecular Biology* 33, 929–935.

Griffitts, J.S. and Aroian, R.V. (2005) Many roads to resistance: how invertebrates adapt to Bt toxins. *Bioessays* 27, 614–624.

Gunning, R.V., Dang, H.T., Kemp, F.C., Nicholson, I.C. and Moores, G.D. (2005) New resistance mechanism in *Helicoverpa armigera* threatens transgenic crops expressing *Bacillus thuringiensis* Cry1Ac toxin. *Applied and Environmental Microbiology* 71, 2558–2563.

Hayakawa, T., Shitomi, Y., Miyamoto, K. and Hori, H. (2004) GalNAc pretreatment inhibits trapping of *Bacillus thuringiensis* Cry1Ac on the peritrophic membrane of *Bombyx mori*. *FEBS Letters* 576, 331–335.

Heckel, D.G. (2012) Learning the ABCs of Bt: ABC transporters and insect resistance to *Bacillus thuringiensis* provide clues to a crucial step in toxin mode of action. *Pesticide Biochemistry and Physiology* 104, 103–110.

Heckel, D.G., Gahan, L.J., Baxter, S.W., Zhao, J.Z., Shelton, A.M. *et al.* (2007) The diversity of Bt resistance genes in species of Lepidoptera. *Journal of Invertebrate Pathology* 95, 192–197.

Herrero, S., Gechev, T., Bakker, P.L., Moar, W.J. and de Maagd, R.A. (2005) *Bacillus thuringiensis* Cry1Ca-resistant *Spodoptera exigua* lacks expression of one of four aminopeptidase N genes. *BMC Genomics* 6:96.

Iracheta, M.M., Pereyra-Alférez, B., Galán-Wong, L. and Ferré, J. (2000) Screening for *Bacillus thuringiensis* crystal proteins active against the cabbage looper, *Trichoplusia ni*. *Journal of Invertebrate Pathology* 76, 70–75.

Janmaat, A.F. and Myers, J. (2003) Rapid evolution and the cost of resistance to *Bacillus thuringiensis* in greenhouse populations of cabbage loopers, *Trichoplusia ni*. *Proceedings of the Royal Society, B: Biological Sciences* 270, 2263–2270.

Janmaat, A.F., Wang, P., Kain, W., Zhao, J.Z. and Myers, J. (2004) Inheritance of resistance to *Bacillus thuringiensis* subsp. *kurstaki* in *Trichoplusia ni*. *Applied and Environmental Microbiology* 70, 5859–5867.

Jurat-Fuentes, J.L. and Adang, M.J. (2004) Characterization of a Cry1Ac-receptor alkaline phosphatase in susceptible and resistant *Heliothis virescens* larvae. *European Journal of Biochemistry* [now *The FEBS Journal*] 271, 3127–3135.

Jurat-Fuentes, J.L., Karumbaiah, L., Jakka, S.R., Ning, C., Liu, C. *et al.* (2011) Reduced levels of membrane-bound alkaline phosphatase are common to lepidopteran strains resistant to Cry toxins from *Bacillus thuringiensis*. *PloS One* 6(3): e17606.

Kain, W.C., Zhao, J.Z., Janmaat, A.F., Myers, J., Shelton, A.M. *et al.* (2004) Inheritance of resistance to *Bacillus thuringiensis* Cry1Ac toxin in a greenhouse-derived strain of cabbage looper (Lepidoptera: Noctuidae). *Journal of Economic Entomology* 97, 2073–2078.

Karumbaiah, L., Oppert, B., Jurat-Fuentes, J.L. and Adang, M.J. (2007) Analysis of midgut proteinases from *Bacillus thuringiensis*-susceptible and -resistant *Heliothis virescens* (Lepidoptera: Noctuidae). *Comparative Biochemistry and Physiology Part B: Biochemistry and Molecular Biology* 146, 139–146.

Knight, P.J., Crickmore, N. and Ellar, D.J. (1994) The receptor for *Bacillus thuringiensis* CryIA(c) delta-endotoxin in the brush border membrane of the lepidopteran *Manduca sexta* is aminopeptidase N. *Molecular Microbiology* 11, 429–436.

Li, C., Song, X., Li, G. and Wang, P. (2009) Midgut cysteine protease-inhibiting activity in *Trichoplusia ni* protects the peritrophic membrane from degradation by plant cysteine proteases. *Insect Biochemistry and Molecular Biology* 39, 726–734.

Li, H., Oppert, B., Higgins, R.A., Huang, F., Zhu, K.Y. *et al.* (2004) Comparative analysis of proteinase activities of *Bacillus thuringiensis*-resistant and -susceptible *Ostrinia nubilalis*

(Lepidoptera: Crambidae). *Insect Biochemistry and Molecular Biology* 34, 753–762.

Lingren, P.D. and Green, G.L. (eds) (1984) *Suppression and Management of Cabbage Looper Populations.* Technical Bulletin No. 1684, US Department of Agriculture Agricultural Research Service, Washington, DC.

Lorence, A., Darszon, A. and Bravo, A. (1997) Aminopeptidase dependent pore formation of *Bacillus thuringiensis* CrylAc toxin on *Trichoplusia ni* membranes *FEBS Letters* 414, 303–307.

Ma, G., Roberts, H., Sarjan, M., Featherstone, N., Lahnstein, J. *et al.* (2005) Is the mature endotoxin Cry1Ac from *Bacillus thuringiensis* inactivated by a coagulation reaction in the gut lumen of resistant *Helicoverpa armigera* larvae? *Insect Biochemistry and Molecular Biology* 35, 729–739.

Matten, S.R., Head, G.P. and Quemada, H.D. (2008) How governmental regulation can help or hinder the integration of Bt crops within IPM programs. In: Romeis, J., Shelton, A.M. and Kennedy, G.G. (eds) *Integration of Insect Resistant Genetically Modified Crops within IPM Programs. Progress in Biological Control, Volume 5.* Springer, Dordrecht, The Netherlands, pp. 27–39.

McGaughey, W.H. (1985) Insect resistance to the biological insecticide *Bacillus thuringiensis. Science* 229, 193–195.

Melander, A.L. (1914) Can insects become resistant to sprays? *Journal of Economic Entomology* 7, 167–173.

Morin, S., Biggs, R.W., Sisterson, M.S., Shriver, L., Ellers-Kirk, C. *et al.* (2003) Three cadherin alleles associated with resistance to *Bacillus thuringiensis* in pink bollworm. *Proceedings of the National Academy of Sciences of the United States of America* 100, 5004–5009.

Mota-Sánchez, D., Whalon, H., Hollingworth, R.M. and Xue, Q. (2008) Documentation of pesticide resistance in arthropods. In: Whalon, M.E., Mota-Sánchez, D. and Hollingworth, R.M. (eds.) *Global Pesticide Resistance in Arthropods.* CAB International, Wallingford, UK, pp. 32–39.

Nagamatsu, Y., Koike, T., Sasaki, K., Yoshimoto, A. and Furukawa, Y. (1999) The cadherin-like protein is essential to specificity determination and cytotoxic action of the *Bacillus thuringiensis* insecticidal CrylAa toxin. *FEBS Letters* 460, 385–390.

Oppert, B., Kramer, K.J., Beeman, R.W., Johnson, D. and McGaughey, W.H. (1997) Proteinase-mediated insect resistance to *Bacillus thuringiensis* toxins. *The Journal of Biological Chemistry* 272, 23473–23476.

Pardo-López, L., Soberón, M. and Bravo, A. (2013) *Bacillus thuringiensis* insecticidal three-domain Cry toxins: mode of action, insect resistance and consequences for crop protection. *FEMS Microbiology Reviews* 37, 3–22.

Pigott, C.R. and Ellar, D.J. (2007) Role of receptors in *Bacillus thuringiensis* crystal toxin activity. *Microbiology and Molecular Biology Reviews* 71, 255–281.

Rahman, M.M., Roberts, H.L., Sarjan, M., Asgari, S. and Schmidt, O. (2004) Induction and transmission of *Bacillus thuringiensis* tolerance in the flour moth *Ephestia kuehniella. Proceedings of the National Academy of Sciences of the United States of America* 101, 2696–2699.

Sangadala, S., Walters, F.S., English, L.H. and Adang, M.J. (1994) A mixture of *Manduca sexta* aminopeptidase and phosphatase enhances *Bacillus thuringiensis* insecticidal CryIA(c) toxin binding and 86Rb(+)-K+ efflux *in vitro. The Journal of Biological Chemistry* 269, 10088–10092.

Schnepf, E., Crickmore, N., Van Rie, J., Lereclus, D., Baum, J. *et al.* (1998) *Bacillus thuringiensis* and its pesticidal crystal proteins. *Microbiology and Molecular Biology Reviews* 62, 775–806.

Schwartz, P.H. (1983) Losses in yield of cotton due to insects. In: Ridgway, R.L., Lloyd, E.P. and Cross, W.H. (eds) *Cotton Insect Management with Special Reference to the Boll Weevil.* Agriculture Handbook, No. 589, US Department of Agriculture Agricultural Research Service, Washington, DC, pp. 329–358.

Shao, Z., Cui, Y., Liu, X., Yi, H., Ji, J. *et al.* (1998) Processing of δ-endotoxin of *Bacillus thuringiensis* subsp. *kurstaki* HD-1 in *Heliothis armigera* midgut juice and the effects of protease inhibitors. *Journal of Invertebrate Pathology* 72, 73–81.

Shelton, A.M., Robertson, J.L., Tang, J.D., Perez, C., Eigenbrode, S.D. *et al.* (1993) Resistance of diamondback moth (Lepidoptera: Plutellidae) to *Bacillus thuringiensis* subspecies in the field. *Journal of Economic Entomology* 86, 697–705.

Storer, N.P., Babcock, J.M., Schlenz, M., Meade, T., Thompson, G.D. *et al.* (2010) Discovery and characterization of field resistance to Bt maize: *Spodoptera frugiperda* (Lepidoptera: Noctuidae) in Puerto Rico. *Journal of Economic Entomology* 103, 1031–1038.

Tabashnik, B.E. (1994) Evolution of resistance to *Bacillus thuringiensis. Annual Review of Entomology* 39, 47–79.

Tabashnik, B.E., Cushing, N.L., Finson, N. and Johnson, M.W. (1990) Field development of resistance to *Bacillus thuringiensis* in diamondback moth (Lepidoptera: Plutellidae). *Journal of Economic Entomology* 83, 1671–1676.

Tabashnik, B.E., Finson, N., Groeters, F.R., Moar, W.J., Johnson, M.W. *et al.* (1994) Reversal of resistance to *Bacillus thuringiensis* in *Plutella xylostella. Proceedings of the National Academy of Sciences of the United States of America* 91, 4120–4124.

Tabashnik, B.E., Liu, Y.B., Malvar, T., Heckel, D.G., Masson, L. *et al.* (1998) Insect resistance to *Bacillus thuringiensis*: uniform or diverse? *Philosophical Transactions of the Royal Society B: Biological Sciences* 353, 1751–1756.

Tabashnik, B.E., Carrière, Y., Dennehy, T.J., Morin, S., Sisterson, M.S. *et al.* (2003) Insect resistance to transgenic Bt crops: lessons from the laboratory and field. *Journal of Economic Entomology* 96, 1031–1038.

Tabashnik, B.E., van Rensburg, J.B. and Carrière, Y. (2009) Field-evolved insect resistance to Bt crops: definition, theory, and data. *Journal of Economic Entomology* 102, 2011–2025.

Tanaka, S., Miyamoto, K., Noda, H., Jurat-Fuentes, J.L., Yoshizawa, Y. *et al.* (2013) The ATP-binding cassette transporter subfamily C member 2 in *Bombyx mori* larvae is a functional receptor for Cry toxins from *Bacillus thuringiensis*. *The FEBS Journal* 280, 1782–1794.

Tiewsiri, K. and Wang, P. (2011) Differential alteration of two aminopeptidases N associated with resistance to *Bacillus thuringiensis* toxin Cry1Ac in cabbage looper. *Proceedings of the National Academy of Sciences of the United States of America* 108, 14037–14042.

van Rensburg, J.B.J. (2007) First report of field resistance by the stem borer, *Busseola fusca* (Fuller) to Bt-transgenic maize. *South African Journal of Plant and Soil* 24, 147–151.

Van Rie, J., McGaughey, W.H., Johnson, D.E., Barnett, B.D. and Van Mellaert, H. (1990) Mechanism of insect resistance to the microbial insecticide *Bacillus thuringiensis*. *Science* 247, 72–74.

Wang, P., Zhang, X. and Zhang, J. (2005) Molecular characterization of four midgut aminopeptidase N isozymes from the cabbage looper, *Trichoplusia ni. Insect Biochemistry and Molecular Biology* 35, 611–620.

Wang, P., Zhao, J.Z., Rodrigo-Simon, A., Kain, W., Janmaat, A.F. *et al.* (2007) Mechanism of resistance to *Bacillus thuringiensis* toxin Cry1Ac in a greenhouse population of the cabbage looper, *Trichoplusia ni. Applied and Environmental Microbiology* 73, 1199–1207.

Wan, P., Huang, Y., Wu, H., Huang, M., Cong, S. *et al.* (2012) Increased frequency of pink bollworm resistance to Bt toxin Cry1Ac in China. *PloS One* 7(1): e29975.

Xu, X., Yu, L. and Wu, Y. (2005) Disruption of a cadherin gene associated with resistance to Cry1Ac δ-endotoxin of *Bacillus thuringiensis* in *Helicoverpa armigera. Applied and Environmental Microbiology* 71, 948–954.

Yang, Y., Chen, H., Wu, S., Yang, Y., Xu, X. *et al.* (2006) Identification and molecular detection of a deletion mutation responsible for a truncated cadherin of *Helicoverpa armigera. Insect Biochemistry and Molecular Biology* 36, 735–740.

Zhang, H., Yin, W., Zhao, J., Jin, L., Yang, Y. *et al.* (2011) Early warning of cotton bollworm resistance associated with intensive planting of Bt cotton in China. *PloS One* 6(8): e22874.

Zhang, H., Tian, W., Zhao, J., Jin, L., Yang, J. *et al.* (2012a) Diverse genetic basis of field-evolved resistance to Bt cotton in cotton bollworm from China. *Proceedings of the National Academy of Sciences of the United States of America* 109, 10275–10280.

Zhang, X., Kain, W. and Wang, P. (2013) Sequence variation and differential splicing of the midgut cadherin gene in *Trichoplusia ni. Insect Biochemistry and Molecular Biology* 43, 712–723.

Zhang, X., Tiewsiri, K., Kain, W., Huang, L. and Wang, P. (2012b) Resistance of *Trichoplusia ni* to *Bacillus thuringiensis* toxin Cry1Ac is independent of alteration of the cadherin-like receptor for Cry toxins. *PloS One* 7(5): e35991.

9

Roles of ABC Proteins in the Mechanism and Management of Bt Resistance

David G. Heckel*

Max Planck Institute for Chemical Ecology, Jena, Germany

Summary

Genetic studies of strains of insects that have developed resistance to pore-forming Cry toxins from *Bacillus thuringiensis* (Bt) have provided useful and unexpected insights into the mode of action of the toxin. Independent approaches in five species of Lepidoptera have converged on the same result: that mutations in a member of the superfamily of ABC transporters confer resistance to Cry toxins. These mutations range from a single amino acid insertion to truncations that delete most of the protein. This result is surprising, because since the first detection of the specific binding of Cry toxins to sites in the lepidopteran midgut in 1988, no studies had documented any sort of interaction between Cry toxins and ABC proteins. It is hypothesized that ABC transporters mediate the critical step of Cry pore insertion into the membrane. Heterologous expression of ABC proteins has recently shown that they can facilitate cell swelling by Cry toxins, which is indirect evidence of pore formation. It is suggested that the 'ATP-switch' mechanism of ABC proteins that drives the transport of small molecules across cell membranes might be exploited to potentiate Bt action and to guide the design of improved Cry toxins.

9.1 Introduction

The Cry family of insecticidal protein toxins from *Bacillus thuringiensis* (Bt) is one of the most potent virulence factors used by this pathogenic bacterium to overcome its insect hosts (Schnepf *et al.*, 1998). These proteins are synthesized before sporulation, and are packed into a crystalline parasporal inclusion, hence the name Cry. When the spore is ingested by a lepidopteran caterpillar, the crystal dissolves in the alkaline milieu of the midgut and the protoxin which is then in solution is attacked by the insect's digestive proteases until it is whittled down to a relatively protease-resistant core toxin. This toxin interacts with several proteins on the surface of the insect's midgut epithelium, undergoing additional processing steps and assembling into an oligomeric pre-pore structure (Soberón *et al.*, 2010). Finally, the pore inserts into the epithelial membrane, disrupting the osmotic balance of the cell and ultimately killing it (Knowles and Ellar, 1987). Death of the midgut epithelium provides a suitable environment for Bt to deploy an arsenal of other virulence factors to invade the rest of the body and exploit its resources (Raymond *et al.*, 2010). Separate quorum-sensing systems govern the expression of these virulence factors and of

* Corresponding author. E-mail address: heckel@ice.mpg.de

additional gene products enabling survival in the dead host and, finally, of genes that promote sporulation and the synthesis of additional Cry toxins (Slamti et al., 2014).

Although pore formation has been widely recognized as the cytotoxic mechanism of the Cry toxins, the details of pore insertion into the membrane have remained obscure. Because Cry toxins must bind to specific receptors on the midgut epithelium before killing the cell, most research has taken a biochemical approach focused on the myriad of binding proteins and the nature of their interactions with the toxin, and on the processing steps that produce a competent pore structure. Only recently have genetic studies of Bt-resistant strains of various species of Lepidoptera pointed to a previously unsuspected role of ABC proteins in the pore insertion step.

ABC proteins comprise a superfamily of transmembrane transporters found in all prokaryotes and eukaryotes. The name ABC comes from the ATP-binding cassette, an intracellular domain that binds and hydrolyses ATP to drive the transport of small molecules across a lipid bilayer membrane. The functional transporter consists of two integral transmembrane domains, each with six membrane-spanning helices, and two of the cytosolic ATP-binding domains. Many ABC transporters play an important role in detoxification by excreting xenobiotics from the cell. These include the multiple drug resistance (Mdr) proteins that were first discovered as a result of their overexpression leading to resistance of cancer cells to chemotherapeutic agents.

Of particular interest is the unique reaction cycle that drives the transport of compounds across the membrane, the so-called ATP-switch mechanism (Higgins and Linton, 2004). The transporter alternates between conformations in which the channel surrounded by the trans-membrane domains is either open or closed to the outside of the cell. The transition between these conformations is driven by the binding and hydrolysis of ATP by the two cytosolic domains. The transported molecule is expelled from the cell when the channel momentarily opens to the outside, and then the channel closes again to receive another molecule from the cytoplasmic side. Continual opening and closing of the channel is required for the continuous efflux of the xenobiotics or other transported compounds. It also provides the basis for a hypothesis about the mechanism of Cry pore insertion into the membrane.

9.2 Genetic Approaches Converge on ABC Proteins

The study of insect strains that had developed resistance to Bt toxins, either by selection in the field or in the laboratory, has yielded much useful information about the mode of action of the toxin (Heckel, 2012). In the YHD2 strain of the tobacco budworm, Heliothis virescens, a positional cloning approach led to the discovery of a mutation that conferred high levels of resistance to Cry1Ab and Cry1Ac, along with a loss of binding of these toxins to the midgut epithelial membrane (Gahan et al., 2010). This mutation did not affect any of the proteins known to interact with Cry toxins at that time. Instead, it was a 22-base deletion in exon 2 of a gene for an ABC protein in the C subfamily, named ABCC2. This protein is one of a cluster of three similar ABCC proteins, the other two being ABCC1 and ABCC3. The frameshift caused by the deletion in the gene was predicted to produce a truncated 99-residue protein instead of the full-length protein of 1339 residues. Thus, absence of the full-length protein from the midgut epithelium was correlated with resistance to Cry1Ab and Cry1Ac, and the loss of epithelial binding to these two toxins. Nevertheless, the ability to bind the Cry1Aa toxin persisted.

Previous work on H. virescens had identified a mutation inactivating a 12-cadherin domain cadherin protein expressed in the midgut that also conferred resistance to Cry1A toxins (Gahan et al., 2001). This mutation abolished binding of Cry1Aa to the membrane, but the binding of Cry1Ab and Cry1Ac still occurred, despite resistance to those two toxins. Subsequent work on the cadherin protein in Manduca *sexta* showed that it played an important role in an additional processing step that accelerated the oligomerization step before pore formation (Bravo *et al.*, 2004). When the cadherin mutation and the ABCC2 mutation were combined into the same strain of *H. virescens*, Cry1Ac resistance levels were much higher than with either mutation alone, and binding to Cry1Aa, Cry1Ab and Cry1Ac was abolished (Gahan *et al.*, 2010). Hence, the cadherin and the ABC protein appeared to play non-redundant roles in the potency of the Cry toxins.

The NO-QA strain of diamondback moth, *Plutella xylostella*, was collected from Hawaii, USA, and represents the first case of the evolution of Bt resistance in field populations of an insect pest. Previous genetic studies had shown that a single gene conferred resistance to Cry1Aa, Cry1Ab, Cry1Ac and Cry1F toxins (Tabashnik *et al.*, 1997a), and that resistance was accompanied by a loss of toxin binding to midgut membranes (Tabashnik *et al.*, 1997b). After a long search in which every protein with a known interaction with Bt toxins was eliminated (Baxter *et al.*, 2005, 2008), a combined mapping and sequencing approach revealed a mutation in the *Plutella* homologue of ABCC2 (Baxter *et al.*, 2011). This was a deletion of ten amino acids in the middle of the twelfth transmembrane helix, which was predicted to flip the second ATP-binding domain to the extracellular space instead of the cytoplasmic side of the membrane.

A Cry1Ac-resistant strain of the cabbage looper, *Trichoplusia ni*, was isolated from commercial greenhouses in British Columbia, Canada, and represents the second case of Bt resistance in open populations of an insect pest (Janmaat and Myers, 2003). One of the two resistance mechanisms in this strain was mapped to a chromosomal region containing the ABCC2 protein, and complete linkage was found between resistance and a marker in the ABCC2 gene among more than 300 backcross progeny (Baxter *et al.*, 2011). However, the entire sequence of the gene has not yet been determined and a resistance-conferring mutation has not yet been identified.

The C2 strain of the domesticated silkworm, *Bombyx mori*, was found to be susceptible to Cry1Aa, the most potent toxin against this species, but resistant to Cry1Ab. A positional cloning approach involving mapping in several thousand backcross progeny eventually led to the ABCC2 protein, although there was no obvious mutation in the sequence of the resistant allele that would incapacitate the protein (Atsumi *et al.*, 2012). In addition to 12 amino acid substitutions, the resistant protein had an insertion of a single tyrosine residue at position 234, at the tip of the second extracellular loop of the first transmembrane domain. Germline transformation was used to prove that this apparently minor difference was actually responsible for resistance. By transforming a resistant line with the susceptible allele lacking tyrosine at position 234 (Tyr234), and expressing that allele specifically in the midgut, susceptibility to Cry1Ab was restored. This transgenic approach provides compelling evidence that even small changes in the ABCC2 protein can have a large impact on Cry toxicity (Atsumi *et al.*, 2012).

9.3 Proposed Role of ABCC2 in Cry Toxin Mode of Action

Based on the findings in *H. virescens*, Gahan *et al.* (2010) proposed an extension of the sequential binding model put forth by Bravo *et al.* (2007). In the first step of the Bravo *et al.* model, reversible toxin binding to the 12-cadherin domain protein enables an additional processing reaction, the cleavage of an N-terminal α-helix, that speeds up the formation of an oligomeric pre-pore structure in solution. In the second step, the oligomers bind various membrane-bound proteins such as aminopeptidase or alkaline phosphatase, resulting in an increased concentration of oligomers at the epithelial surface. It has been suggested that the oligomers have a higher affinity for aminopeptidase than does the monomer (Bravo *et al.*, 2007).

Gahan *et al.* (2010) proposed a third step in which the oligomeric pre-pore structure is inserted into the membrane. This is presumed to happen when the channel of the ABC protein is open to the outside of the cell, momentarily exposing hydrophobic patches lining the channel to binding by the extended helix α-4/α-5 hairpins of the pre-pore structure. When the channel closes again in the ATP-switch cycle, the α-helix hairpins of the pre-pore are forced out of the channel. At this point, the pre-pore may be ejected back out into the extracellular space. This toxin–protein binding interaction was therefore reversible and the pre-pore has another chance for insertion. Alternatively, when the channel closes, the pre-pore may slide sideways into the lipid bilayer, where it starts to function as a pore. In this case, the transient interaction with the ABC protein has resulted in irreversible binding of a different kind – the stable insertion of the pore into the membrane (Heckel, 2012).

The fleeting nature of the proposed binding interaction between the Cry toxin and the ABC protein makes it very difficult to demonstrate directly. However, the hypothesis makes some testable predictions. If the fraction of time that the transporter channel is open to the outside limits the opportunities for toxin insertion, then compounds that stimulate the ATP-switch cycle should potentiate Cry toxicity. In addition, the loss of a functional ABC transporter due to Bt resistance may have fitness costs resulting from a lower ability to export other toxic substances such as host-plant secondary compounds or insecticides. Moreover, heterologous expression of the right ABC protein should confer toxin susceptibility on otherwise resistant cells. Alternatively, RNA inhibition of the right ABC protein *in vivo* should reduce toxin susceptibility. Two recent studies have made progress in testing some of these assumptions.

9.4 Heterologous Expression Studies

An important contribution was made by Tanaka *et al.* (2013), who explored toxin–receptor interactions by expressing *B. mori* ABCC2 (BmABCC2) in the Sf9 insect cell line using a baculovirus as transfer agent. EGFP

(enhanced green fluorescent protein) was included in the expression constructs so that fluorescence could be used as a marker for virus-infected cells. Pore formation was indirectly measured by microscopic observation of cell swelling as cells were incubated with different amounts of toxin. When BmABCC2 from a Cry1Ab-susceptible strain was expressed, all three Cry1A toxins induced swelling, but to differing degrees. Half of the cells swelled in the presence of 0.8 nM Cry1Aa, 8.0 nM Cry1Ac or 80 nM Cry1Ab. When, instead, the BmABCC2 from a Cry1Ab-resistant strain was expressed, the cells showed a similar response to Cry1Aa, but neither Cry1Ab nor Cry1Ac induced any observable swelling up to concentrations of 1 μM. Thus, BmABCC2 facilitates pore formation by all three toxins, but the amino acid sequence differences in the resistant protein hinder this for Cry1Ab and Cry1Ac only. This corresponds to the fact that the Cry1Ab-resistant strain was still susceptible to Cry1Aa.

The ABCC2 proteins from the Cry1Ab-resistant and susceptible strains differed in 12 amino acid substitutions as well as in the presence or absence of the tyrosine insertion at position 234. Mutagenized constructs with Tyr234 present in the susceptible protein (BmABCC2 + Tyr234) or Tyr234 absent from the resistant protein (BmABCC2 – Tyr234) were also tested, confirming that the Tyr234 insertion alone was sufficient to abolish Cry1Ab- and Cry1Ac-induced swelling, while still allowing Cry1Aa-induced swelling.

An antibody against the toxin was used to detect Cry1A toxin binding to cells expressing BmABCC2. More Cry1Ab binding was observed to cells expressing BmABCC2 – Tyr234 than to cells expressing BmABCC2 + Tyr234; while Cry1Aa bound similarly to both types of cells. However, specific binding was not demonstrated, and homologous binding competition assays were not performed so that irreversible pore insertion could not be distinguished from a strong reversible binding to a receptor ectodomain in this experiment.

As well as expressing BmABCC2 in Sf9 cells, Tanaka *et al.* (2013) also expressed

BtR175-TBR, the toxin-binding region of the 12-cadherin domain protein BtR175, in Sf9 cells and repeated the swelling assay in the presence of Cry1A toxins. This receptor was much less effective than the ABC protein in facilitating swelling; 600 nM of Cry1Aa or Cry1Ab were required for 20% swelling; 600 nM of Cry1Ac produced only 5% swelling. Whether an endogenous ABC protein expressed by the Sf9 cells played any role in the swelling was not investigated. Any such protein would be likely to be expressed at a far lower level than the BmABCC2 produced by the baculovirus-infected cells, which could account for the higher concentrations of toxin required.

To examine the sequential action hypothesis that Cry1A toxins must interact with the cadherin before pore insertion, cells expressing BtR175-TBR and BmABCC2 – Tyr234 either separately or together were examined. For all three Cry1A toxins, a significant degree of synergism was observed, i.e. there was more swelling when both proteins were expressed than was expected from their independent actions. When, instead, BmABCC2 + Tyr234 was co-expressed with BtR175-TBR, the results for Cry1Aa were unchanged, but synergism was now absent for Cry1Ab and Cry1Ac because the amount of swelling was similar to that induced when BtR175-TBR was expressed alone. Therefore, the effect of the Tyr234 insertion seems to be magnified according to the sequential binding model in this case, producing a higher level of Cry1Ab resistance in larvae than would be anticipated from consideration of its effect on ABCC2–toxin interactions alone.

Comparable synergism of BtR175-TBR and BmABCC2 – Tyr234 was also observed with Cry1Fa, which was approximately as toxic to susceptible larvae as the Cry1A toxins (but was not tested against resistant larvae in this study). Cross-resistance between Cry1A and Cry1F toxins had previously been observed in the NO-QA strain of P. xylostella, which is now known to have a mutated ABCC2 (Tabashnik et al., 1997a). Surprisingly, a slight amount of swelling and synergism was even observed for Cry8Ca, which is active against Coleoptera but hardly toxic against B. mori larvae (but another beetle-active toxin, Cry3Bb, produced no swelling at all). This results hints that the sequential binding model, first developed for Cry1A toxins, may apply to other toxins as well.

9.5 ABCC2 Interactions with an Additional Toxin

Another recent publication (Park et al., 2014) used an innovative genetic mapping strategy to implicate the ABCC2 protein in resistance to a different lepidopteran-active toxin, Cry1Ca. This toxin is much more potent against Spodoptera species than are the Cry1A toxins. A strain of the beet armyworm, S. exigua, was selected for resistance to Xentari, a commercial formulation containing both Cry1Ac and Cry1Ca toxins. The resistant strain was crossed with a susceptible strain and the hybrids backcrossed to the resistant strain. Backcross progeny were bioassayed using a high concentration of Xentari that was chosen to kill all but resistant homozygotes. RNA from the survivors was pooled and sequenced, and the frequencies of single nucleotide polymorphisms (SNPs) in the transcriptome were compared with the resistant and susceptible grandparents to identify alleles from the resistant strain that were over-represented in the survivors. The Spodoptera transcripts were mapped on to the genome sequence of B. mori to facilitate the comparison. The method is, then, a variant of the classical approach of bulked segregant analysis in quantitative trait locus (QTL) mapping (Michelmore et al., 1991), which uses polymorphisms scored from the transcriptome rather than the genome and borrows the genetic map from a related organism. This procedure indicated two genomic regions that were strongly associated with resistance, one corresponding to a region of B. mori chromosome 15 known to contain a cluster of three ABCC proteins, including ABCC2.

Comparison of the sequences of these proteins revealed that resistant individuals had a deletion of 82 amino acids in the second ATP-binding region of ABCC2. In susceptible individuals, the ABCC2 gene was highly expressed in the midgut, as was the neighbouring ABCC3 gene. To examine the role of these two genes in toxin susceptibility, each one was suppressed using RNA interference (RNAi) by feeding double-stranded (ds) RNA to larvae. Insects were then exposed to Cry1Ac or Cry1Ca protoxin separately. Suppression of ABCC2 transcripts decreased the susceptibility to both toxins, especially Cry1Ca; suppression of ABCC3 transcripts also decreased susceptibility but to a lesser extent. This is the first evidence that ABC proteins other than ABCC2 may mediate Cry toxicity.

Competitive binding assays with iodinated Cry1Ca were conducted on membrane preparations from resistant and susceptible larvae. Equilibrium binding parameters did not differ, but the resistant insects showed a lower irreversible component of the specific binding than did susceptible insects. This outcome would be expected if Cry1Ca formed fewer pores in the membranes of resistant insects. The 82-amino acid deletion in one of the cytosolic ATP-binding domains most likely hinders or blocks the ATP-switch mechanism, and the model predicts that this should produce pore insertion at a lower rate. These results are thus not only consistent with the theory that the active ABC protein transport cycle facilitates pore insertion, but also extend the theory from the Cry1A family to the Cry1C family of toxins as well, at least for *Spodoptera*. This extension appears to be taxon specific, however, because the NO-QA strain of *P. xylostella* is not resistant to Cry1C (Tabashnik *et al.*, 1996). Other Cry1C-resistant strains of this species have a resistance mechanism that is genetically independent of Cry1A resistance (Zhao *et al.*, 2001).

9.6 Summary and Prospects for Future Research

The toxin–ABCC2 interactions described above are represented schematically in Fig.

9.1 for the four species with known ABCC2 mutations associated with Bt resistance. The subtlest interactions occur in *B. mori*, where a full-length, functional ABC transporter is apparently present in both susceptible and resistant insects. The single resistance-conferring Tyr insertion in the second extracellular loop is unlikely to have any effect on the operation of the ATP-switch mechanism, because it permits the insertion of Cry1Aa, yet apparently has a toxin-specific effect in hindering the insertion of Cry1Ab and Cry1Ac into the membrane. Its effect on Cry1F insertion remains unknown. At the other extreme, the complete absence of ABCC2 in resistant *H. virescens* blocks Cry1Ab and Cry1Ac but not Cry1Aa; an independent mechanism must be invoked for the insertion of the latter. In *P. xylostella*, the deletion of the twelfth transmembrane domain drastically alters the membrane topology, flipping the normally intracellular second ATP-binding domain to the outside, which certainly blocks the ATP-switch mechanism and probably results in the degradation and removal of the protein from the membrane. This lesion blocks the insertion of the three Cry1A toxins as well as that of Cry1F, but not that of Cry1C and several other toxins. In *S. exigua*, a large deletion within the second ATP-binding domain could allow the ABCC2 protein to maintain its correct membrane topology but probably compromises the ATP-switch mechanism, resulting in reduced pore insertion of Cry1Ac and Cry1Ca.

The current data are incomplete, because not all toxins have been tested for toxicity on resistant and susceptible strains of all species, and subjected to binding studies that can distinguish between non-specific, specific, reversible and irreversible components of binding. A comprehensive comparative study in which all the available resistant ABCC2 proteins and their resistant counterparts are heterologously expressed in the same cell system, and subjected to the same swelling and binding assays with all of the available toxins, would be extremely helpful. Such a study should also include *M. sexta*, because studies in this species have been crucial to the formation of the

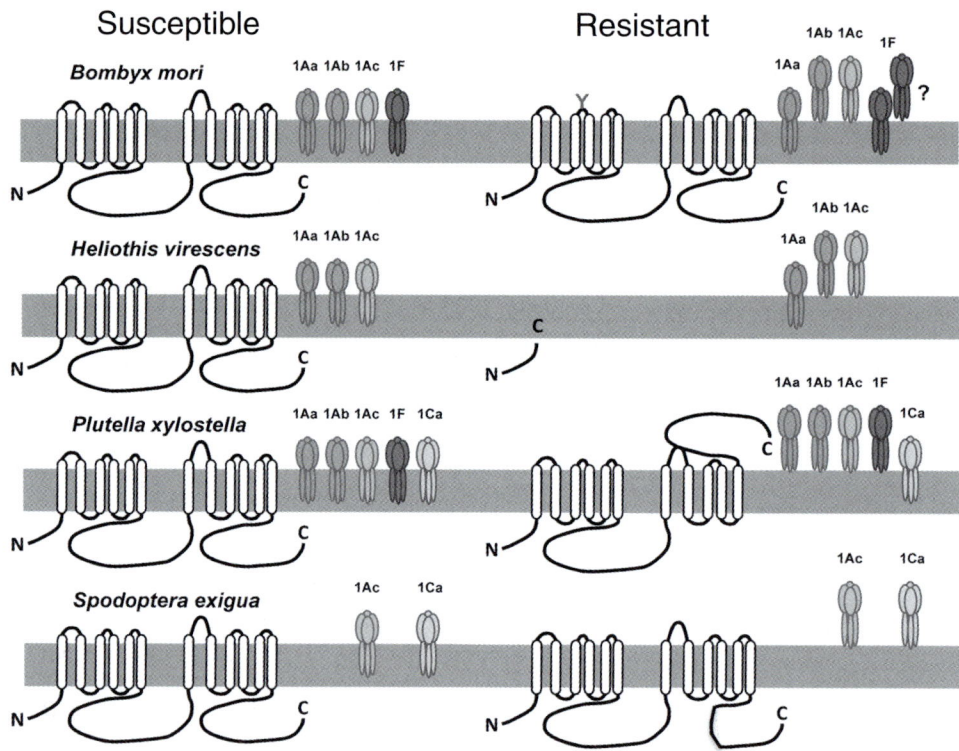

Fig. 9.1. Diagrammatic representation of *Bacillus thuringiensis* (Bt)-resistance-conferring mutations in the ABCC2 transporter proteins (N- and C-terminals indicated) of four lepidopteran species, and their consequences for pore insertion of various Cry toxins (1Aa, 1Ab, 1Ac, 1F, 1Ca). Drawing is not to scale. In the top row, Y represents the tyrosine insertion at position 234 (Tyr234) in resistant *B. mori*. The 99-amino acid truncated *H. virescens* protein is probably not associated with the membrane. *P. xylostella* and *S. exigua* proteins are depicted as if the mutations did not affect the membrane topology, but evidence for physical presence of these forms is lacking.

sequential binding model, even though no resistance due to ABCC2 protein modification is known in *Manduca* so far.

It would also be useful to interfere with the ATP-switch mechanism in such a comparative study, independently of the protein sequence differences conferring resistance. Although the dynamic aspect of the transport cycle has been hypothesized to drive pre-pore insertion, other possibilities can be envisioned. For example, an alteration in protein structure that blocks the normal transport cycle but favours a longer lasting opening of the channel to the outside might also facilitate pore insertion by increasing the time available for the pre-pore to interact with the hydrophobic interior.

Even such a comparative expression study would not reveal all because of the species-specific differences in the types and affinities of other Cry toxin binding targets in the epithelial membrane. The toxin must run the gauntlet of these interactions as it is further processed and oligomerizes and makes its way to the final goal. The challenge remains of connecting this well-studied binding network, which is different for each insect species and toxin, to the rapid, transient interaction with the ABC protein that is almost immediately interrupted when the toxin pore finally and irreversibly inserts into the membrane. When we understand all of this, we will finally understand how Cry toxins kill lepidopteran caterpillars – and perhaps other organisms as well.

This knowledge has long-term but significant potential for understanding and managing resistance. Understanding the transient interactions between toxins and ABC proteins may enable the design of improved toxins that form pores faster, or that target additional members of the ABC superfamily of transporters. Working out the precise role of the dynamic ATP-switch mechanism in pore insertion opens up the possibility of manipulating the transport cycle, by administering chemical stimulants or relying on plant-derived chemical toxins that are normally excreted by these transporters. Also, appreciation of the fitness benefits of Phase III detoxification mechanisms, including efflux by ABC transporters, can be used to exploit the fitness costs imposed by gaining resistance by losing a functional transporter to an inactivating mutation. Although these benefits are still a long way off, the need to sustain the enormous agricultural benefits of transgenic Bt-expressing crops provides a strong motivation for starting to pursue them now.

Acknowledgement

Preparation of this review was supported by the Max-Planck-Gesellschaft.

References

Atsumi, S., Miyamoto, K., Yamamoto, K., Narukawa, J., Kawai, S. et al. (2012) A single amino acid mutation in an ABC transporter causes resistance to Bt toxin Cry1Ab in the silkworm, *Bombyx mori*. *Proceedings of the National Academy of Sciences of the United States of America* 109, E1591–E1598.

Baxter, S.W., Zhao, J.Z., Gahan, L.J., Shelton, A.M., Tabashnik, B.E. et al. (2005) Novel genetic basis of field-evolved resistance to Bt toxins in *Plutella xylostella*. *Insect Molecular Biology* 14, 327–334.

Baxter, S.W., Zhao, J.Z., Shelton, A.M., Vogel, H. and Heckel, D.G. (2008) Genetic mapping of Bt-toxin binding proteins in a Cry1A-toxin resistant strain of diamondback moth *Plutella*

xylostella. *Insect Biochemistry and Molecular Biology* 38, 125–135.

Baxter, S.W., Badenes-Pérez, F.R., Morrison, A., Vogel, H., Crickmore, N. et al. (2011) Parallel evolution of *Bacillus thuringiensis* toxin resistance in Lepidoptera. *Genetics* 189, 675–679.

Bravo, A., Gómez, I., Conde, J., Muñoz-Garay, C., Sánchez, J. et al. (2004) Oligomerization triggers binding of a *Bacillus thuringiensis* Cry1Ab pore-forming toxin to aminopeptidase N receptor leading to insertion into membrane microdomains. *Biochimica et Biophysica Acta (BBA) – Biomembranes* 1667, 38–46.

Bravo, A., Gill, S.S. and Soberón, M. (2007) Mode of action of *Bacillus thuringiensis* Cry and Cyt toxins and their potential for insect control. *Toxicon* 49, 423–435.

Gahan, L.J., Gould, F. and Heckel, D.G. (2001) Identification of a gene associated with Bt resistance in *Heliothis virescens*. *Science* 293, 857–860.

Gahan, L.J., Pauchet, Y., Vogel, H. and Heckel, D.G. (2010) An ABC transporter mutation is correlated with insect resistance to *Bacillus thuringiensis* Cry1Ac toxin. *PLoS Genetics* 6(12): e1001248.

Heckel, D.G. (2012) Learning the ABCs of Bt: ABC transporters and insect resistance to *Bacillus thuringiensis* provide clues to a crucial step in toxin mode of action. *Pesticide Biochemistry and Physiology* 104, 103–110.

Higgins, C.F. and Linton, K.J. (2004) The ATP switch model for ABC transporters. *Nature Structural and Molecular Biology* 11, 918–926.

Janmaat, A.F. and Myers, J. (2003) Rapid evolution and the cost of resistance to *Bacillus thuringiensis* in greenhouse populations of cabbage loopers, *Trichoplusia ni*. *Proceedings of the Royal Society B: Biological Sciences* 270, 2263–2270.

Knowles, B.H. and Ellar, D.J. (1987) Colloid-osmotic lysis is a general feature of the mechanism of action of *Bacillus thuringiensis* δ-endotoxins with different insect specificity. *Biochimica et Biophysica Acta* 924, 509–518.

Michelmore, R.W., Paran, I. and Kesseli, R.V. (1991) Identification of markers linked to disease-resistance genes by bulked segregant analysis: a rapid method to detect markers in specific genomic regions by using segregating populations. *Proceedings of the National Academy of Sciences of the United States of America* 88, 9828–9832.

Park, Y., González-Martínez, R.M., Navarro-Cerrillo, G., Chakroun, M., Kim, Y. et al. (2014) ABCC transporters mediate insect resistance to multiple Bt toxins revealed by bulk segregant analysis. *BMC Biology* 12:46.

Raymond, B., Johnston, P.R., Nielsen-LeRoux, C., Lereclus, D. and Crickmore, N. (2010) *Bacillus thuringiensis*: an impotent pathogen? *Trends in Microbiology* 18, 189–194.

Schnepf, E., Crickmore, N., Van Rie, J., Lereclus, D., Baum, J. *et al.* (1998) *Bacillus thuringiensis* and its pesticidal crystal proteins. *Microbiology and Molecular Biology Reviews* 62, 775–806.

Slamti, L., Perchat, S., Huillet, E. and Lereclus, D. (2014) Quorum sensing in *Bacillus thuringiensis* is required for completion of a full infectious cycle in the insect. *Toxins* 6, 2239–2255.

Soberón, M., Pardo, L., Muñóz-Garay, C., Sánchez, J., Gómez, I. *et al.* (2010) Pore formation by Cry toxins. In: Anderluh, G. and Lakey, J., (eds.) *Proteins: Membrane Binding and Pore Formation. Advances in Experimental Medicine and Biology Volume 677.* Landes Bioscience/Springer, Austin, Texas, pp. 127–142.

Tabashnik, B.E., Malvar, T., Liu, Y.B., Finson, N., Borthakur, D. *et al.* (1996) Cross-resistance of the diamondback moth indicates altered interactions with domain II of *Bacillus thuringiensis* toxins. *Applied and Environmental Microbiology* 62, 2839–2844.

Tabashnik, B.T., Liu, Y.-B., Finson, N., Masson, L. and Heckel, D.G. (1997a) One gene in diamondback moth confers resistance to four *Bacillus thuringiensis* toxins. *Proceedings of the National Academy of Sciences of the United States of America* 94, 1640–1644.

Tabashnik, B.E., Liu, Y.B., Malvar, T., Heckel, D.G., Masson, L. *et al.* (1997b) Global variation in the genetic and biochemical basis of diamondback moth resistance to *Bacillus thuringiensis*. *Proceedings of the National Academy of Sciences of the United States of America* 94, 12780–12785.

Tanaka, S., Miyamoto, K., Noda, H., Jurat-Fuentes, J.L., Yoshizawa, Y. *et al.* (2013) The ATP-binding cassette transporter subfamily C member 2 in *Bombyx mori* larvae is a functional receptor for Cry toxins from *Bacillus thuringiensis*. *The FEBS Journal* 280, 1782–1794.

Zhao, J.Z., Li, Y.X., Collins, H.L., Cao, J., Earle, E.D. and Shelton, A.M. (2001) Different cross-resistance patterns in the diamondback moth (Lepidoptera : Plutellidae) resistant to *Bacillus thuringiensis* toxin CryIC. *Journal of Economic Entomology* 94, 1547–1552.

10

The Role of Proteolysis in the Biological Activity of Bt Insecticidal Crystal Proteins

Igor A. Zalunin,[1] Elena N. Elpidina[2] and Brenda Oppert[3]*

[1]The State Research Institute for Genetics and Selection of Industrial Microorganisms, Moscow, Russia; [2]A.N. Belozersky Institute of Physico-Chemical Biology, Moscow State University, Moscow, Russia; [3]USDA Agricultural Research Service, Center for Grain and Animal Health Research, Manhattan, Kansas, USA

Summary

The crystal toxins (Cry) produced by the bacterium *Bacillus thuringiensis* (Bt) have been successfully used in both spray formulations and transgenic crops to control some of the most problematic insect pests, as has been discussed in previous chapters. The δ-endotoxins of Bt are functionally active in the insect gut and interact with and are processed by proteolytic enzymes. The structure of Cry proteins has specific features that not only permit them to retain their biological activity in the hostile environment of the insect gut, but also to use the process of proteolysis in the solubilization and activation of Cry protoxins. Because the proteolysis of Cry proteins is critical to their biological activity, we review the literature on studies related to insect and mammalian proteases and their effects on toxin structure and toxicity.

10.1 Introduction

Cry proteins are found in parasporal inclusions within Bt. They are grouped by amino acid sequence identity, with over 70 Cry proteins described so far (Crickmore *et al.*, 2014). Many Cry proteins belong to a group of toxins referred to as 3d-Cry toxins (Pardo-López *et al.*, 2013). The term '3d' refers to the three domains connected by extended loops that are found in all 3d-Cry proteins. Within the 3d-Cry toxins, there are 130–145 kDa Cry protoxins (Cry1, Cry4, Cry9) that contain an extended C-terminus that is hydrolysed by gut proteases, and 65–73 kDa Cry protoxins (Cry2, Cry3, Cry11) that lack an extended C-terminus and therefore do not require this processing for activity. All 3d-Cry proteins are hydrolysed at the N-terminus by gut proteases to produce an active toxin. Hence, proteases are essential to the functionality of Cry insecticidal toxins, and thus significant research efforts have been made to understand the integral role of proteases in Cry toxicity.

10.2 Proteolysis of Cry Proteins by Model Proteases

Even before the three-dimensional structure of the active portion of the 130–145 kDa Cry proteins was known, it was determined that the C-terminus contains 15 kDa domains that are sequentially removed

---

* Corresponding author: E-mail address: brenda.oppert@ars.usda.gov or bso@k-state.edu

during proteolysis (Chestukhina *et al.*, 1982). Apparently, loops that connect these domains are readily accessible to proteolytic enzymes, and so the process is largely independent of specific proteases. After separation from the parent molecule, the hydrolysis products are easily subjected to continued hydrolysis to ten amino acid peptides (Choma *et al.*, 1990). As a rule, the activation of protoxins is also associated with the cleavage of a small N-terminus fragment of about 30 amino acid residues from the molecule (Nagamatsu *et al.*, 1984; Chestukhina *et al.*, 1994; Zalunin *et al.*, 1998). During the hydrolysis, the toxic 65–70 kDa core protein is formed. Generally, the toxic core adopts a conformation that is resistant to continued proteolysis; however, some exceptions to this rule have been observed, as will be discussed in the next section.

The 65–70 kDa Cry proteins lack the C-terminus portion of the 130–145 kDa Cry proteins, but they are homologous to the core toxins in primary structure and spatial organization (Pardo-López *et al.*, 2013). As with the 130–145 kDa Cry proteins, a short fragment is proteolytically cleaved from the N-terminus (Carroll *et al.*, 1989; Audtho *et al.*, 1999), but in the 65–70 kDa Cry proteins, the cleavage of peptide bonds in the core region is more frequent than in the 130–145 Cry proteins.

10.3 Limited Proteolysis of Core Toxins

The resistance of the core toxins to additional degradation is ensured by the integrity of their molecular structure, namely, an intimate connection of the domains by multiple non-covalent bonds, i.e. the loops by which the domains are linked are effectively protected from proteolytic enzymes (Li *et al.*, 1991; Grochulski *et al.*, 1995; Boonserm *et al.*, 2005). However, a few core toxins do undergo continued proteolysis under certain conditions, more frequently in the 65–70 kDa than in the 130–145 kDa Cry proteins. The hydrolysed peptide bond is located in one of the loops

that bind various elements of the molecular secondary structure. Examples can be divided into two groups: the cleavage occurs in one of the loops that link the α-helices of the N-terminus domain; or it takes place in the loops that bind the β-strands of the second and third domains.

10.3.1 Proteolytic cleavage of peptide bonds in loops linking the N-terminal domain α-helices

The processing of core toxins in the α-helices of the N-terminal domain is observed in Cry4A and Cry4B mosquitocidal proteins (Angsuthanasombat *et al.*, 1991, 1992, 1993). Trypsin hydrolyses these proteins to approximately 50 and 20 kDa fragments (Zalunin *et al.*, 1998; Yamagiwa *et al.*, 1999), at Arg-203–Ser-204 (Cry4B (Chungjatupornchai *et al.*, 1988) or Arg-235–Gln-236 (Cry4A) (Ward and Ellar, 1987) in the C-terminal portion of the loop that links helices α-5 and α-6. The 20 kDa fragment corresponds to the first five helices of the N-terminal portion of the core toxin, and the approximately 50 kDa fragment includes the α-6 and α-7 helices and both β-structure domains (Yamagiwa *et al.*, 1999). Chymotrypsin is incapable of hydrolysing peptide bonds in these Cry4A and Cry4B loops or in any other segment within the core region (Zalunin *et al.*, 1998). Neither trypsin nor chymotrypsin hydrolyse bonds in the loops α-5 to α-6 in the Cry1A subclass endotoxins.

These selective hydrolyses are explained by the structures of the similar segments in Cry1Aa and Cry4B (Grochulski *et al.*, 1995; Boonserm *et al.*, 2005). By X-ray diffraction analysis, the loop α-5 to α-6 is buried between helices α-5 and α-6. In fact, only the C-terminus of the loop is in contact with the solvent and is accessible to proteases. In Cry4A and Cry4B, this segment contains an Arg residue that is consistent with trypsin specificity, and this Arg is lacking in the C-terminus of the appropriate loop in Cry1Aa.

The proteolysis of Cry3A endotoxin, which is toxic to the larvae of some beetle species, has been well investigated. The protein is synthesized as a 73 kDa polypeptide, but is processed in the bacterium to a 67 kDa protein (Höfte et al., 1987; McPherson et al., 1988; Carroll et al., 1989). The processing starts from the N-terminus with the cleavage of 57 amino acid residues, so that the core toxin starts with Asp-58. Trypsin or gut extracts of some coleopteran insects convert this protein to a 55 kDa core toxin with insecticidal activity (Carroll et al., 1989). The hydrolysis occurs at the Lys-158–Asn-159 bond located in the loop α-3 to α-4 (Carroll et al., 1989). The reaction products contain an 11 kDa fragment corresponding to the N-terminal portion of the 67 kDa protein and an 8 kDa fragment resulting from continued hydrolysis. Chymotrypsin degrades the 67 kDa Cry3a protoxin to 49, 11, 8 and 6 kDa fragments. The formation of a 49 kDa fragment is accompanied by the hydrolysis of the His-161–Ser-162 peptide bond and also the Tyr-587–Tyr-588 bond at the end of β-strand 19 in domain III. The additional cleavage of the C-terminal fragment of the third domain by chymotrypsin sharply increases the solubility of the protein at pH 5–9, which makes the 49 kDa peptide more accessible to binding with the gut epithelium of the coleopteran target.

Therefore, the C-terminus of the loop α-3 to α-4 in the Cry3Aa protoxin is amenable to protease activity, and the precise place and rate of the process depend on the specificity of the protease (Carroll et al., 1997). The proteolytic cleavage of the bond inside the loop α-3 to α-4 has also been reported for Cry2Aa (Audtho et al., 1999; Morse et al., 2001; Ohsawa et al., 2012), Cry9Ca (Lambert et al., 1996; Brunet et al., 2010), Cry3Ba and Cry3Ca (Rausell et al., 2004).

10.3.2 Proteolytic cleavage in loops that link β-structural elements in domains II and III

The proteolysis of Cry11A endotoxin differs from that of other mosquitocidal toxins (Dai and Gill, 1993). This 70 kDa endotoxin is hydrolysed by trypsin, chymotrypsin and thermolysin to fragments of 40 and 34 kDa that are converted to 36 and 33 kDa proteins (Yamagiwa et al., 1999, 2002). The 36 kDa fragment derives from the N-terminus and is a result of hydrolysis of the Ile-28–Ala-29 bond. The 33 kDa fragment is a product of the hydrolysis of the Arg-360–Asp-361 bond that occurs in the loop linking β4 and β5 strands in domain II (Gutierrez et al., 2001), a loop that is sensitive to other proteases (Dai and Gill, 1993; Revina et al., 2004). Apparently, a major portion of the β4–β5 loop occurs on the surface of the toxin molecule. The proteolysis scheme for Cry1A and Cry11 endotoxins is similar. A limited proteolysis of Cry1Ab by chymotrypsin, thermolysin, pronase, subtilisin and papain ensures the cleavage of the peptide bond in loops that link the β strands in domain II (Chestukhina et al., 1990; Choma et al., 1990; Convents et al., 1991). However, the disruption of the bond in the second domain of Cry1A proteins occurs with much less efficiency than in Cry11A. With Cry1A, the disruption requires either partial denaturing of the protein (Chestukhina et al., 1990; Choma et al., 1990; Convents et al., 1991), or a relatively large amount of the proteolytic enzyme and long-term incubation (Pang et al., 1999; Miranda et al., 2001). Therefore, the data suggest that the sensitivity of the β-strand linking loops to protease activity is determined by the length of the loop and the accessibility to proteolytic enzymes (Revina et al., 2004), which is similar to proteolysis within the N-terminal domain.

10.4 Biological Activity of Cry Protein Hydrolysates

Almost all hydrolysates of Bt Cry protoxins obtained with commercial preparations of purified mammalian proteases either retain or increase the toxicity of the original protein (Dai and Gill, 1993; Carroll et al., 1997; Yamagiwa et al., 2002; Revina et al., 2004). This tenet demonstrates the inherent functionality of the proteolysis of Cry proteins and the resiliency of the hydrolysed protein (i.e. the toxin). For example, tryptic hydrolysates of Cry4B protoxin were approximately twice as active against *Aedes aegypti* larvae as the protoxin (Zalunin *et al.*, 1998). Cry2Aa9 (60 kDa) and its chymotryptic hydrolysate (containing a 50 kDa fragment) have similar activity towards *Bombyx mori* larvae, whereas the chymotryptic hydrolysate was twofold more active than the protoxin in *Lymantria dispar* (Ohsawa *et al.*, 2012).

In fact, hydrolysates were used to confirm the mechanism of protease-mediated resistance in the Indianmeal moth, *Plodia interpunctella* (Herrero *et al.*, 2001) and the European corn borer, *Ostrinia nubilalis* (Li *et al.*, 2004). In both resistant insect populations, resistance to Cry1Ab was reduced when insects were fed the protoxin activated with bovine pancreatic α-chymotrypsin or gut extracts. Some of the information summarized in this review can be found in a compilation of bioassay data (van Frankenhuyzen and Nystrom, 2002) that is useful in assessing the response of insects to a number of Cry proteins.

10.5 Proteolysis Products of the Bt Endotoxin Core Form an Intermolecular Complex

The separation of hydrolysis products by ion exchange chromatography or gel filtration is difficult because of their simultaneous elution from columns in the forms of putative intermolecular complexes. These complexes have been observed in the hydrolysis of various Cry proteins: Cry4A by the gut extract of *Culex pipiens* larvae (Yamagiwa *et al.*, 1999); Cry4B by *Ae. aegypti* proteases (Angsuthanasombat *et al.*, 1993); Cry3Aa by different proteases (Carroll *et al.*, 1997); and Cry11A by trypsin (Yamagiwa *et al.*, 2002). The molecular mass of this complex and the core protein were similar, as demonstrated by gel filtration (Yamagiwa *et al.*, 2002). Together with data on the toxicity of hydrolysates (see Section 10.6.4), these results suggest that the complex of proteolysis products

retains the active conformation of the initial molecule.

This hypothesis is supported by the fact that, in most cases, Cry toxin fragments that are separated by denaturation reagents are non-toxic. For example, 36 and 30 kDa fragments from a Cry11A trypsin hydrolysate that were separated by chromatography in the presence of 2 M urea both lacked toxicity, but toxicity was restored when the fragments were combined (Revina *et al.*, 2004; Yamagiwa *et al.*, 2004).

Various data suggest that the fragments of Cry proteins corresponding to the α-helical domain have membranotropic and pore-forming effects (Walters *et al.*, 1993; Puntheeranurak *et al.*, 2004), although despite this, these fragments did not bind to midgut receptor(s) and were non-toxic to neonate *Heliothis virescens* larvae (Walters *et al.*, 1993). Cry1Aa activity against *B. mori* larvae was decreased as a result of the degradation of the core toxin to 31.8 kDa and 29.6 kDa fragments by pronase or subtilisin in high concentrations, although a complex of the proteolysis products was formed (Pang *et al.*, 1999). Apparently, retaining association between the proteolysis products does not always guarantee their biological activity.

A question arises: can proteolysis in the core part of Bt toxins have a functional significance, such as increasing the flexibility of the molecule and its ability to influence molecular dynamics? The data suggest otherwise, for the elimination of the proteolytic site in Cry4A, Cry4B, Cry9Ca and Cry11A has no effect on toxicity (Angsuthanosombat *et al.*, 1991, 1992, 1993; Dai and Gill, 1993; Lambert *et al.*, 1996; Yamagiwa *et al.*, 1999).

10.6 Specific Traits of the Effect of Insect Digestive Proteases on Bt Cry Proteins

Since the publication of pioneering works of Lecadet *et al.* (Lecadet and Martouret, 1965, 1967; Lecadet and Dedonder, 1967), numerous investigations have demonstrated that insect gut extracts and proteolytic enzymes convert Cry protoxins to active toxins (reviewed in Oppert, 1999). An insect gut extract contains many endoproteases and exoproteases that complicate the study of the effect of protoxin hydrolysis, but analysis of the composition of the hydrolysate can contribute to an understanding of toxin activation and specificities. Frequently, the products of hydrolysis by gut extracts are similar to those of commercial proteases (e.g. Carroll et al., 1997; Bulushova et al., 2011). However, the action of midgut proteases in a complex can be more variable and profound in comparison with that of individual enzymes, as considered in the following paragraphs.

10.6.1 The activity of insect gut enzymes can alter Cry toxicity

In early studies, the treatment of Cry1Ab protoxin from an IC1 strain of B. t. subsp. aizawai (Bta) by mosquito larvae gut extract induced toxicity to dipteran cell lines (Haider et al., 1986). Cry1B and Cry7A proteins were active against some coleopteran species only after preliminary in vitro dissolution and activation, an indication that conditions in the gut must be favourable for biological activity (Lambert et al., 1992; Bradley et al., 1995).

In some cases, the treatment of Cry protoxins with gut extract reduces biological activity. For example, the 48 and 47 kDa fragments from the hydrolysis of the Cry4B protoxin by Ae. aegypti gut extract are much less toxic to this mosquito than a 50 kDa fragment from the tryptic hydrolysate of the protein (Angsuthanasombat et al., 1991; Zalunin et al., 1998). Hydrolysis of Cry1A to 30 kDa fragments by gut proteases of either Spodoptera frugiperda or Manduca sexta renders them inactive (Miranda et al., 2001).

10.6.2 Proteases of midgut extracts differ in their effect on Cry protoxin processing

Because of the complexity of proteases in the insect gut, both in function and temporal expression, the effect of insect proteases on Bt δ-endotoxins can be difficult to predict. For example, hydrolysis of the 130 kDa Cry1Ab protoxin by two trypsin enzymes from Sesamia nonagrioides provided a single 69 kDa fragment, whereas a third trypsin degraded these fragments further (Díaz-Mendoza et al., 2007). Similarly, although five trypsin-like proteases were found in the gut extract of Plutella xylostella, only one was found to activate Cry1Aa and Cry1Ab protoxins to active toxins (Mohan and Gujar, 2002).

Attempts have been made to define a connection between sensitivity to Bt δ-endotoxins and the proteolytic activation or inactivation of Cry toxicity. A comparison of the effects of gut extracts of protoxin-sensitive (Pieris brassicae) and protoxin-insensitive (Mamestra brassicae) insects on CryA1c protoxin demonstrated that a rapid proteolysis of the N-terminus occurred in both cases (Lightwood et al., 2000), but that the processes of limited proteolysis in these insects were different. Nonetheless, a mutant toxin incapable of being hydrolysed to smaller fragments did not increase the toxicity to M. brassicae.

Dai and Gill (1993) demonstrated that enzymes with trypsin, chymotrypsin and thermolysin activities were involved in cleaving Cry11A by C. quinquefasciatus midgut proteases. Interestingly, the products of midgut proteolysis differ, depending on whether Cry11A occurs in the dissolved state or in the crystal form. Moreover, the researchers showed that the in vivo proteolysis products differed from those observed in vitro. Therefore, physiologically relevant experiments should be compared with actual molecular events in the insect gut.

It is apparent that the relative expression of proteases can affect the differential sensitivity to toxins in different larval instars. The Cry1C δ-endotoxin is toxic to S. littoralis larvae, but the first instar is more sensitive than later instars, which have increased proteolytic activity that can degrade the toxin completely (Keller et al., 1996).

10.6.3 Proteolysis is not a rate-limiting process for Cry toxicity in sensitive insects

In a *Galleria mellonella* larval gut extract, five and two enzymes were described with trypsin-like and chymotrypsin-like activities, respectively (Bulushova *et al.*, 2011). Trypsins or chymotrypsins from this insect are equally involved in the processing of the endotoxins from *B. t.* subsp. *galleria* (Btg) crystals. The following characteristics were critical for Cry toxicity in *G. mellonella*: (i) the level of protease activity in the anterior midgut of a larva (0.314 U of tryptic activity); (ii) the concentration of midgut extract that is required for a complete activation of the Btg protein crystals (0.00625 U of tryptic activity mg^{-1} protein h^{-1}); and (iii) the concentration of the activated toxin that induces a complete cessation of feeding (0.6 µg per insect). These data demonstrate that the level of the proteolytic activity in anterior midgut was 3000 times more than is needed for the activation of a lethal dose of the protoxin for this insect. Consequently, the activation of the endotoxin in a sensitive insect is not a rate-limiting stage in its toxic effect.

10.6.4 Can the properties of an insect proteolytic complex underlie species specificity of Cry toxicity?

Attempts have been made to define a connection between sensitivity to Bt δ-endotoxins and the proteolytic activation or inactivation of Cry toxicity. As previously discussed, gut extracts of Cry1Ac protoxin-sensitive (*P. brassicae*) and protoxin-insensitive (*M. brassicae*) insects could rapidly hydrolyse the N-terminus of Cry1Ac protoxin (Lightwood *et al.*, 2000). However, in the insensitive *M. brassicae*, 60, 58, 40 and 20 kDa proteins were the primary products of the proteolysis *in vitro*. Chymotrypsin cleaved the peptide bonds at the N-terminus of the molecule, whereas trypsin was responsible for the hydrolysis within the second domain. In the sensitive *P. brassicae*, the proteolysis was limited to 60 and 56 kDa fragments. A mutant toxin that could not be hydrolysed to 40 and 20 kDa fragments did not show increased toxicity to *M. brassicae*.

Notwithstanding, the expression of midgut proteases can relate to differential sensitivity to toxins in different larval instars. The Cry1C δ-endotoxin is toxic to *S. littoralis* larvae, but the first instar was more sensitive than later instars (Keller *et al.*, 1996). In the incubation of Cry1C with gut extract from larvae of the first and second instars, the toxic portion of Cry1C was found. In contrast, gut extract from fifth instar larvae degraded the protein completely. The role of larval age in decreased toxin sensitivity was confirmed by a protease inhibitor analysis.

10.6.5 Qualitative and quantitative changes in the composition of the digestive complex of the sensitive insect can lead to Cry resistance

Changes in the processing of the Cry protein can be one of the reasons for the emergence of toxin resistance in insects, as was first demonstrated in a series of papers from Oppert *et al.* (1994, 1996, 1997). A strain of *P. interpunctella*, selected in the laboratory as resistant to *B. t.* subsp. *entomocidus* (Bte), had reduced sensitivity to Cry1Ab, Cry1Ac and Cry1C protoxins. The activation of Cry1Ac protoxin by the gut extract of this resistant *P. interpunctella* was considerably slower than that of the susceptible strain, which was attributed to a fivefold decrease in the tryptic activity of the gut extract of this insect (Oppert *et al.*, 1994). The decrease in trypsin activity was found to be related to the loss of a major trypsin enzyme (Oppert *et al.*, 1996) and was genetically linked to resistance (Oppert *et al.*, 1997). This Bte-resistant strain of *P. interpunctella* had a slight delay in development (about a day) but suffered approximately 50% mortality from egg to adult when reared on either Bt-treated or untreated diets (Oppert *et al.*, 2000). The high mortality in this strain was due mostly to a low percentage of egg hatch (unpublished data), but it is unknown whether a loss of trypsin activity was involved.

A comparative analysis of *H. virescens* sensitive to Cry1Ac and Cry2Aa δ-endotoxins and three resistant populations demonstrated that the mechanisms of toxin resistance in different strains were not the same (Karumbaiah *et al.*, 2007). Protease-mediated resistance was found in a strain with reduced protoxin processing, but both processing defects and toxin-receptor binding were found in another, and multifactorial resistance mechanisms in yet another. These differences were attributed to the collective rate of protoxin hydrolysis, and the intermediate and final toxin products were the same among the strains. Nevertheless, zymographic analysis revealed that the decrease in rate of activation of protoxins by the gut extract of one of the resistant strains correlated with a reduction in trypsin hydrolysis due to the loss of a 32 kDa trypsin-like protease, while a decrease in protoxin proteolysis in another strain correlated with a reduction in a 36 kDa chymotrypsin-like enzyme activity. The targeted reduction of serine protease activity and its relation to activation of the Cry1Ab protoxin was also associated with resistance to Cry1Ab in *Ostrinia nubilalis* (Li *et al.*, 2004) and with a 75-fold enhanced resistance to Cry1A in a laboratory population of *Helicoverpa armigera* (Rajagopal *et al.*, 2009). Resistance mediated by changes in proteolytic activity in the insect gut is mostly associated with lower resistance levels, but is seemingly widespread among insects faced with increased selection pressure from Cry toxins.

10.7 Emergence of Cytolytic Activity during Cry Protein Proteolysis

In some cases, extended proteolysis of Cry proteins by gut extracts leads to the appearance of cytolytic activity. Gut extracts from *Ae. aegypti* produce 46–48 kDa Cry4B fragments that exhibit cytolytic activity against *Ae. aegypti* cell cultures, but the cell cultures derived from other mosquito species were resistant to these fragments (Angsuthanasombat *et al.*, 1991). The Cry4A endotoxin treated by extracts of *Ae. aegypti*,

Anopheles gambiae or *C. quinquefasciatus* midgut extracts (but not by trypsin) acquired cytolytic activity against *An. gambiae* and *C. quinquefasciatus* cells; however, these hydrolysis products were inactive against *Ae. aegypti* cells (Angsuthanasombat *et al.*, 1992). The cytolytic activity is probably associated with the cleavage of a small fragment of the core toxin molecule from the C-terminus, and results in changes in the conformational flexibility of the molecule that is responsible for its binding to cell membrane receptors.

10.8 Proteolysis of Cry Proteins by Membrane Proteases

10.8.1 Cleavage of the α-1 helix

The incubation of the Cry1Ab endotoxin with preparations from *M. sexta* midgut epithelium apical (brush border) membrane vesicles (BBMVs) provides the hydrolysis of the Phe-50–Val-51 peptide bond located in the loop that links helices α-1 and α-2, apparently by membrane proteases (Gómez *et al.*, 2002). In turn, the removal of the first α-helix results in oligomerization of the toxin molecules, an important stage preceding the formation of transmembrane pores. The proteolysis and oligomerization are associated with toxin binding to a cadherin-like receptor in the epithelial cell membranes (Gómez *et al.*, 2002; Muñoz-Garay *et al.*, 2009).

10.8.2 ADAM proteases

The incubation of Cry3Aa endotoxin with BBMVs from *Leptinotarsa decemlineata* results in the hydrolysis of the insecticidal protein into three fragments (Rausell *et al.*, 2007). The membrane-bound protease from BBMVs of *L. decemlineata* was identified as an ADAM10 (a disintegrin and metallo-proteinase domain-containing protein 10)-like metalloprotease (Ochoa-Campuzano *et al.*, 2007). An ADAM10 recognition sequence (342FHTRFQPG349) was found in loop 1 of domain II in the Cry3Aa endotoxin. This finding provided evidence that the enzyme binds to the toxin at the second domain and hydrolyses the peptide bond in the third domain, which is consistent with the mechanism of ADAM activity.

The fact that ADAMs are involved in numerous signal pathways may indicate that this enzyme is also a toxin receptor. Competing with cytokines and some messenger molecules for protease binding, the toxin could inhibit various vital signal processes in the cell, resulting in indirect cell death mechanisms (Ochoa-Campuzano *et al.*, 2007). *L. decemlineata* cysteine proteases (intestains) may also interact with the Cry3Aa protoxin, as was suggested by inhibitor analysis and the binding of intestains to Cry3Aa (García-Robles *et al.*, 2013).

Interestingly, the food source of *L. decemlineata* larvae affects the intensity of proteolysis of Cry3Aa protoxin by BBMVs (García-Robles *et al.*, 2013). The proteolytic system of the insect is, of course, plastic and responds to the nature of food supply by modulating proteolytic activity; this, in turn, affects its ability to activate and/or degrade the toxin.

10.9 Conclusion

The data discussed in this review illustrate how Bt insecticidal toxicity is interrelated with insect proteolytic enzymes and dependent on this enzyme activity for toxicity. Undoubtedly, the adaptation of entomocidal proteins to insect gut conditions was achieved during prolonged evolution. The data on mechanisms of the resistance to δ-endotoxins suggest the co-adaptive evolution of insect digestive systems and δ-endotoxins (Raymond *et al.*, 2010), in a manner similar to the co-adaptive evolution of angiosperm plant inhibitors of proteases and insect proteolytic enzymes (Lopes *et al.*, 2006; Srinivasan *et al.*, 2006; Sato, *et al.*, 2008). It is likely that the large expansion of protease genes in many insects is due to adaptive pressure from plant inhibitors and microbial toxins. Continued research into the native interactions of microbial toxins in the insect gut, in addition to a complete characterization of the complex of digestive enzymes in target pests, will contribute to improved microbial insect control products.

Acknowledgements

Mention of trade names or commercial products in this publication is solely for the purpose of providing specific information and does not imply recommendation or endorsement by the US Department of Agriculture (USDA). The USDA is an equal opportunity provider and employer.

This chapter is dedicated to Professor Galina G. Chestukhina, whose fundamental research on the proteolysis of Cry toxins has provided much of our understanding of the mode of action and specificity of these toxins in the early stages of insect intoxication. Her name will forever remain a part of the glorious galaxy of the pioneers of the study of *Bacillus thuringiensis* δ-endotoxins. She will be missed.

References

Angsuthanasombat, C., Crickmore N. and Ellar, D.J. (1991) Cytotoxicity of a cloned *Bacillus thuringiensis* subsp. *israelensis* CryIVB toxin to an *Aedes aegypti* cell line. *FEMS Microbiology Letters* 83, 273–276.

Angsuthanasombat, C., Crickmore N. and Ellar, D.J. (1992) Comparison of *Bacillus thuringiensis* subsp. *israelensis* Cry IVA and CryIVB cloned toxins reveals synergism *in vivo*. *FEMS Microbiology Letters* 94, 63–68.

Angsuthanasombat, C., Crickmore, N. and Ellar, D.J. (1993) Effects on toxicity of eliminating a cleavage site in a predicted interhelical loop in *Bacillus thuringiensis* CryIVB δ-endotoxin. *FEMS Microbiology Letters* 111, 255–262.

Audtho, M., Valaitis, A.P., Alzate, O. and Dean, D.H. (1999) Production of chymotrypsin-resistant *Bacillus thuringiensis* Cry2Aa1 δ-endotoxin by protein engineering. *Applied and Environmental Microbiology* 65, 4601–4605.

Boonserm, P., Davis, P., Ellar, D.J. and Li, J. (2005) Crystal structure of the mosquito-larvicidal toxin Cry4Ba and its biological implications. *Journal of Molecular Biology* 348, 363–382.

Bradley, D., Harkey, M.A., Kim, M.K., Biever, K.D. and Bauer, L.S. (1995) The insecticidal Cry1B crystal protein of *Bacillus thuringiensis* ssp. *thuringiensis* has dual specificity to coleopteran and lepidopteran larvae. *Journal of Invertebrate Pathology* 65, 162–173.

Brunet, J.-F., Vachon, V., Marsolais, M., Van Rie, J., Schwartz, J.L. *et al.* (2010) Midgut juice components affect pore formation by the *Bacillus thuringiensis* insecticidal toxin Cry9Ca. *Journal of Invertebrate Pathology* 104, 203–208.

Bulushova, N.V., Elpidina, E.N., Zhuzhikov, D.P., Lyutikova, L.I., Ortego, RF. *et al.* (2011) Complex of digestive proteinases of *Galleria mellonella* caterpillars. Composition, properties, and limited proteolysis of *Bacillus thuringiensis* endotoxins. *Biochemistry (Moscow)* 76, 581–589.

Carroll, J., Li, J. and Ellar, D.J. (1989) Proteolytic processing of a coleopteran-specific δ-endotoxin produced by *Bacillus thuringiensis* var. *tenebrionis*. *Biochemical Journal* 261, 99–105.

Carroll, J., Convents, D., Van Damme, J., Boets, A., Van Rie, J. *et al.* (1997) Intramolecular proteolytic cleavage of *Bacillus thuringiensis* Cry3A δ-endotoxin may facilitate its coleopteran toxicity. *Journal of Invertebrate Pathology* 70, 41–49.

Chestukhina, G.G., Kostina, L.I., Mikhailova, A.L., Tyurin, S.A., Klepikova F.S. *et al.* (1982) The main features of *Bacillus thuringiensis* δ-endotoxin molecular structure. *Archives of Microbiology* 132, 159–162.

Chestukhina, G.G., Tyurin, S.A., Kostina, L.I., Osterman, A.L., Zalunin, I.A. *et al.* (1990) Subdomain organization of *Bacillus thuringiensis* entomocidal protein's N-terminal domains. *Journal of Protein Chemistry*, 9, 501–507.

Chestukhina, G.G., Kostina, L.I., Zalunin, I.A., Revina, L.P., Mikhailova, A.L. *et al.* (1994) Production of multiple δ-endotoxins by *Bacillus thuringiensis*: δ-endotoxins produced by strains of the subspecies *galleriae* and *wuhanensis*. *Canadian Journal of Microbiology* 40, 1026–1034.

Choma, C.T., Surewicz, W.K., Carey, P.R., Pozsgay, M., Raynor, T. *et al.* (1990) Unusual proteolysis of the protoxin and toxin from *Bacillus thuringiensis* structural implications. *European Journal of Biochemistry* 189, 523–527.

Chungjatupornchai, W., Höfte, H., Seurinck, J., Angsuthanasombat, C. and Vaeck, M. (1988) Common features of *Bacillus thuringiensis* toxins specific for Diptera and Lepidoptera. *European Journal of Biochemistry* 173, 9–16.

Convents, D., Cherlet, M., Van Damme, J., Lasters, I. and Lauwereys, M. (1991) Two structural domains as a general fold of the toxic fragment of the *Bacillus thuringiensis* δ-endotoxins. *European Journal of Biochemistry* 195, 631–635.

Crickmore, N., Baum, J., Bravo, A., Lereclus, D., Narva, K. *et al.* (2014) *Bacillus thuringiensis* toxin nomenclature. Available at: *http://www.btnomenclature.info/* (accessed 22 October 2014).

Dai, S.-M. and Gill, S.S. (1993) *In vitro* and *in vivo* proteolysis of the *Bacillus thuringiensis* subsp. *israelensis* Cry IVD protein by *Culex quinquefasciatus* larval midgut proteases. *Insect Biochemistry and Molecular Biology* 23, 273–283.

Díaz-Mendoza, M., Farinós, G.P., Castañera, P., Hernández-Crespo, P. and Ortego, F. (2007) Proteolytic processing of native Cry1Ab toxin by midgut extracts and purified trypsins from the Mediterranean corn borer *Sesamia nonagrioides*. *Journal of Insect Physiology* 53, 428–435.

García-Robles, I., Ochoa-Campuzano, C., Fernández-Crespo, E., Camañes, G., Martínez-Ramírez, A.C. *et al.* (2013) Combining hexanoic acid plant priming with *Bacillus thuringiensis* insecticidal activity against Colorado potato beetle. *International Journal of Molecular Sciences* 14, 12138–12156.

Gómez, I., Sánchez, J., Miranda, R., Bravo, A. and Soberón, M. (2002) Cadherin-like receptor binding facilitates proteolytic cleavage of helix α-1 in domain I and oligomer pre-pore formation of *Bacillus thuringiensis* CryAb toxin. *FEBS Letters* 513, 242–246.

Grochulski, P., Masson, L., Borisova, S., Pusztai-Carey, M., Schwartz, J.L. *et al.* (1995) *Bacillus thuringiensis* CryIA(a) insecticidal toxin: crystal structure and channel formation. *Journal of Molecular Biology* 254, 447–464.

Gutierrez, P., Alzate, O. and Orduz, S. (2001) A theoretical model of the tridimensional structure of *Bacillus thuringiensis* subsp. *medellin* Cry 11Bb toxin deduced by homology modelling. *Memórias do Instituto Oswaldo Cruz* 96, 357–364.

Haider, M. Z., Knowles, B.H. and Ellar, D.J. (1986) Specificity of *Bacillus thuringiensis* var. *colmeri* insecticidal δ-endotoxin is determined by differential proteolytic processing of the protoxin by larval gut proteases. *European Journal of Biochemistry* 156, 531–540.

Herrero, S., Oppert, B. and Ferré, J. (2001) Different mechanisms of resistance to *Bacillus thuringiensis* in the Indianmeal moth. *Applied and Environmental Microbiology* 67, 1085–1089.

Höfte, H., Seurinck, J., Van Houkven, A. and Vaeck, M. (1987) Nucleotide sequence of a gene encoding an insecticidal protein of *Bacillus thuringiensis* var. *tenebrionis* toxic against Coleoptera. *Nucleic Acid Research* 15(17): 7183.

Karumbaiah, L., Oppert, B., Jurat-Fuentes, J.L. and Adang, M.J. (2007) Analysis of midgut proteinases from *Bacillus thuringiensis* susceptible and resistant *Heliothis virescens* (Lepidoptera: Noctuidae). *Comparative Biochemistry and Physiology Part B: Biochemistry and Molecular Biology* 146, 139–146.

Keller, M., Sneh, B., Strizhov, N., Prudovsky, E., Regev, A. *et al.* (1996) Digestion of δ-endotoxin by gut proteases may explain reduced sensitivity of advanced instar larvae of *Spodoptera littoralis* to CryIC. *Insect Biochemistry and Molecular Biology* 26, 365–373.

Lambert, B., Höfte, H., Annys, K., Jansens, S., Soetraert, P. *et al.* (1992) Novel *Bacillus thuringiensis* insecticidal crystal protein with a silent activity against coleopteran larvae. *Applied and Environmental Microbiology* 58, 2536–2542.

Lambert, B., Buysse, L., Decock, C., Jansens, S., Piens, C. *et al.* (1996) A *Bacillus thuringiensis* insecticidal crystal protein with a high activity against members of the family Noctuidae. *Applied and Environmental Microbiology* 62, 80–86.

Lecadet, M.M. and Dedonder, R. (1967) Enzymatic hydrolysis of the crystals of *Bacillus thuringiensis* by the proteases of *Pieris brassicae* I. Preparation and fractionation of the lysates. *Journal of Invertebrate Pathology* 9, 310–321.

Lecadet, M.M. and Martouret, D. (1965) The enzymic hydrolysis of *Bacillus thuringiensis* Berliner crystals and the liberation of the toxic fragments of bacterial origin by the chyle of *Pieris brassicae* (Linnaeus). *Journal of Invertebrate Pathology* 7, 105–108.

Lecadet, M.M. and Martouret, D. (1967) Enzymatic hydrolysis of the crystals of *Bacillus thuringiensis* by the proteases of *Pieris brassicae* II. Toxicity of the different fractions of the hydrolyzate for larvae of *Pieris brassicae*. *Journal of Invertebrate Pathology* 9, 322–330.

Li, H., Oppert, B., Higgins, R.A., Huang, F., Zhu, K.Y, *et al.* (2004) Comparative analysis of proteinase activities of *Bacillus thuringiensis* resistant and susceptible *Ostrinia nubilalis* (Lepidoptera: Crambidae). *Insect Biochemistry and Molecular Biology* 34, 753–762.

Li, J., Carroll, J. and Ellar, D.J. (1991) Crystal structure of insecticidal δ-endotoxin from *Bacillus thuringiensis* at 2,5 Å resolution. *Nature* 353, 815–821.

Lightwood, D.J., Ellar, D.J. and Jarrett, P. (2000) Role of proteolysis in determining potency of *Bacillus thuringiensis* Cry1Ac δ-endotoxin. *Applied and Environmental Microbiology* 66, 5174–5181.

Lopes, A.R., Juliano, M.A., Juliano, L., Marana, S.R. and Terra, W.R. (2006) Substrate specificity of insect trypsins and the role of their subsites in catalysis. *Insect Biochemistry and Molecular Biology* 36, 130–140.

McPherson, S.A., Perlak, F.J., Fuchs, R.L., Marrone, P.G., Larvik, P.B. *et al.* (1988) Characterization of the protein gene of *Bacillus thuringiensis* var. *tenebrionis*. *Bio/Technology* 6, 61–66.

Miranda, R., Zamudio, F.Z. and Bravo, A. (2001) Processing of Cry1Ab δ-endotoxin from *Bacillus thuringiensis* by *Manduca sexta* and *Spodoptera frugiperda* midgut proteases: role in protoxin activation and toxin inactivation. *Insect Biochemistry and Molecular Biology* 31, 1155–1163.

Mohan, M. and Gujar, G.T. (2002) Geographical variation in larval susceptibility of the diamondback moth, *Plutella xylostella* (Lepidoptera: Plutellidae) to *Bacillus thuringiensis* spore–crystal mixtures and purified crystal proteins and associated resistance development in India. *Bulletin of Entomological Research* 92, 489–498.

Morse, R.J., Yamamoto, T. and Stroud, R.M. (2001) Structure of Cry2Aa suggests an unexpected receptor binding epitope. *Structure* 9, 409–417.

Muñóz-Garay [Muñoz-Garay], C., Portugal, L., Pardo-López, L., Jiménez-Juárez, N., Arenas, I. *et al.* (2009) Characterization of the mechanism of action of the genetically modified Cry1AbMod toxin that is active against Cry1Ab-resistant insects. *Biochimica et Biophysica Acta* 1788, 2229–2237.

Nagamatsu, Y., Itai, Y., Hatanaka, G., Funatsu, G. and Hayashi, K. (1984) A toxic fragment from the entomocidal crystal protein of *Bacillus thuringiensis*. *Agricultural and Biological Chemistry* 48, 611–619.

Ochoa-Campuzano, C., Real, M.D., Martínez-Ramírez, A.C., Bravo, A. and Rausell, C. (2007) An ADAM metalloprotease is a Cry3Aa *Bacillus thuringiensis* toxin receptor. *Biochemical and Biophysical Research Communications* 362, 437–442.

Ohsawa, M., Tanaka, M., Morijama, K., Shimazu,

M., Asano, S.-I. *et al.* (2012). A 50-kilodalton Cry 2A peptide is lethal to *Bombyx mori* and *Lymantria dispar*. *Applied and Environmental Microbiology* 78, 4755–4757.

Oppert, B. (1999) Protease interactions with *Bacillus thuringiensis* insecticidal toxins. *Archives of Insect Biochemistry and Physiology* 42, 1–12.

Oppert, B., Kramer, K.J., Johnson, D.E., MacIntosh, S.C. and McGaughey, W.H. (1994) Altered protoxin activation by midgut enzymes from a *Bacillus thuringiensis* resistant strain of *Plodia interpunctella*. *Biochemical and Biophysical Research Communications* 198, 940–947.

Oppert, B., Kramer, K.J., Johnson, D., Upton, S.J. and McGaughey, W.H. (1996) Luminal proteinases from *Plodia interpunctella* and the hydrolysis of *Bacillus thuringiensis* CryIA(c) protoxin. *Insect Biochemistry and Molecular Biology* 26, 571–583.

Oppert, B., Kramer, K.J., Beeman, R.W., Johnson, D. and McGaughey, W.H. (1997) Proteinase-mediated insect resistance to *Bacillus thuringiensis* insecticidal toxins. *The Journal of Biological Chemistry* 272, 23473–23476.

Oppert, B., Hammel, R., Throne, J. E. and Kramer, K.J. (2000) Fitness costs of resistance to *Bacillus thuringiensis* in the Indianmeal moth, *Plodia interpunctella* (Lepidoptera: Pyralidae). *Entomologia Experimentalis et Applicata* 96, 281–287.

Pang, A.S., Gringorten, J.L. and Bai, C. (1999) Activation and fragmentation of *Bacillus thuringiensis* δ-endotoxin by high concentrations of proteolytic enzymes. *Canadian Journal of Microbiology* 45, 816–825.

Pardo-López, L., Soberón, M. and Bravo, A. (2013) *Bacillus thuringiensis* insecticidal three-domain Cry toxins: mode of action, insect resistance and consequences for crop protection. *FEMS Microbiology Reviews* 37, 3–22.

Puntheeranurak, T., Uawithya, P., Potvin, L., Angsuthanasombat, C. and Schwartz, J.L. (2004) Ion channels formed in planar lipid bilayers by the dipteran-specific Cry4B *Bacillus thuringiensis* toxin and its α1–α5 fragment. *Molecular Membrane Biology* 21, 67–74.

Rajagopal, R., Arora, N., Sivakumar, S., Rao, N.G., Nimbalkar, S.A. *et al.* (2009) Resistance of *Helicoverpa armigera* to Cry1Ac toxin from *Bacillus thuringiensis* is due to improper processing of the protoxin. *Biochemical Journal* 419, 309–316.

Rausell, C., García-Robles, I., Sánchez, J., Muñoz-Garay, C., Martínez-Ramírez, A.C. *et al.* (2004) Role of toxin activation on binding and pore formation activity of the *Bacillus thuringiensis*

Cry3 toxins in membranes of *Leptinotarsa decemlineata* (Say). *Biochimica et Biophysica Acta* 1660, 99–105.

Rausell, C., Ochoa-Campuzano, C., Martínez-Ramírez, A.C., Bravo, A. and Real, M.D. (2007) A membrane associated metalloprotease cleaves Cry3Aa *Bacillus thuringiensis* toxin reducing pore formation in Colorado potato beetle brush border membrane vesicles. *Biochimica et Biophysica Acta* 1768, 2293–2299.

Raymond, B., Johnston, P.R., Nielsen-LeRoux, C., Lerecluse, D. and Crickmore, N. (2010) *Bacillus thuringiensis*: an impotent pathogen? *Trends in Microbiology* 18, 189–194.

Revina, L.P., Kostina, L.I., Ganushkina, L.A., Mikhailova, A.L., Zalunin, I.A. *et al.* (2004) Reconstruction of *Bacillus thuringiensis* ssp. *israelensis* Cry11A endotoxin from fragments corresponding to its N- and C-moieties restores its original biological activity. *Biochemistry (Moscow)* 69, 181–187.

Sato, P.M., Lopes, A.R., Juliano, L., Juliano, M.A. and Terra, W.R. (2008) Subsite substrate specificity of midgut insect chymotrypsins. *Insect Biochemistry and Molecular Biology* 38, 628–633.

Srinivasan, A., Giri, A.P. and Gupta, V.S. (2006) Structural and functional diversities in lepidopteran serine proteinases. *Cellular and Molecular Biology Letters* 11, 132–154.

van Frankenhuyzen, K. and Nystrom, C. (2002) The *Bacillus thuringiensis* toxin specificity database (2002). Available at: http://www.glfc.cfs.nrcan.gc.ca/bacillus (accessed 22 October 2014).

Walters, F.S., Slatin, S.L., Kulesza, C.A. and English, L.H. (1993) Ion channel activity of N-terminal fragments from Cry1A(c) delta-endotoxin. *Biochemical and Biophysical Research Communications* 196, 921–926.

Ward, E.S. and Ellar, D.J. (1987) Nucleotide sequence of a *Bacillus thuringiensis* var. *israelensis* gene encoding a 130 kDa delta-endotoxin. *Nucleic Acids Research* 15(17): 7195.

Yamagiwa, M., Esaki, M., Otake, K., Inagaki, M., Komano, T. *et al.* (1999) Activation process of dipteran-specific insecticidal protein produced by *Bacillus thuringiensis* subsp. *israelensis*. *Applied and Environmental Microbiology* 65, 3464–3459.

Yamagiwa, M., Ogawa, R., Yasuda, K., Natsuyama, H., Sen, K. *et al.* (2002) Active form of dipteran-specific insecticidal protein Cry11A produced by *Bacillus thuringiensis* subsp. *israelensis*. *Bioscience, Biotechnology, and Biochemistry* 66, 516–522.

Yamagiwa, M., Sakagawa, K. and Sakai, H. (2004) Functional analysis of two processed fragments of *Bacillus thuringiensis* Cry11A toxin. *Bioscience, Biotechnology, and Biochemistry* 68, 523–528.

Zalunin, I.A., Revina, L.P., Kostina, L.I., Chestuk-hina, G.G. and Stepanov, V.M. (1998) Limited proteolysis of *Bacillus thuringiensis* CryIG and CryIVB δ-endotoxins leads to formation of active fragments that do not coincide with the structural domains. *Journal of Protein Chemistry* 17, 463–471.

11 The Lessons that *Caenorhabditis elegans* Has Taught Us About the Mechanism of Action of Crystal Proteins

Anand Sitaram and Raffi V. Aroian*

Program in Molecular Medicine, University of Massachusetts Medical School, Worcester, Massachusetts, USA

Summary

Caenorhabditis elegans is susceptible to three domain crystal proteins similar to those that intoxicate insects. Investigations of this organism have several important strengths, for example, ease of forward genetic screens, range of molecular genetic tools available and ease of carrying out RNAi (RNA interference) studies. These have been exploited to study cellular responses known as cellular non-immune defences (CNIDs) to the crystal proteins of *Bacillus thuringiensis* (Bt) and pore-forming toxins in general. We will discuss what we have learned through genetics and RNAi (including genome-wide RNAi) to elucidate the pathways that allow cells to respond productively to crystal protein attack. In addition, key results will be discussed from investigations with mammalian pore-forming proteins to highlight the conservation of cellular responses to crystal proteins with cellular responses to pore-forming proteins in general.

11.1 Introduction

Pore-forming toxins (PFTs) constitute the single largest class (25–30%) of bacterial protein toxins. They are critical virulence factors for many important human pathogens, including *Vibrio cholerae*, *Streptococcus pneumoniae* and *Staphylococcus aureus* (Los *et al.*, 2013). PFT monomers – some of which require processing by host proteases – bind to receptors on the host cell surface and oligomerize into a pre-pore conformation that is inserted into the host membrane, subsequently forming an ion-permeable pore. The PFTs can be categorized by their structure or their pore sizes into several different families. For example, *S. aureus* alpha-toxin and *V. cholerae* cytolysin (Füssle *et al.*, 1981; Zitzer *et al.*, 1997) fall into the small-diameter category (0.5–5 nm), while streptolysin O from *Streptococcus pyogenes* is a cholesterol-dependent cytolysin family member with a pore size of around 30 nm (Bhakdi and Tranum-Jensen, 1985).

While exposure to high concentrations of PFTs can cause cell lysis – reflected in the '-lysin' ending of many PFT names – it has been increasingly appreciated that at low concentrations of PFT exposure, many cell types have innate mechanisms (so-called cellular non-immune defences, or CNIDs (Aroian and van der Goot, 2007) for detecting and limiting the damage caused by PFT pores, thereby ensuring cell survival. Many of these defences were first characterized in the roundworm *Caenorhabditis elegans* when

* Corresponding author. E-mail address: raffi.aroian@umassmed.edu

it was exposed to a group of PFTs called crystal (Cry) proteins made by the (invertebrate) pathogenic bacterium *Bacillus thuringiensis* (Bt). Despite the fact that Cry protein host receptors are specific to invertebrates, the mechanisms that cells employ to resist the toxic effects of sublytic doses of PFTs are nevertheless remarkably conserved in nematodes, insects and mammalian cells, as discussed below. Thus, the genetically tractable host *C. elegans* and nematicidal PFTs such as Cry5B present a useful model for illumination of the general mechanisms of action of CNIDs against PFTs. The purpose of this article is to review the conservation of the CNIDs elicited by Cry proteins, along with *bona fide* PFTs, from hosts such as *C. elegans* to other systems. Figure 11.1 depicts the pathways that are discussed.

11.2 Mechanism of Action of Cry Toxins Active Against Nematodes

The Cry proteins, which are produced as parasporal crystalline inclusions by the soil bacterium Bt during sporulation, form a large and diverse group of bacterial proteins. While many Cry proteins have been utilized for years to control insect pests in agriculture, it was subsequently shown that a different subset of Cry proteins specifically target nematodes such as *C. elegans* (Marroquin *et al.*, 2000; Wei *et al.*, 2003). The toxic effects of nematicidal Cry proteins include developmental delay, reduction in brood sizes and increased mortality. In accordance with the target tissue in insects being the gut, the Cry proteins affect the nematode intestine (Griffitts *et al.*, 2001, 2003). Ingested Cry proteins are solubilized

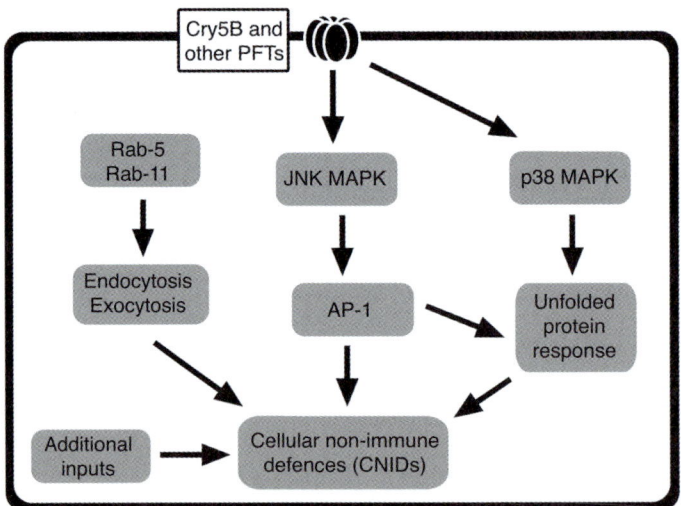

Fig. 11.1. Attack by bacterial pore-forming toxins (PFTs) such as Cry5B activates several cellular non-immune defence (CNID) responses. The pathways shown here and discussed in this review include signalling through the p38 and JNK (c-Jun N-terminal kinase) MAPK (mitogen-activated protein kinase) pathways, initiation of the unfolded protein response, and endocytosis and exocytosis to remove pores from the cell surface. Loss of any one of these pathways, all but one of which were first demonstrated in *Caenorhabditis elegans* and later shown in other systems, leads to compromised animal/cellular survival during PFT attack. All of these pathways are conserved among nematodes, insects and mammalian cells, demonstrating the ancient, complex and cell-autonomous programme of defence against membrane perforation. RAB-5 and RAB-11 are endosomal trafficking regulators: the early endosomal RAB-5 and the recycling endosomal RAB-11. AP-1 is the activator protein 1 transcription factor.

and, like other PFTs, bind host receptors (in the case of Cry5B, invertebrate-specific glycolipids; see Griffitts *et al.*, 2005 and Barrows *et al.*, 2007), oligomerize and insert into the apical surface of the intestinal epithelium. The Cry proteins have a pore diameter of approximately 1–2 nm, placing them in the α-helical, small-pore category of PFTs (Carroll and Ellar, 1997; Peyronnet *et al.*, 2002). The nematicidal crystal protein Cry5B is clearly a pore-forming protein as it allows current to pass when incubated with lipid bilayers (Kao *et al.*, 2011) and cell-impermeable dyes to access the cytoplasm of intestinal epithelia following ingestion by *C. elegans* (Los *et al.*, 2011).

11.3 Cellular Non-immune Defences to Cry Toxin Action

11.3.1 Role of p38 mitogen-activated protein kinase

As already mentioned, cells employ CNIDs to resist the effects of sublytic doses of PFTs. The first of these defences was identified using the *C. elegans*-Cry5B system. Micro-array screening of *C. elegans* fed Cry5B versus unexposed *C. elegans* identified genes that were upregulated in response to Cry5B PFT. Notably, these included one member of the p38 mitogen-activated protein kinase (MAPK) signalling axis, an important stress-activated signalling pathway that is conserved in humans (Huffman *et al.*, 2004). The screening also identified the c-Jun N-terminal kinase (JNK)-like MAPK pathway as important for defence. The availability of null mutants of *C. elegans* in the genes for each of these pathways allowed simple validation of such hits from the screen, and additional testing showed that the mutants were also hypersensitive to a second nematicidal Bt Cry protein, Cry21A. The mutants were hypersensitive even to very short toxin exposure times, confirming that the MAPK pathways mediate critical early responses to PFTs. Importantly, in the same paper, Huffman *et al.* (2004) showed that the p38 defence is conserved in mammalian cells, as baby hamster kidney

cells treated with a p38 inhibitor were hypersensitive to the PFT proaerolysin from *Aeromonas hydrophila*. This was the first report that p38 MAPK pathway was required for a defence against a specific virulence factor. Indeed, the subsequent incubation of cultured mammalian epithelial cells with isogenic bacterial strains that did or did not express PFTs confirmed that defensive MAPK activation required the presence of the PFT (Ratner *et al.*, 2006).

Since the *C. elegans* report by Huffman *et al.* (2004), the activation of p38 MAPK has been shown to be conserved as a response to exposure to several other PFTs, including *Clostridium perfringens* beta-toxin (Nagahama *et al.*, 2013), *S. aureus* alpha-toxin, *Escherichia coli* haemolysin, *V. cholerae* cytolysin, *S. pyogenes* streptolysin O (SLO) (Kloft *et al.*, 2009), *S. pneumoniae* pneumoly-sin (Ratner *et al.*, 2006), *Bacillus anthracis* anthrolysin O (Popova *et al.*, 2006) and *Lactobacillus iners* inerolysin (Rampersaud *et al.*, 2010) in mammalian cells, as well as insecticidal Cry proteins Cry1Ab in larvae of the lepidopteran crop pest *Manduca sexta* and Cry11Aa in the dipteran disease vector *Aedes aegypti* (Cancino-Rodezno *et al.*, 2010).

With p38 MAPK mutants available, a second microarray experiment in toxin-exposed *C. elegans* further identified p38 target genes whose expression was up-regulated by toxin exposure and that were needed for PFT defence, such as ttm-1 and ttm-2. The ttm-1 predicted product shows homology to mammalian cation efflux channels, suggesting that a conserved response of intoxicated cells may involve trying to counter the dysregulation of ion balances (Huffman *et al.*, 2004). Numerous reports with PFTs in mammalian cells have correlated K+ efflux with MAPK activation (Kloft *et al.*, 2009; Gonzalez *et al.*, 2011; Nagahama *et al.*, 2013). Whether this will hold true for the Cry proteins remains to be seen. Although p38 MAPK induction was not required for the protection of mammalian cells following exposure to SLO (Husmann *et al.*, 2006), the JNK MAPK pathway was needed for *C. elegans* defence against SLO (Kao *et al.*, 2011), suggesting that MAPK signalling broadly assists in cellular survival. MAPK activity is not universally protective, however; for example, the pharmacological inhibition of p38 was required to rescue a human neuronal cell line from death following exposure to pneumolysin (Stringaris et al., 2002).

11.3.2 Role of the unfolded protein response

An additional PFT defence pathway first discovered via the C. elegans-Cry system was the unfolded protein response (UPR). Various cellular stresses can have a negative impact on the protein-folding and glycosylation activities of the secretory system, and the cell has a three-pronged response to deal with such a situation. Activation of the PERK (perturbed folding-activated kinase) pathway leads to a general suppression of protein translation. Activation of the ATF6 (activating transcription factor) pathway leads to the proteolytic release of ATF6 from the membrane, allowing the liberated protein to translocate to the nucleus and upregulate UPR target genes. Activation of the IRE1 RNase (and protein kinase) enzyme causes cytoplasmic splicing of the mRNA xbp-1 (for X-box binding protein 1), leading to the translation of a mature transcription factor that activates the expression of UPR target genes. Cry5B exposure induced expression from an IRE-1 target promoter in C. elegans (Bischof et al., 2008). UPR activation is a protective defence against the toxin, as xbp-1 mutants as well as worms knocked down for xbp-1 were hypersensitive to Cry5B. The ATF6 pathway was also partially required for defence, but not the PERK arm of the UPR. Loss of xbp-1 did not lead to general sickness, as mutant animals showed normal sensitivity to the heavy metal compound $CuSO_4$, the oxidative stressor H_2O_2 and the non-PFT-expressing C. elegans bacterial pathogen Pseudomonas aeruginosa. Furthermore, the UPR is needed directly in the intoxicated cells, as intestine-specific re-expression of xbp-1 in a mutant strain partially rescued the hypersensitivity. The toxin-induced UPR response receives

inputs from both the p38 and JNK-like MAPK pathways: splicing of xbp-1 is blocked in mutants of the p38 MAPK activator sek-1 or in mutants of jun-1, which constitutes one half of the AP-1 (activator protein 1) transcription factor activated downstream of the JNK-like MAPK kgb-1 (Kao et al., 2011). Consistently, the hypersensitivity phenotype is stronger in p38-deficient animals than in UPR-deficient animals, suggesting that the UPR is just one component of a p38-controlled PFT defence programme. Strikingly, induction of the UPR by heat or the glycosylation inhibitor tunicamycin did not require intact MAPK signalling, revealing a level of specificity toward PFT defence (Bischof et al., 2008). A transcriptomic assay of C. elegans exposed to another PFT, V. cholerae cytolysin, found that the induction of several genes is required for a non-canonical UPR (Urano et al., 2002; Sahu et al., 2012). The PFT-induced activation of the UPR is conserved in insects as well. Exposure of A. aegypti larvae to tunicamycin or Cry11Aa induced splicing of xbp-1 (Bedoya-Pérez et al., 2013). As in C. elegans, the UPR is protective, because RNAi knockdown of xbp-1 led to a significant increase in the lethality of Cry11Aa. Also as in C. elegans, p38-silenced larvae are more sensitive to toxin than UPR-silenced larvae. The results suggest that the UPR may also be downstream of p38 in insect PFT defences, though this has not yet been demonstrated (Cancino-Rodezno et al., 2010). The induction of the UPR after toxin exposure is conserved in mammalian cells too: HeLa cells exposed to aerolysin also increased XBP1 splicing (Bischof et al., 2008).

11.3.3 Endocytosis

Another cellular response to PFTs observed in C. elegans was an increase in endocytosis from the plasma membrane (Los et al., 2011). Within 2 h of feeding on E. coli expressing recombinant Cry5B or Cry21A, there is a significant increase in the number of internalized endosomal vesicles seen in the cytoplasm of intestinal epithelial cells. The activation of endocytosis was specific to

PFTs, as feeding on cytolysin-positive *V. cholerae* also induced endocytosis, but cytolysin-negative bacteria or exposure to heat, heavy metals or increased salt did not. RNAi and genetic mutants were used to show that the endocytic effect depends on the presence of two endosomal trafficking regulators, the early endosomal RAB-5 and the recycling endosomal RAB-11. While previous papers had demonstrated endo-cytosis as a defensive PFT response in mammalian cells (Idone *et al.*, 2008; Husmann *et al.*, 2009; Thiery *et al.*, 2010), this was the first paper that showed that endocytosis had a positive role for organismal survival. The dipteran toxin Cry11Aa and the lepidopteran toxin Cry1Ab were also endocytosed in an *A. aegypti* cell line using both clathrin-dependent and -independent routes, and internalized toxin co-localized with RAB5 and RAB11 (Vega-Cabrera *et al.*, 2014).

Previous work had suggested that *C. elegans* intestinal epithelia internalize fluorescently labelled Cry5B into auto-fluorescent gut granules (Griffitts *et al.*, 2001, 2003), which are lysosome-related organelles (Hermann *et al.*, 2005). Similarly, mammalian cells treated with SLO increased dynamin-independent endocytosis of the toxin (Idone *et al.*, 2008) and delivery through the conventional endocytic pathway (including ubiquitination and the engage-ment of the ESCRT – endosomal sorting complexes required for transport – proteins) to lysosomes (Corrotte *et al.*, 2012). The reduction in toxin levels over time could be blocked by lysosomal inhibition, suggesting that the toxin is degraded in the lysosomes. In contrast to their response to treatment with SLO, mammalian cells treated with *S. aureus* alpha-toxin also initiate a protective endocytic response, but it is dynamin-dependent, and the subsequent reduction in the level of cell-associated toxin is unaffected by lysosomal inhibitors, leading to the finding that the toxin is expelled from the cell in an ESCRT-dependent process on exosome-like structures dubbed toxosomes (Husmann *et al.*, 2009). Similarly, the endocytosed Cry11Aa and Cry1Ab in-secticidal toxins did not co-localize with a lysosomal marker in insect cells (Vega-

Cabrera *et al.*, 2014). What distinguishes the decision between lysosomal and non-lysosomal trafficking routes for internalized toxin is unclear, but a correlative difference in the responses to the two mammalian toxins is the requirement for extracellular calcium to stimulate repair. SLO-treated mammalian cells require extracellular calcium to activate the restoration of plasma membrane integrity (Idone *et al.*, 2008). A parallel result was found in *C. elegans*, where it was seen that SLO intoxication is reduced by the presence of calcium (Kao *et al.*, 2011). Cry5B, in contrast, may better model the similarly sized alpha-toxin, as survival against both of these small-pore toxins is calcium independent. Furthermore, Cry5B exposure caused shedding of microvilli into the lumen of the *C. elegans* intestine (Los *et al.*, 2011). One possible interpretation of this response is as a mechanistically distinct but functionally similar process to toxosome production as a method of removing toxin-bearing membranes from the cell.

11.4 Conclusion

As we have seen, several studies in *C. elegans*, insects and mammalian cells demonstrate numerous parallels and strong conservation in how cells respond to PFTs. Because of these parallels, the studies also reinforce the generally accepted idea that *B. thuringiensis* Cry proteins act on cells as PFTs and highlight the strength of *C. elegans* in illuminating PFT–host interactions. The studies likewise demonstrate the great complexity of how cells respond to Cry proteins and PFTs and belie the simple model that cellular exposure to PFTs simply results in osmotic lysis. Indeed, subsequent research will focus on determining how the cellular responses intersect and elucidating both the PFT-induced pathways that protect the cell (CNIDs) as well as those that result in cellular and organismal intoxication. Indeed, insulin signalling, negative regulation of hypoxia, magnesium signalling and cyclic AMP signalling have all been implicated in Cry protein-induced intoxi-cation (Zhang *et al.*, 2005, 2006; Bellier *et* al., 2009; Chen et al., 2010), all of which could involve PFT responses.

Although several important CNIDs have been identified through use of the C. elegans system, many important questions remain. It is still not clear what is the most proximal event that activates the cellular defences, whether it is a signalling pathway activated by the toxin interacting with a host factor or a sensing of ion fluxes, or something else entirely. While transcriptomics has revealed some of the changes in gene expression that result from toxin exposure, how these changes are reflected in the post-exposure proteome and drive the defence pathways must still be determined. The transcriptomics data from C. elegans show that the pattern of genes activated by exposure to PFTs is distinct from those activated by many other common stressors (Kao et al., 2011), thus further reinforcing the idea that with its wealth of genetic tools, the C. elegans system is poised to give us many new insights into a fundamental genetic programme devoted to maintaining cellular membrane integrity.

References

Aroian, R. and van der Goot, F.G. (2007) Pore-forming toxins and cellular non-immune defenses (CNIDs). Current Opinion in Microbiology 10, 57–61.

Barrows, B.D., Griffitts, J.S. and Aroian, R.V. (2007) Resistance is non-futile: resistance to Cry5B in the nematode Caenorhabditis elegans. Journal of Invertebrate Pathology 95, 198–200.

Bedoya-Pérez, L.P., Cancino-Rodezno, A., Flores-Escobar, B., Soberón, M. and Bravo, A. (2013) Role of UPR pathway in defense response of Aedes aegypti against Cry11Aa toxin from Bacillus thuringiensis. International Journal of Molecular Sciences 14, 8467–8478.

Bellier, A., Chen, C.S., Kao, C.Y., Cinar, H.N. and Aroian, R.V. (2009) Hypoxia and the hypoxic response pathway protect against pore-forming toxins in C. elegans. PLoS Pathogens 5(12): e1000689.

Bhakdi, S. and Tranum-Jensen, J. (1985) Formation of protein channels in target membranes. Advances in Experimental Medicine and Biology 184, 3–21.

Bischof, L.J., Kao, C.Y., Los, F.C., Gonzalez, M.R., Shen, Z. et al. (2008) Activation of the unfolded protein response is required for defenses

against bacterial pore-forming toxin in vivo. PLoS Pathogens 4(10): e1000176.

Cancino-Rodezno, A., Alexander, C., Villaseñor, R., Pacheco, S., Porta, H. et al. (2010) The mitogen-activated protein kinase p38 is involved in insect defense against Cry toxins from Bacillus thuringiensis. Insect Biochemistry and Molecular Biology 40, 58–63.

Carroll, J. and Ellar, D.J. (1997) Analysis of the large aqueous pores produced by a Bacillus thuringiensis protein insecticide in Manduca sexta midgut-brush-border-membrane vesicles. European Journal of Biochemistry 245, 797–804.

Chen, C.S., Bellier, A., Kao, C.Y., Yang, Y.L., Chen, H.D. et al. (2010) WWP-1 is a novel modulator of the DAF-2 insulin-like signaling network involved in pore-forming toxin cellular defenses in Caenorhabditis elegans. PLoS One 5(3): e9494.

Corrotte, M., Fernandes, M.C., Tam, C. and Andrews, N.W. (2012) Toxin pores endocytosed during plasma membrane repair traffic into the lumen of MVBs for degradation. Traffic 13, 483–494.

Füssle, R., Bhakdi, S., Sziegoleit, A., Tranum-Jensen, J., Kranz, T. et al. (1981) On the mechanism of membrane damage by Staphylococcus aureus α-toxin. Journal of Cell Biology 91, 83–94.

Gonzalez, M.R., Bischofberger M., Frêche, B., Ho, S., Parton, R.G. et al. (2011) Pore-forming toxins induce multiple cellular responses promoting survival. Cellular Microbiology 13, 1026-1043.

Griffitts, J.S., Whitacre, J.L., Stevens, D.E. and Aroian, R.V. (2001) Bt toxin resistance from loss of a putative carbohydrate-modifying enzyme. Science 293, 860–864.

Griffitts, J.S., Huffman, D.L., Whitacre, J.L., Barrows, B.D., Marroquin, L.D. et al. (2003) Resistance to a bacterial toxin is mediated by removal of a conserved glycosylation pathway required for toxin–host interactions. The Journal of Biological Chemistry 278, 45594–45602.

Griffitts, J.S., Haslam, S.M., Yang, T., Garczynski, S.F., Mulloy, B. et al. (2005) Glycolipids as receptors for Bacillus thuringiensis crystal toxin. Science 307, 922–925.

Hermann, G.J., Schroeder, L.K., Hieb, C.A., Kershner, A.M., Rabbitts, B.M. et al. (2005) Genetic analysis of lysosomal trafficking in Caenorhabditis elegans. Molecular Biology of the Cell 16, 3273–3288.

Huffman, D.L., Abrami, L., Sasik, R., Corbeil, J., van der Goot, F.G. et al. (2004) Mitogen-activated protein kinase pathways defend against bacterial pore-forming toxins. Proceedings of the National Academy of Sciences of the United States of America 101, 10995–11000.

Husmann, M., Dersch, K., Bobkiewicz, W., Beckmann, E., Veerachato, G. *et al.* (2006) Differential role of p38 mitogen activated protein kinase for cellular recovery from attack by pore-forming *S. aureus* α-toxin or streptolysin O. *Biochemical and Biophysical Research Communications* 344, 1128–1134.

Husmann, M., Beckmann, E., Boller, K., Kloft, N., Tenzer, S. *et al.* (2009) Elimination of a bacterial pore-forming toxin by sequential endocytosis and exocytosis. *FEBS Letters* 583, 337–344.

Idone, V., Tam, C., Goss, J.W., Toomre, D., Pypaert, M. *et al.* (2008) Repair of injured plasma membrane by rapid Ca^{2+}-dependent endocytosis. *Journal of Cell Biology* 180, 905–914.

Kao, C.Y., Los, F.C., Huffman, D.L., Wachi, S., Kloft, N. *et al.* (2011) Global functional analyses of cellular responses to pore-forming toxins. *PLoS Pathogens* 7(3): e1001314.

Kloft, N., Busch, T., Neukirch, C., Weis, S., Boukhallouk, F. *et al.* (2009) Pore-forming toxins activate MAPK p38 by causing loss of cellular potassium. *Biochemical and Biophysical Research Communications* 385, 503–506.

Los, F.C., Kao, C.Y., Smitham, J., Mcdonald, K.L., Ha, C. *et al.* (2011) RAB-5- and RAB-11-dependent vesicle-trafficking pathways are required for plasma membrane repair after attack by bacterial pore-forming toxin. *Cell Host and Microbe* 9, 147–157.

Los, F.C., Randis, T.M., Aroian, R.V. and Ratner, A.J. (2013) Role of pore-forming toxins in bacterial infectious diseases. *Microbiology and Molecular Biology Reviews* 77, 173–207.

Marroquin, L.D., Elyassnia, D., Griffitts, J.S., Feitelson, J.S. and Aroian, R.V. (2000) *Bacillus thuringiensis* (Bt) toxin susceptibility and isolation of resistance mutants in the nematode *Caenorhabditis elegans*. *Genetics* 155, 1693–1699.

Nagahama, M., Shibutani, M., Seike, S., Yonezaki, M., Takagishi, T. *et al.* (2013) The p38 MAPK and JNK pathways protect host cells against *Clostridium perfringens* beta-toxin. *Infection and Immunity* 81, 3703–3708.

Peyronnet, O., Nieman, B., Généreux, F., Vachon, V., Laprade, R. *et al.* (2002) Estimation of the radius of the pores formed by the *Bacillus thuringiensis* Cry1C δ-endotoxin in planar lipid bilayers. *Biochimica et Biophysica Acta* 1567, 113–122.

Popova, T.G., Millis, B., Bradburne, C., Nazarenko, S., Bailey, C. *et al.* (2006) Acceleration of epithelial cell syndecan-1 shedding by anthrax hemolytic virulence factors. *BMC Microbiology* 6:8.

Rampersaud, R., Planet, P.J., Randis, T.M., Kulkarni, R., Aguilar, J.L. *et al.* (2010) Inerolysin, a cholesterol-dependent cytolysin produced by *Lactobacillus iners*. *Journal of Bacteriology* 193, 1034–1041.

Ratner, A.J., Hippe, K.R., Aguilar, J.L., Bender, M.H., Nelson, A.L. *et al.* (2006) Epithelial cells are sensitive detectors of bacterial pore-forming toxins. *The Journal of Biological Chemistry* 281, 12994–12998.

Sahu, S.N., Lewis, J., Patel, I., Bozdag, S., Lee, J.H. *et al.* (2012) Genomic analysis of immune response against *Vibrio cholerae* hemolysin in *Caenorhabditis elegans*. *PLoS One* 7(5), e38200.

Stringaris, A.K., Geisenhainer, J., Bergmann, F., Balshusemann, C., Lee, U. *et al.* (2002) Neurotoxicity of pneumolysin, a major pneumococcal virulence factor, involves calcium influx and depends on activation of p38 mitogen-activated protein kinase. *Neurobiology of Disease* 11, 355–368.

Thiery, J., Keefe, D., Saffarian, S., Martinvalet, D., Walch, M. *et al.* (2010) Perforin activates clathrin- and dynamin-dependent endocytosis, which is required for plasma membrane repair and delivery of granzyme B for granzyme-mediated apoptosis. *Blood* 115, 1582–1593.

Urano, F., Calfon, M., Yoneda, T., Yun, C., Kiraly, M. *et al.* (2002) A survival pathway for *Caenorhabditis elegans* with a blocked unfolded protein response. *Journal of Cell Biology* 158, 639–646.

Vega-Cabrera, A., Cancino-Rodezno, A., Porta, H. and Pardo-López, L. (2014) *Aedes aegypti* Mos20 cells internalizes [*sic*] Cry toxins by endocytosis, and actin has a role in the defense against Cry11Aa toxin. *Toxins* 6, 464–487.

Wei, J.Z., Hale, K., Carta, L., Platzer, E., Wong, C. *et al.* (2003) *Bacillus thuringiensis* crystal proteins that target nematodes. *Proceedings of the National Academy of Sciences of the United States of America* 100, 2760–2765.

Zhang, X., Candas, M., Griko, N.B., Rose-Young, L. and Bulla, L.A. Jr (2005) Cytotoxicity of *Bacillus thuringiensis* Cry1Ab toxin depends on specific binding of the toxin to the cadherin receptor BT-R_1 expressed in insect cells. *Cell Death and Differentiation* 12, 1407–1416.

Zhang, X., Candas, M., Griko, N.B., Taussig, R. and Bulla, L.A. Jr (2006) A mechanism of cell death involving an adenylyl cyclase/PKA signaling pathway is induced by the Cry1Ab toxin of *Bacillus thuringiensis*. *Proceedings of the National Academy of Sciences of the United States of America* 103, 9897–9902.

Zitzer, A., Palmer, M., Weller, U., Wassenaar, T., Biermann, C. *et al.* (1997) Mode of primary binding to target membranes and pore formation induced by *Vibrio cholerae* cytolysin (hemolysin). *European Journal of Biochemistry* 247, 209–216.

12 The Development and Prospect of Discovery of Bt Toxin Genes

Jie Zhang,* Changlong Shu and Zeyu Wang

State Key Laboratory for Biology of Plant Diseases and Insect Pests, Institute of Plant Protection, Chinese Academy of Agricultural Sciences, Beijing, People's Republic of China

Summary

Bacillus thuringiensis (Bt) Cry protein-based insect control has proven to be effective in reducing the use of chemical insecticides and in increasing crop yields. However, Bt crops increase selection pressure for resistant insects and accelerate their succession. The discovery and application of different Bt toxins that have no cross-resistance with known toxins has been proposed as a strategy for the management of resistant insects. Additionally, the discovery of novel Bt toxins with a new insecticidal spectrum will control insect succession. In fact, the discovery of new Bt toxins is one of the most important areas in Bt research, and the most advanced molecular biology methods available have been applied to this task. In this chapter, we summarize the published methods for Bt toxin discovery.

12.1 Introduction

Bacillus thuringiensis (Bt) is an insect pathogen that can produce insecticidal crystal (Cry) proteins (ICPs) within a crystalline inclusion in the mother cell during sporulation, and it can also secrete vegetative insecticidal proteins (Vips) into the extracellular medium during vegetative growth (Schnepf *et al.*, 1998; Raymond *et al.*,

2010). Bt was first isolated by the Japanese scientist S. Ishiwata from silkworm larvae exhibiting the symptoms of Sotto disease (Ishiwata, 1901); the species was named *Bacillus thuringiensis* in 1911, after the province of Thuringia in Germany where an infected moth, *Anagasta kuehniella*, was found (Berliner, 1911).

Because of its insecticidal properties, Bt was originally considered to be a risk for silkworm rearing but, later on, its success in biologically controlling insect pests caused it to become an important biological insecticide. The insecticidal activity of Bt largely results from the ICP and Vip insecticidal proteins. Currently, the genes for these insecticidal proteins are being used to develop insect-resistant crops. Bt pesticides and insect-resistant crops not only reduce the use of chemical insecticides but also bring enormous commercial benefits to biotechnology companies. Because the insecticidal proteins were demonstrated to be the primary active substances in Bt, the identification and cloning of Bt insecticidal genes has become an important research component. Currently, nearly 800 toxin genes have been cloned and named (see Crickmore *et al.*, 2014).

In Bt insecticide research, the identification of the genes that are responsible for the production of insecticidal proteins not only identifies active substance-coding genes in commercially important strains,

* Corresponding author. E-mail address: jzhang@ippcaas.cn

but can also be used to predict activities of Bt isolates to reduce bioassay-screening labour. Moreover, because some commercial strains tend to lose toxin genes during subculturing and fermentation, the identification of insecticidal genes is an important way to ensure product quality and stability across different batches. Furthermore, to search for new strains or new Cry proteins with toxic effects against different organisms or with broader specificities, or to search for strains/Cry proteins that can be used with different insecticidal models to provide alternatives after insect resistance has appeared, researchers must continuously search for new experimental approaches to identify and clone Bt insecticidal genes. Currently, the reported methods include: (i) the cloning of DNA libraries; (ii) PCR; and (iii) hybridization and microarray analysis. Historically, each method has its own advantages, and, currently, researchers combine the methods for the efficient cloning and identification of toxin genes. In this chapter, we describe the major features of each method, focusing on definitions and demonstrating how each method accomplishes particular research goals. We also discuss the recent development of next-generation sequencing (NGS) and its application to finding Bt toxin genes.

12.2 Cloning of DNA Libraries

The development of molecular cloning changed Bt research, allowed for investigations of Bt toxins at the gene level and laid the foundation for the subsequent development of insect-resistant genetically modified organisms. Since the successful cloning and expression of the Bt crystal protein gene in *Escherichia coli* (Schnepf and Whiteley, 1981), a variety of DNA library cloning technologies has been established for cloning novel Bt toxin genes. Typically, the two main steps in these technologies include: (i) construction of a Bt DNA library; and (ii) screening of the library.

12.2.1 Library construction

The first step of DNA library construction is the formation of DNA fragments from genomic/plasmid DNA by mechanical cutting or restriction endonuclease cleavage. With restriction endonucleases, researchers have performed both complete and partial digestions. The partial digestion stratagem is more random, and a full-length gene is more easily obtained with this method. In contrast, with the complete digestion stratagem, only partial genes can be obtained if restriction endonuclease sites are within gene sequences.

Once the Bt DNA fragments are prepared, the vector needs to be determined. Bt DNA libraries are mainly constructed and expressed in *E. coli*. To construct a library in *E. coli*, an *E. coli* plasmid vector, such as pBR322 (Schnepf and Whiteley, 1981) or pBlueScript II SK(+) (Yu *et al.*, 2006) is needed. If a toxin gene is silent or under special regulation, it will not be easy to purify enough protein for antibody generation or to facilitate amino acid sequencing for nucleic acid probe preparation for library screening; in these cases, researchers can express libraries in a Bt acrystalliferous mutant and can screen for positive clones by examining Bt crystal morphology. To construct and express a library in a Bt acrystalliferous mutant, an *E. coli*–Bt shuttle vector, such as pHT315 (Guo *et al.*, 2008), is needed.

Once the type of library to be created is determined and the vector selected, genomic DNA fragments are ligated into vectors and the ligation mixture is transformed into *E. coli*. If the library is to be expressed in Bt, after amplification in *E. coli*, plasmids (shuttle vectors) present in the recombinant *E. coli* pool are extracted and transferred into Bt by electroporation (Guo *et al.*, 2008).

12.2.2 Library Screening

Once a library is constructed successfully, a method(s) is needed to screen it. As noted above, libraries can be expressed in a Bt acrystalliferous mutant, which can then be screened for positive clones by observing crystal shape (Guo *et al.*, 2008). For toxin genes that can be expressed in *E. coli*, Western blots are suitable for library screening. Antibodies can be prepared from rabbits or mice using protein purified from original Bt strains. For toxin genes that cannot be expressed in *E. coli*, Southern blots are a good alternative (Sekar *et al.*, 1987; Chambers *et al.*, 1991). For new genes, probes are synthesized based on the amino acid sequencing of proteins purified from original Bt strains. For variants of homologous genes, probes can be prepared by amplifying conserved fragments with universal primers. Furthermore, for libraries with partial target gene sequences/fragments available, PCR techniques can also be used for efficient to detection and screening (Yu *et al.*, 2006).

12.3 PCR-based Methods

Since PCR was proposed by Kleppe *et al.* (1971) and independently conceived and developed by Mullis and co-workers (Saiki *et al.*, 1985), it has been used widely to amplify specific DNA fragments and to determine the presence or absence of target genes. Its development has also been a milestone in Bt toxin gene research because it is highly sensitive, requires very little DNA for analysis and allows for large numbers of samples to be processed in a relatively short time. PCR-based methods cannot only be used to identify previously reported genes, they can also be used to discover novel genes, as discussed in Sections 12.3.1 and 12.3.2 below.

12.3.1 Prediction of the spectrum of insecticidal activity by toxin gene identification

Bt insecticidal proteins have specific insecticidal activity spectra, which are usually restricted to a few species within one particular order of insects. The combination of toxins within a given strain, therefore, defines the activity spectrum of that strain. To date, Bt insecticidal proteins (Cry and Vip) against insect species in the orders Lepidoptera, Diptera, Coleoptera and Hymenoptera have been identified (Schnepf *et al.*, 1998; Raymond *et al.*, 2010). Additionally, a small minority of crystal toxins have activity against non-insect species, such as nematodes. Consequently, it is possible to predict the activity spectra of novel Bt strains by identification of their insecticidal protein genes; Carozzi *et al.* (1991), for example, showed that the electrophoretic profiles of insecticidal gene PCR products corresponded with the insecticidal activities of new and known isolates. In fact, all of the methods used for identifying genes in Bt strains can also be used to predict insecticidal activity spectra.

12.3.2 Screening for novel toxin genes

Currently, the reported Bt insecticidal proteins can be classified into approximately eight groups according to their phylogenetic relationships. The groups include four insecticidal crystal protein groups and four vegetative insecticidal protein (Vip) groups: the three-domain Cry toxin (3d) family, the mosquitocidal Cry toxin (Mtx) family, the binary-like (Bin) family, the Cyt (cytolytic) family of toxins (Bravo *et al.*, 2011), Vip1/Vip2 (a binary toxin) and Vip3 and Vip4 (Crickmore *et al.*, 1998).

There are conserved sequences present in each group, for example, in the largest 3d-Cry toxin group, sequence alignments revealed five conserved blocks of sequence common to a large majority of the 3d-Cry proteins, even though sequence identities were lower than 45% between group members (Crickmore *et al.*, 1998; Schnepf *et al.*, 1998). Using these conserved primers, it is possible to amplify novel genes. Various PCR-based methods have now been developed to screen and clone novel genes. In the rest of Section 12.3, we introduce the theories behind these reported methods. Typically, their first step is to amplify novel gene fragments using universal primers designed from conserved regions, and then to use suitable stratagems to detect new genes.

12.3.3 Multiplex PCR

Multiplex PCR is a variation on PCR developed by Chamberlain *et al.* (1988), and it enables simultaneous amplification of many targets of interest in one reaction by using more than one pair of primers. In Bt research, it is applied to *cry* gene typing and toxicity predictions. Additionally, it can be applied to novel toxin gene discovery, especially to screens for homologous novel toxin genes. For screening novel genes, multiplex primers are designed to a known reference toxin gene and are used to screen Bt strains by multiplex PCR. If a Bt strain contains the reference gene, all of the predicted fragments will be amplified, whereas if a Bt strain contains gene variants with only partially conserved sequences that

only partially match the primers, some fragments will not be amplified and the difference will be seen by gel analysis and reveal the existence of a novel homologous toxin gene (Fig. 12.1). Using this method, Kalman *et al.* (1993) cloned *cry*1Cb1 and Fang *et al.* (2007) cloned *vip*3Ac1. Although these researchers did not intend to design conserved primers, by designing multiplex primers to known reference toxin genes, some of the primers are probes to the conserved regions. These conserved primers can produce PCR products from the novel homologue genes.

12.3.4 Exclusive PCR

Exclusive PCR was first developed by Juárez-Pérez *et al.* (1997) to discover novel toxin genes. In this method, one primer pair is designed from the DNA sequence of a conserved region of a class of genes and the family primer is used to amplify products

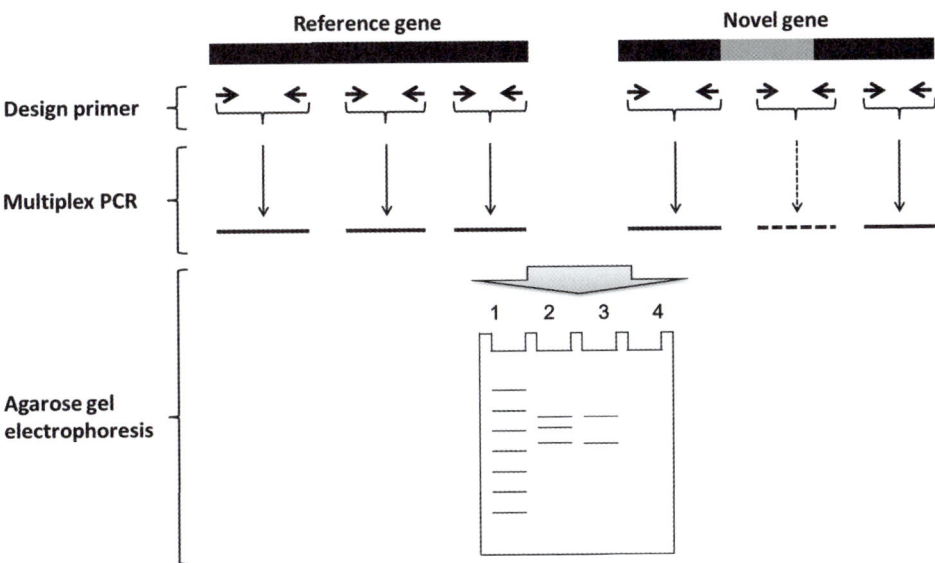

Fig. 12.1. The multiplex PCR method. Primers were designed to a known reference gene (left). Novel gene (right) indicates a possible novel gene with sequence variation (in grey). Primer pairs are designed to several regions of the reference gene. During multiplex PCR amplification, three fragments are generated if the reference gene is present. The novel gene may be missing some regions of the reference gene, thus, not all of the reference primers will bind to the novel gene and fewer PCR products will be generated and detected by agarose gel electrophoresis.

from the same family of genes; this amplified fragment is always the larger (Fig. 12.2, PCR1). The other primer pairs are designed to match sequences within unique regions of known genes in the family; these are the type primers that amplify specific gene's fragments from specific subclasses; the amplified fragments are always shorter than the fragment generated with the family primers (Fig. 12.2, PCR2). When both family primers and type primers are used in one PCR reaction, the known gene generates shorter fragments, whereas the novel gene will only be amplified by the family primers and will generate the larger fragment (Fig. 12.2, PCR3). The presence of the novel gene can be detected by agarose gel electrophoresis. Using this method, Juárez-Pérez *et al.* (1997) cloned a novel *cry*1B gene.

12.3.5 PCR-restriction fragment length polymorphism (RFLP)

The RFLP method, previously called the 'DNA marker loci' method, was developed by Botstein *et al.* (1980). This method reveals sequence variation and polymorphisms between genomic DNA samples by gel analysis. When PCR technology was developed, RFLP analysis was applied to detect differences among amplified frag-

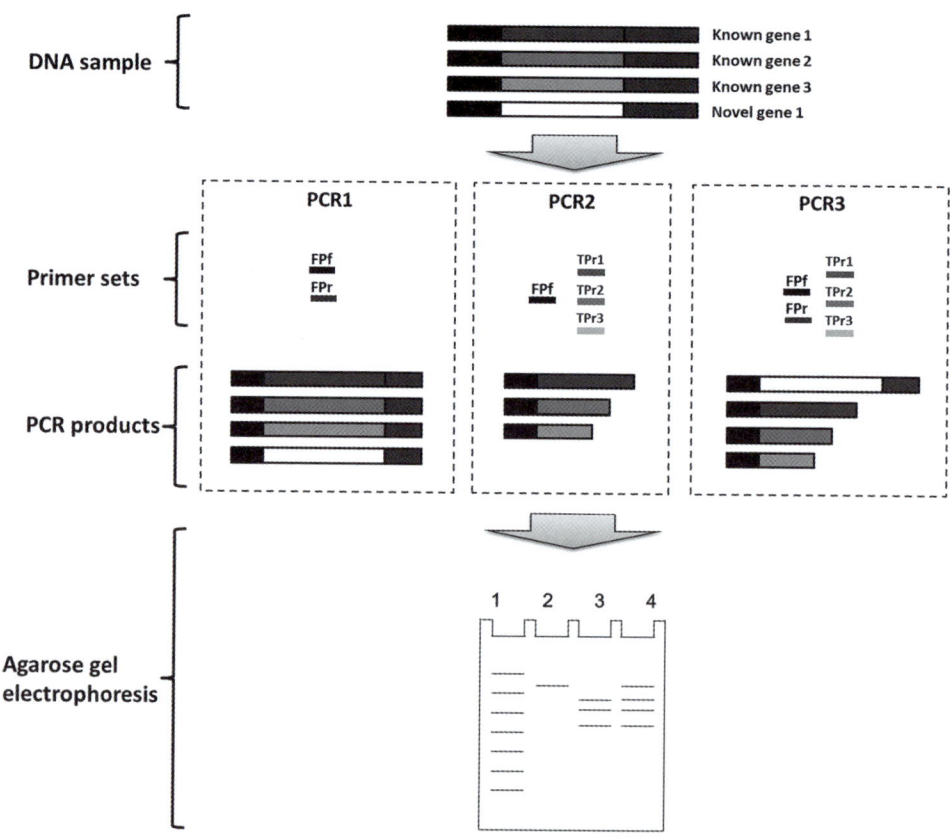

Fig. 12.2. The exclusive PCR method. This example shows a sample containing four types of genes. Four unique sequences are shown in four different combinations of black/grey/white shading. Three of the genes are known and one is novel. FPf and FPr indicate family primers. TPr1, TPr2 and TPr3 indicate type primers. See text for further explanation of the method.

ments. The PCR-RFLP method was first applied to Bt insecticidal gene typing by Kuo and Chak (1996). The method involves two main steps: in the first step, universal primers are used to amplify all possible toxin genes; in the second, RFLP analysis is employed to identify the gene types present by comparing gel profiles along with RFLP predictions (Fig. 12.3). Using this method, Kuo and Chak (1996) identified 14 distinct *cry*-type genes from Bt strains. Among them, six *cry*-type genes were found to have sequences distinct from the corresponding published *cry* gene sequences. These results demonstrated that the PCR-RFLP typing system is easy to use for detecting both known and novel *cry* genes in Bt strains. Subsequently, numerous novel *cry* genes

were cloned by the method (Song *et al.*, 2003; Yu *et al.*, 2006; Shu *et al.*, 2009, 2013a; Hernández-Rodríguez *et al.*, 2009; Patel and Ingle, 2012).

12.3.6 PCR-high-resolution melting analysis (HRMA)

The melting curve of a DNA fragment is determined by its DNA sequence composition and can be used to scan for unknown mutations. To improve resolution ratios and to meet the need for single nucleotide polymorphism (SNP) analysis, advances in DNA melting techniques have been developed; these include instrumentation that allows for highly controlled

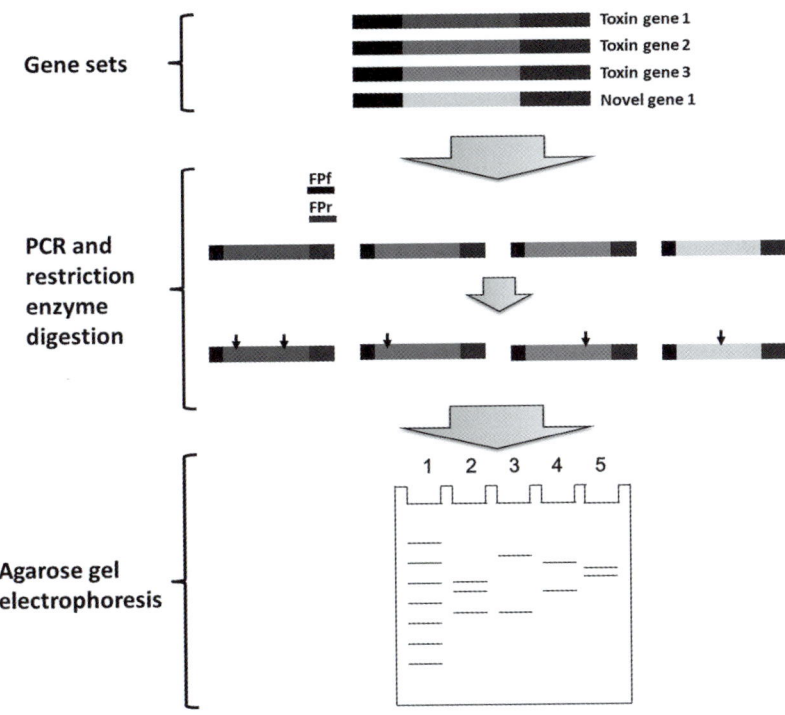

Fig. 12.3. The PCR-restriction fragment length polymorphism (RFLP) method. The gene set shown is a gene family containing four members: three known genes (toxin genes 1 to 3) and one novel gene. All four genes can be amplified with universal (family) primers (FPf and FPr) during the PCR step. Arrows indicate restriction endonuclease cleavage sites that are cut during the enzyme digestion step. After agarose gel electrophoresis, each gene is found to produce a unique RFLP profile (lanes 2–5). When comparing profiles with RFLP predictions, RFLP profile (lane 5) can be detected.

temperature transitions and data acquisition (Gundry *et al.*, 2003), and the development of fluorescent DNA-binding dyes with improved saturation properties. The new dyes, called LCGreen, exhibit minimal redistribution during melting and do not inhibit PCR, which were two common difficulties associated with the use of earlier DNA binding dyes in PCR (Wittwer *et al.*, 2003). HRMA of PCR products amplified in the presence of LCGreen can identify both heterozygous and homozygous sequence variants. Therefore, it can also be used to identify Bt toxin gene variants among PCR products. Li *et al.* (2012a) and Shu *et al.* (2013b) applied HRMA to the identification of *vip* and *cry* genes. They designed a pair of conserved primers to amplify a short DNA region containing nucleotide polymorphisms. Then, by comparing their

melting curves with those of a reference gene, they identified gene types (Fig. 12.4). Finally, several *vip3* variants and novel *cry9* genes (e.g. *cry9Ee1*) were successfully cloned (Li *et al.*, 2012a; Shu *et al.*, 2013b).

12.3.7 PCR-high-throughput sequencing (HTS)

DNA sequencing is the process of determining the precise sequence of nucleotides within a DNA molecule. Because it provides accurate genetic information, it is useful for identifying gene types in Bt strains. Before the development of HTS, unique PCR fragments had to be cloned for sequencing, and this process was time-consuming and inefficient. However, recently developed HTS methods can be used to generate libraries

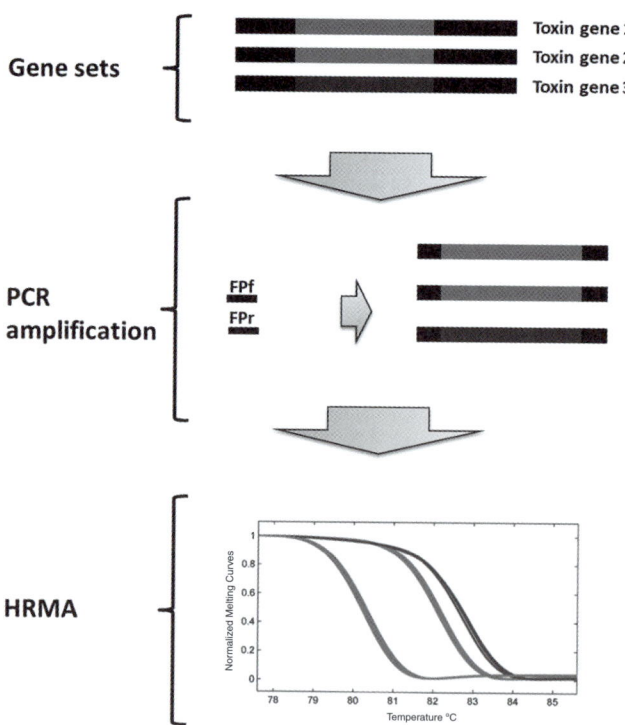

Fig. 12.4. The PCR-high-resolution melting analysis (HRMA) method. The gene set shown is a gene family containing three members. All three genes can be amplified with universal (family) primers (FPf and FPr) during the PCR step. The HRMA profile shows that, because of sequence variations between the amplified fragments, their melting curves are different and can be used to distinguish genes.

from single DNA molecules for sequencing; they can also be used to determine the sequence of each PCR-amplified DNA molecule. Chen *et al.* (2014) used 454 sequencing technology to determine the sequences of PCR products amplified from a template pool containing DNA from 2000 strains. Four pairs of universal primers were designed to amplify overlapping fragments (of less than 750 bp) within a gene family (Fig. 12.5a). Each fragment was then amplified separately and sequenced using 454 sequencing technology (Fig. 12.5b), which produces read lengths of approximately 450 bp, so that reads can cover a whole gene region (Fig. 12.5a). Using this method, Chen *et al.* (2014) cloned a novel *cry2* gene and determined the distribution of every *cry2* gene in the collection.

12.3.8 Metagenomic PCR cloning

As Bt is widely distributed in soil, the metagenomic PCR cloning method was mainly developed for identifying and cloning genes direct from soil samples, and it eliminates the strain isolation steps required for cloning Bt toxin genes. Shu *et al.* (2013c) combined a metagenomic method with PCR to clone *cry2* from a pooled soil sample. They randomly collected 235 soil samples from the Beijing Botanical Garden; 2 g of each sample was mixed in 500 ml of sterile water and heated at 75°C for 30 min to kill vegetative bacteria. The mixture was filtered to remove large soil particles and then centrifuged. The spore-containing pellet was resuspended in 100 ml LB medium and incubated at 30°C overnight with shaking at

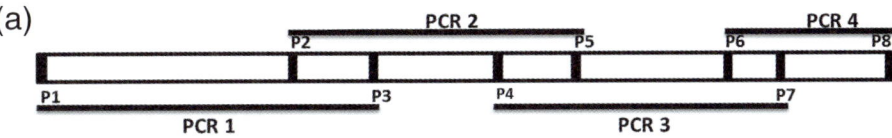

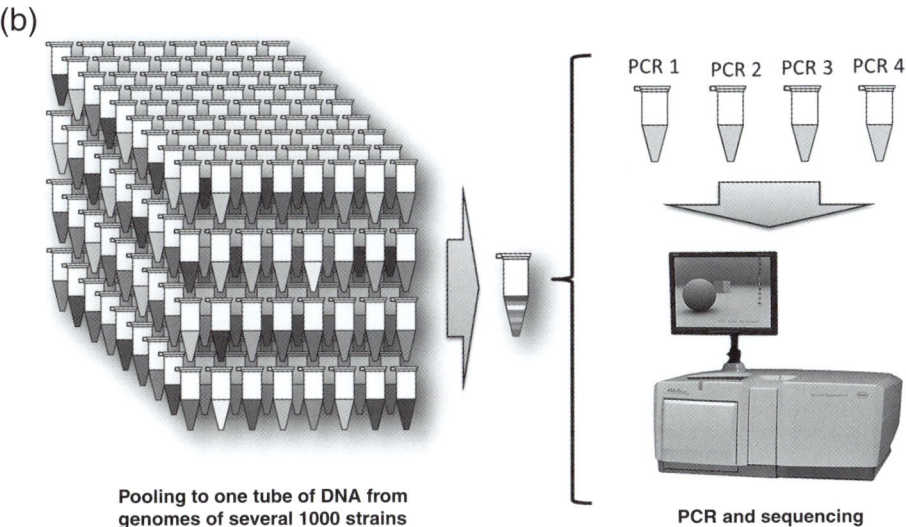

Fig. 12.5. The PCR high-throughput sequencing (HTS) method. Four pairs of universal primers were designed to amplify overlapping fragments (<750 bp) within a gene family (a). Each fragment was then amplified separately using a template pool containing DNA pooled from several 1000 strains and sequenced using 454 sequencing technology (b), which produces read lengths of approximately 450 bp, so that reads can cover a whole gene region (a). Reproduced with permission, from authors Chen *et al.*, 2014, (*Chinese Journal of Biological Control*, 30, 610–617).

220 rpm to allow for spore germination. Cells were collected by centrifugation and plasmid DNA was prepared for PCR amplification. Conserved *cry2* primers to conserved N- and C-terminal sequences of the Cry2A toxin family were used to amplify *cry2* genes from this pooled plasmid sample. The resulting PCR products were purified, cloned and sequenced. Using this method, the distribution of *cry2* gene types was revealed and a novel gene was cloned (Shu *et al.*, 2013c).

12.3.9 Genome sequencing

Second-generation sequencing technologies provide a great deal of genomic data efficiently and at low cost (Table 12.1) (Liu *et al.*, 2012; Quail *et al.*, 2012). Some laboratories have adopted this technology for toxin gene mining (Ye *et al.*, 2012). For gene mining, researchers construct a library of small fragments, for example, 500 bp, and sequence them. Once the short reads are produced, they are assembled into contiguous units or scaffolds, and coding genes are predicted and annotated. Toxin genes will be identified during the annotation process.

Although the high efficiency and low cost of second-generation DNA sequencing technologies has significantly improved the efficiency of the toxin gene discovery process, some problems still need to be solved, for example, the read lengths are short (Table 12.1). In addition, many Bt strains contain multiple homologous *cry* genes that share identical domains (for example, *cry*1Aa, *cry*1Ab and *cry*1Ac). In most of the available whole-genome assembly programs, to reasonably and accurately join reads into contiguous sequence contigs, the elegant computational de Bruijn graph is used (Li *et al.*, 2012b). During genome assembly, once erroneous connections are removed or corrected, the assembly programs break repeat connections on the graph and output the linear sequences as contigs. Identical sequences between homologous *cry* genes are considered to be repeated connections in the de Bruijn graph

Table 12.1. Comparison of sequencing systems.

	1st generation sequencing	2nd generation sequencing		3rd generation sequencing		
	Chain termination (Sanger sequencing)	Ion semiconductor (ion torrent sequencing)	Pyro-sequencing (454 sequencing)	Sequencing by synthesis (Illumina)	Sequencing by ligation (SOLiD[a] sequencing)	Single-molecule real-time sequencing (Pacific Bio)
Read length	400–900 bp	Up to 400 bp	700 bp	50–300 bp	50 + 35 or 50 + 50 bp	5500–8500 bp av. (10,000 bp N50[b]); maximum read length >30,000 bp
Accuracy	99.90%	98%	99.90%	98%	99.90%	87% single read
Reads per run	N/A	Up to 80 million	1 million	Up to 3 billion	1.2–1.4 billion	50,000 per SMRT[c] cell, or ~400 mb
Time per run	20 min–3 h	2 h	24 h	1–10 days	1–2 weeks	30 min–2 h
Cost per 1 million bases	US$2400	US$1	US$10	US$0.0–0.15	US$0.13	US$0.33–$1.00

[a]Sequencing by oligonucleotide ligation and detection.
[b]The N50 statistic, N50 is the length of the contig that just covers the 50th percentile.
[c]Single molecule real time sequencing.

and are output as incomplete insecticidal gene segments.

The third generation of DNA sequencing technology has overcome several of the shortcomings of second-generation sequencing technologies. For example, the read lengths are longer (Table 12.1), thereby significantly reducing the probability of splicing mistakes in assembling the full-length *cry* gene. However, the cost is still relatively high (Table 12.1).

12.3.10 Hybridization and microarray analysis

Southern blot hybridization has been used to identify homologous DNA sequences using designed DNA probes. The percentage identity between two *cry* gene sequences is variable, with some regions as high as 95% and other regions as low as 45%. Because of this, putative new *cry* genes can be detected and characterized using hybridization techniques with specific and conserved probes. The use of entire gene sequences or probe mixtures containing only conserved regions within a subfamily in combination with hybridization allows for the detection of many distantly related (unknown) *cry* genes (Porcar and Juárez-Pérez, 2003). Beard *et al.* (2001) used a modified hybridization method to screen gene libraries derived from novel Bt strains, and their results suggested that this method could rapidly and sensitively detect novel genes that were not detected by PCR. The development of microarray technology improved the throughput of hybridizations because small microarray chips contain thousands of probes and, thus, thousands of target genes can be detected in one hybridization (Letowski *et al.*, 2005).

With hybridization and microarray technology, researchers can design more probes to detect different gene regions to gain more information on the genes contained in samples. Nevertheless, because the technology is costly, time-consuming, and requires specialized equipment, it has not been a popular method for Bt toxin gene discovery.

12.4 Discussion and Future Directions

Looking back on the history of Bt insecticidal protein and gene research, although the initial work was focused on characterizing and analysing insecticidal proteins, the subsequent insecticidal gene discovery work was based on techniques that involved DNA homology, such as PCR, hybridization, and high-throughput sequencing-based technologies. The DNA homology-based methods have limited ability to find new insecticidal genes that are not homologous with reported genes. Even though some strategies, such as library-based gene cloning, have been developed to detect and clone new genes, they are time-consuming when studying numerous isolates. Therefore, new methods are still needed for the efficient detection and cloning of new insecticidal genes.

To search for new insecticidal genes, the purification and characterization of insecticidal proteins is the only road to take. The recent development of high-throughput sequencing and mass spectrum analysis technologies has increased the characterization efficiency of insecticidal proteins. Both Bt genome sequencing and protein mass spectrum analyses can be performed in high throughput once the Bt genomic data have been sequenced and accumulated, and mass spectrum analyses and search techniques can then be used to bridge the information gaps between purified insecticidal proteins and coding genes. Thus, combining mass spectrum analysis and genome sequencing stratagems is an alternative method for efficiently detecting and cloning new insecticidal genes, especially when the costs decrease.

References

Beard, C.E., Ranasinghe, C. and Akhurst, R.J. (2001) Screening for novel *cry* genes by hybridization. *Letters in Applied Microbiology* 33, 241–245.

Berliner, E. (1911) Über die Schalffsucht der Mehlmottentraupe. *Zeitschrift für das Gesamte Getreidewesen* 3, 63–70.

Botstein, D., White, R.L., Skolnick, M. and Davis, R.W. (1980) Construction of a genetic linkage map in man using restriction fragment length polymorphisms. *American Journal of Human Genetics* 32, 314–331.

Bravo, A., Likitvivatanavong, S., Gill, S.S. and Soberón, M. (2011) *Bacillus thuringiensis*: a story of a successful bioinsecticide. *Insect Biochemistry and Molecular Biology* 41, 423–431.

Carozzi, N.B., Kramer, V.C., Warren, G.W., Evola, S. and Koziel, M.G. (1991) Prediction of insecticidal activity of *Bacillus thuringiensis* strains by polymerase chain reaction product profiles. *Applied and Environmental Microbiology* 57, 3057–3061.

Chamberlain, J.S., Gibbs, R.A., Rainer, J.E., Nguyen, P.N. and Thomas, C. (1988) Deletion screening of the Duchenne muscular dystrophy locus via multiplex DNA amplification. *Nucleic Acids Research* 16, 11141–11156.

Chambers, J.A., Jelen, A., Gilbert, M.P., Jany, C.S., Johnson, T.B. *et al.* (1991) Isolation and characterization of a novel insecticidal crystal protein gene from *Bacillus thuringiensis* subsp. *aizawai*. *Journal of Bacteriology* 173, 3966–3976.

Chen, G., Shu, C., Li, Y., Song F., Guo, Y., Li G. and Zhang, J. (2014) Identification of *cry*2 gene based on polymerase chain reaction-high throughput sequencing (PCR-HTS). *Chinese Journal of Biological Control* 30, 696–702. Available at: http://www.zgswfz.com.cn/qikan/manage/wenzhang/20140520.pdf (accessed 4 November 2014).

Crickmore, N., Zeigler, D.R., Feitelson, J., Schnepf, E., Van Rie, J. *et al.* (1998) Revision of the nomenclature for the *Bacillus thuringiensis* pesticidal crystal proteins. *Microbiology and Molecular Biology Reviews* 62, 807–813.

Crickmore, N., Baum, J., Bravo, A., Lereclus, D., Narva, K. *et al.* (2014) *Bacillus thuringiensis* toxin nomenclature. Full list of delta-endotoxins. Available at: http://www.lifesci.sussex.ac.uk/home/Neil_Crickmore/Bt/toxins2.html/ (accessed 23 October 2014).

Fang, J., Xu, X., Wang, P., Zhao, J.Z., Shelton, A.M. *et al.* (2007) Characterization of chimeric *Bacillus thuringiensis* Vip3 toxins. *Applied and Environmental Microbiology* 73, 956–961.

Gundry, C.N., Vandersteen, J.G., Reed, G.H., Pryor, R.J., Chen, J. *et al.* (2003) Amplicon melting analysis with labeled primers: a closed-tube method for differentiating homozygotes and heterozygotes. *Clinical Chemistry* 49, 396–406.

Guo, S., Liu, M., Peng, D., Ji, S., Wang, P. *et al.* (2008) New strategy for isolating novel nematicidal crystal protein genes from *Bacillus thuringiensis* strain YBT-1518. *Applied and Environmental Microbiology* 74, 6997–7001.

Hernández-Rodríguez, C.S., Boets, A., Van Rie, J. and Ferré, J. (2009) Screening and identification of *vip* genes in *Bacillus thuringiensis* strains. *Journal of Applied Microbiology* 107, 219–225.

Ishiwata, S. (1901) On a kind of severe flacherie (Sotto disease). *Dainihon Sanshi Kaiho* 114, 1–5. [In Japanese.]

Juárez-Pérez, V.M., Ferrandis, M.D. and Frutos, R. (1997) PCR-based approach for detection of novel *Bacillus thuringiensis cry* genes. *Applied and Environmental Microbiology* 63, 2997–3002.

Kalman, S., Kiehne, K.L., Libs, J.L. and Yamamoto, T. (1993) Cloning of a novel *cry*IC-type gene from a strain of *Bacillus thuringiensis* subsp. *galleriae*. *Applied and Environmental Microbiology* 59, 1131–1137.

Kleppe, K., Ohtsuka, E., Kleppe, R., Molineux, I. and Khorana, H. (1971) Studies on polynucleotides: XCVI. Repair replication of short synthetic DNA's [DNAs] as catalyzed by DNA polymerases. *Journal of Molecular Biology* 56, 341–361.

Kuo, W.S. and Chak, K.F. (1996) Identification of novel *cry*-type genes from *Bacillus thuringiensis* strains on the basis of restriction fragment length polymorphism of the PCR-amplified DNA. *Applied and Environmental Microbiology* 62, 1369–1377.

Letowski, J., Bravo, A., Brousseau, R. and Masson, L. (2005) Assessment of *cry*1 gene contents of *Bacillus thuringiensis* strains by use of DNA microarrays. *Applied and Environmental Microbiology* 71, 5391–5398.

Li, H., Shu, C., He, X., Gao, J., Liu, R. *et al.* (2012a) Detection and identification of vegetative insecticidal proteins *vip*3 genes of *Bacillus thuringiensis* strains using polymerase chain reaction-high resolution melt analysis. *Current Microbiology* 64, 463–468.

Li, Z., Chen, Y., Mu, D., Yuan, J., Shi, Y. *et al.* (2012b) Comparison of the two major classes of assembly algorithms: overlap–layout–consensus and de-bruijn-graph. *Briefings in Functional Genomics* 11, 25–37.

Liu, L., Li, Y., Li, S., Hu, N., He, Y. *et al.* (2012) Comparison of next-generation sequencing systems. *Journal of Biomedicine and Biotechnology* 2012, Article ID 251364.

Patel, K.D. and Ingle, S.S. (2012) RFLP analysis of *cry*1 and *cry*2 genes of *Bacillus thuringiensis* isolates from India. *Journal of Microbiology and Biotechnology* 22, 729–735.

Porcar, M. and Juárez-Pérez, V. (2003) PCR-based identification of *Bacillus thuringiensis* pesticidal crystal genes. *FEMS Microbiology Reviews* 26, 419–432.

Quail, M. A., Smith, M., Coupland, P., Otto, T.D., Harris, S.R. *et al.* (2012) A tale of three next generation sequencing platforms: comparison of Ion Torrent, Pacific Biosciences and Illumina MiSeq sequencers. *BMC Genomics* 13:341.

Raymond, B., Johnston, P.R., Nielsen-LeRoux, C., Lereclus, D. and Crickmore, N. (2010) *Bacillus thuringiensis*: an impotent pathogen? *Trends in Microbiology* 18, 189–194.

Saiki, R.K., Scharf, S., Faloona, F., Mullis, K.B., Horn, G.T. *et al.* (1985) Enzymatic amplification of beta-globin genomic sequences and restriction site analysis for diagnosis of sickle cell anemia. *Science* 230, 1350–1354.

Schnepf, E., Crickmore, N., Van Rie, J., Lereclus, D., Baum, J. *et al.* (1998) *Bacillus thuringiensis* and its pesticidal crystal proteins. *Microbiology and Molecular Biology Reviews* 62, 775–806.

Schnepf, H.E. and Whiteley, H.R. (1981) Cloning and expression of the *Bacillus thuringiensis* crystal protein gene in *Escherichia coli*. *Proceedings of the National Academy of Sciences of the United States of America* 78, 2893–2897.

Sekar, V., Thompson, D.V., Maroney, M.J., Bookland, R.G. and Adang, M.J. (1987) Molecular cloning and characterization of the insecticidal crystal protein gene of *Bacillus thuringiensis* var. *tenebrionis*. *Proceedings of the National Academy of Sciences of the United States of America* 84, 7036–7040.

Shu, C., Yan, G., Wang, R., Zhang, J., Feng, S. *et al.* (2009) Characterization of a novel *cry8* gene specific to Melolonthidae pests: *Holotrichia oblita* and *Holotrichia parallela*. *Applied Microbiology and Biotechnology* 84, 701–707.

Shu, C., Liu, D., Zhou, Z., Cai, J., Peng, Q. *et al.* (2013a) An improved PCR-restriction fragment length polymorphism (RFLP) method for the identification of *cry1*-type genes. *Applied and Environmental Microbiology* 79, 6706–6711.

Shu, C., Su, H., Zhang, J., He, K., Huang, D. *et al.* (2013b) Characterization of *cry9Da4*, *cry9Eb2*, and *cry9Ee1* genes from *Bacillus thuringiensis* strain T03B001. *Applied Microbiology and Biotechnology* 97, 9705–9713.

Shu, C., Zhang, J., Chen, G., Liang, G., He, K. *et al.* (2013c) Use of a pooled clone method to isolate a novel *Bacillus thuringiensis* Cry2A toxin with activity against *Ostrinia furnacalis*. *Journal of Invertebrate Pathology* 114, 31–33.

Song, F., Zhang, J., Gu, A., Wu, Y., Han, L., H. *et al.* (2003) Identification of *cry1I*-type genes from *Bacillus thuringiensis* strains and characterization of a novel *cry1I*-type gene. *Applied and Environmental Microbiology* 69, 5207–5211.

Wittwer, C.T., Reed, G.H., Gundry, C.N., Vandersteen, J.G. and Pryor, R.J. (2003) High-resolution genotyping by amplicon melting analysis using LCGreen. *Clinical Chemistry* 49, 853–860.

Ye, W., Zhu, L., Liu, Y., Crickmore, N., Peng, D. *et al.* (2012) Mining new crystal protein genes from *Bacillus thuringiensis* on the basis of mixed plasmid-enriched genome sequencing and a computational pipeline. *Applied and Environmental Microbiology* 78, 4795–4801.

Yu, H., Zhang, J., Huang, D., Gao, J. and Song, F. (2006) Characterization of *Bacillus thuringiensis* strain Bt185 toxic to the Asian cockchafer: *Holotrichia parallela*. *Current Microbiology* 53, 13–17.

13 Cry Toxin Binding Site Models and their Use in Strategies to Delay Resistance Evolution

Siva Jakka,[1] Juan Ferré[2] and Juan Luis Jurat-Fuentes[1]*

[1]*Department of Entomology and Plant Pathology, University of Tennessee, Knoxville, Tennessee, USA;* [2]*Departament de Genètica, Universitat de València, Burjassot, Spain*

Summary

The binding of Cry toxins to receptors on the midgut brush border membrane is a necessary step in the Cry intoxication process. Alterations in this binding step are commonly associated with high levels of resistance to Cry toxins in laboratory and field pest populations. When alterations affect binding to a site shared by distinct Cry toxins, cross-resistance is observed. Consequently, pyramiding of Cry toxins recognizing distinct sites has been used to delay the evolution of resistance against transgenic crops expressing *cry* toxin genes for pest control. The availability of accurate binding site models describing binding sites for diverse Cry toxins in targeted pests is critical to identifying genes that are amenable to pyramiding. Many of these models have been proposed and refined in the past 25 years from quantitative binding competition data. In this chapter, we review and revise Cry toxin binding site models for the most relevant pest species targeted by Bt crops and for *Plutella xylostella* as the insect pest with the most binding data available. Our motivation is to provide binding site models as a tool to allow predictions of resistance and cross-resistance risks in order to optimize combinations of diverse Cry toxins for delaying resistance evolution to Bt crops.

13.1 Introduction

Bacillus thuringiensis (Bt) is a Gram-positive soil bacterium that produces insecticidal proteins during the vegetative (Vip toxins) and sporulation (Cry toxins) phases of growth (Jurat-Fuentes and Jackson, 2012). Currently, the Cry family of toxins contains more than 290 described holotypes (Crickmore *et al.*, 2014), which are active against a number of pest species in the orders Lepidoptera, Coleoptera, Diptera, Hymenoptera, Hemiptera and Blattaria (van Frankenhuyzen, 2009). Transgenic crops expressing Bt toxin genes (Bt crops) are a highly effective yet environmentally safe alternative to synthetic pesticides owing to their high specificity and unique mode of action (Sanchis, 2011).

Although still a matter of debate (Vachon *et al.*, 2012), the mode of action of Cry toxins involves the processing of a protoxin form to an active toxin core by host midgut proteases, followed by binding to receptors and insertion on the enterocyte membrane

---

* Corresponding author. E-mail address: jurat@utk.edu

of the insect larva to form a pore that leads to osmotic cell death. Once the gut epithelial barrier is compromised, bacteria in the lumen invade the haemocoel to cause septicaemia and death of the larva (Raymond et al., 2010). The Cry toxin binding step is necessary (Hofmann et al., 1988b), though not sufficient (Wolfersberger, 1990; Escriche et al., 1997), for specificity. Alterations in midgut proteins resulting in reduced toxin binding are associated with high levels of resistance to Cry toxins (Ferré and van Rie, 2002). Cross-resistance, defined as resistance to a Cry toxin not present in the environment of selection, is typically observed among Cry toxins sharing binding sites. Based on these observations, toxin binding models describing unique and shared binding sites among diverse Cry toxins in the host midgut epithelium can predict combinations of Cry toxins resulting in increased risk of cross-resistance. Alternatively, the use of toxins that do not share binding sites dramatically reduces the probability of resistance evolution, as simultaneous alterations in two distinct binding sites are needed (Moar and Anilkumar, 2007). In this chapter, we review available information and develop or revise binding site models for the most economically relevant pests targeted by Bt crops and for the insect with the most published binding data available (*Plutella xylostella*).

13.2 Methods Used for Analysis of Cry Toxin Binding

Binding assays typically involve monitoring interactions between a labelled ligand (in this case a Cry toxin) and a preparation of receptors, which for Cry toxin binding is usually represented by brush border membrane vesicles (BBMVs) prepared from dissected larval midguts (Wolfersberger et al., 1987). The most common methods to measure Cry toxin binding utilize radiolabelled toxins and BBMVs in solution (Hofmann et al., 1988b), BBMVs immobilized on surface plasmon resonance (SPR) chips (Masson et al., 1994) or histological sections (Bravo et al., 1992).

Analyses of Cry toxin binding to purified putative receptors are performed using SPR (de Maagd et al., 1999), ELISA (Kaur et al., 2007) or ligand/dot blots (Hua et al., 2008). The use of non-radioactive and label-free methods is limited to the qualitative detection of Cry toxin binding (except for SPR), in contrast to solution assays with radiolabelled toxins, which are quantitative. Radiolabelling of Cry toxins is commonly achieved by the incorporation of iodine-125 isotope into tyrosine residues through oxidation involving chloramine-T as the oxidizing agent (Hofmann et al., 1988a,b), although the importance of tyrosine residues for Cry toxin binding (Cummings and Ellar, 1994) may lead to inactivated Cry toxins (Luo et al., 1999). Labelling of Cry toxins with alternative methods and/or isotopes, such as iodination beads or *in vivo* methionine labelling, have also been reported (Sanchis et al., 1994). Reducing the ratio of iodine to toxin during labelling has been shown to overcome any adverse labelling effects (Hernández-Rodríguez et al., 2012). Purification under reducing conditions (Gouffon et al., 2011) and the elimination of contaminating peptides by dithiothreitol (DTT) treatment (van Rie et al., 1990; Luo and Adang, 1994) have been reported to reduce non-specific binding in detecting specific Cry toxin–BBMV interactions. Binding assays with radiolabelled toxins detect both reversible and irreversible interactions, yet most of the detected binding represents irreversible toxin insertion into the BBMV membrane (Liang et al., 1995). Consequently, reduction of the amount of bound labelled toxin by competitors suggests that toxins remain bound to these sites after insertion or that they become unavailable for binding. The unique properties of Cry toxin binding, including irreversible binding (Liang et al., 1995), the formation of toxin oligomers (Walters et al., 1994) and sequential binding interactions with diverse sites (Gómez et al., 2007), complicate the interpretation of results from competition assays.

In SPR assays, binding is detected as changes in polarized light reflected from a sensor chip on which the toxin or BBMVs are attached. The attachment of BBMV proteins is preferentially used because amine coupling to the chip may affect toxin function (Luo *et al.*, 1997); it also allows the performance of binding competition experiments (Masson *et al.*, 1995; Hua *et al.*, 2001; Li *et al.*, 2004). The use of antisera or Cry toxins labelled with biotin (Denolf *et al.*, 1993a) or fluorophores (Higuchi *et al.*, 2007; Sharma *et al.*, 2010) has usually been limited to semi-quantitative binding competition assays and identifying BBMV proteins interacting with Cry toxins on ligand blots (Oddou *et al.*, 1991; Jurat-Fuentes and Adang, 2001). However, the denaturing conditions employed during ligand blotting may limit the biological relevance of the Cry toxin binding proteins identified (Lee and Dean, 1996).

The models proposed in this review are derived from synthesizing the available information from published competition assays. We emphasize the use of 'binding site' as opposed to 'receptor', because even high-affinity binding to some sites may not be conducive to toxicity (Wolfersberger, 1990; Garczynski *et al.*, 1991). We also occasionally refer to individual populations of binding sites as 'binding site' owing to limitations intrinsic to binding assays that hinder the detection of distinct binding sites with similar binding affinities. As a result, the proposed models probably underestimate the number of diverse binding sites present in BBMVs.

13.3 Binding Site Models in Lepidoptera

13.3.1 *Plutella xylostella*

The diamondback moth (*P. xylostella*) is a major pest of cruciferous crops worldwide and is reported to have developed resistance to many groups of insecticides, including Bt pesticides (Ferré *et al.*, 1991; Tabashnik *et al.*, 1997b; Sayyed and Wright, 2001). The Bt specificity database identifies several Cry toxins, including Cry1Aa, Cry1Ab, Cry1Ac, Cry1B, Cry1Ca, Cry1Fa, Cry1Ja and Cry9Ca, as being active against larvae of *P. xylostella*

(van Frankenhuyzen, 2009). Early binding studies with radiolabelled Cry1A, Cry1B and Cry1Ca toxins confirmed the existence of specific binding to BBMVs from *P. xylostella* (Ferré *et al.*, 1991; Tabashnik *et al.*, 1997b), and specific Cry1Fa binding has been reported more recently (Hernández-Rodríguez *et al.*, 2012). Specific Cry1A toxin binding has also been confirmed using fluorescently labelled Cry1A toxins (Higuchi *et al.*, 2007).

A previously proposed binding site model for *P. xylostella* BBMVs included four binding sites, one unique for Cry1Aa, one shared by all Cry1A toxins and Cry1Fa, and two unique sites for Cry1Ba and Cry1Ca (Ballester *et al.*, 1999). However, this model did not clearly explain data from binding assays with BBMVs from resistant *P. xylostella* strains. For instance, a field Cry1Ab-resistant strain from the Philippines presented reduced Cry1Ab binding, but cross-resistance or reduced binding of other Cry1A toxins was not detected (Ballester *et al.*, 1994, 1999).

Based on the available quantitative binding data, we propose an expanded Cry toxin binding site model for *P. xylostella* BBMVs that includes six binding sites (Fig. 13.1). Heterologous competition assays using radiolabelled Cry1A toxins showed that Cry1Aa binds with highest affinity to a site recognized only by this toxin (site A), and that all Cry1A toxins share a binding site (site B) (Ballester *et al.*, 1994). Site B is also recognized by Cry1Fa (Granero *et al.*, 1996), although this toxin does not reduce Cry1Aa binding, probably because site B is a low-affinity site for Cry1Aa, as revealed by homologous Cry1Aa competition data best fitting to a two-site model (Ballester *et al.*, 1999). Alteration of site B confers multiple resistance to all of these toxins and reduced Cry1Ab and Cry1Ac (but not Cry1Aa) binding (Tabashnik *et al.*, 1997a,b), which is probably explained by differences in the binding affinity of the individual toxins for this site (Tabashnik *et al.*, 1997b; Ballester *et al.*, 1999). Based on heterologous incomplete competition, both Cry1Ab and Cry1Ac must have an additional binding site (site C in the model), which is not recognized by Cry1Aa or Cry1Fa (Ballester *et al.*, 1999). The fact

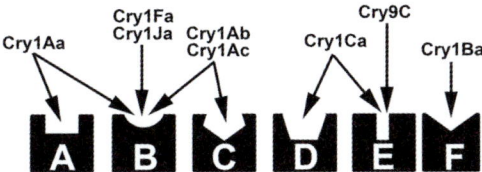

Fig. 13.1. Proposed binding site model with six binding sites (A–F) for Cry toxins in brush border membrane vesicles of *Plutella xylostella*.

that Cry1Ab and Cry1Ac bind to more than one binding site, whereas the homologous competition curve fits a single-site model can be explained by the toxins having similar binding affinity to all sites, which could not be detected from analysis of the competition curve. Heterologous competition with ^{125}I-Cry1Ca revealed that this toxin recognizes a population of binding sites not recognized by Cry1A, Cry1B or Cry1Fa toxins (Ferré *et al.*, 1991; Granero *et al.*, 1996), which is named site D in the model. SPR confirmed that Cry1Ac and Cry1Ca do not share binding sites in *P. xylostella* BBMVs (Masson *et al.*, 1995), and while assays with biotinylated toxins showed that Cry1Ca binds to Cry9Ca binding sites with low affinity (site E), Cry9Ca does not compete for site D (Lambert *et al.*, 1996). Heterologous competition assays with radiolabelled Cry1Ab and Cry1Ac revealed that Cry1Ja binds with high and low affinity to shared Cry1Ab-Cry1Ac binding sites (Herrero *et al.*, 2001). Which of the sites in the model represents this Cry1Ab–Cry1Ac–Cry1Ja site is currently unknown, but based on cross-resistance patterns and homology in the toxin domains involved in binding (Tabashnik *et al.*, 1996), we hypothesize that Cry1Ja and Cry1A toxins share binding site B in *P. xylostella* BBMVs (Fig. 13.1). Binding site F, specific for Cry1Ba, is proposed based on lack of Cry1Ba competition with other toxins (Ferré *et al.*, 1991).

13.3.2 Heliothines

The Heliothinae subfamily of Lepidoptera contains some of the most damaging insect

pests worldwide, including *Heliothis virescens*, *Helicoverpa zea* and *Helicoverpa armigera*. The high biological similarity between these species is reflected in similar susceptibilities to Bt toxins and Cry toxin binding patterns, which have been mostly studied in BBMVs from *H. virescens*, *H. armigera* and, to a lesser extent, *H. zea*. The Bt toxin specificity database identifies Cry1A, Cry1F, Cry1J, Cry2A and Cry9 as the most effective Cry protein families against Heliothinae larvae (van Frankenhuyzen, 2009). In BBMVs from these larvae, Cry1A and Cry1Fa toxins display high affinity and saturable binding (van Rie *et al.*, 1989; Karim *et al.*, 1999; Hernández-Rodríguez *et al.*, 2012), while Cry2A toxin binding is also saturable but of a lower affinity (Hernández-Rodríguez *et al.*, 2008; Caccia *et al.*, 2010). Toxins with low relative activity, such as Cry1B and Cry1C, bind specifically, albeit with low affinity, to BBMVs from *H. armigera* (Lu *et al.*, 2013).

Heterologous competition with Cry1A toxins and Heliothinae BBMVs reveals three Cry1A binding sites (Fig. 13.2). Site A is shared by Cry1Aa, Cry1Ab and Cry1Ac toxins; site B is recognized by Cry1Ab and Cry1Ac; and site C is unique for Cry1Ac (van Rie *et al.*, 1989; Estela *et al.*, 2004). Competition assays labelled Cry1A, Cry1Fa and Cry1Ja toxins demonstrate that Cry1Fa and Cry1Ja recognize site A (Jurat-Fuentes and Adang, 2001; Hernández-Rodríguez and Ferré, 2005). In contrast, Cry2A and Vip3A toxins bind to unique sites (D and E, respectively) for each toxin (Lee *et al.*, 2006; Luo *et al.*, 2007; Gouffon *et al.*, 2011). Site D is shared among Cry2Aa, Cry2Ab and Cry2Ae toxins and is not recognized by Cry1A toxins (Hernández-Rodríguez *et al.*, 2008).

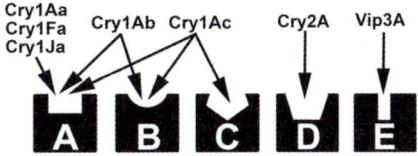

Fig. 13.2. Proposed binding site model with five binding sites (A–E) for Cry and Vip toxins in Heliothinae brush border membrane vesicles.

13.3.3 *Spodoptera*

The genus *Spodoptera* includes highly relevant pest species of many crops, and their high polyphagy also makes them damaging secondary pests on alternative plant hosts. The most relevant species in the group include *S. frugiperda*, *S. exigua*, *S. littoralis* and *S. litura*. Bioassay data identify Cry1Bb, Cry1Ca, Cry1Da, Cry1Fa and Cry2A toxins as active against most *Spodoptera* spp., whereas Cry1A toxins are ineffective (Luo *et al.*, 1999; van Frankenhuyzen, 2009; Lu *et al.*, 2013). Quantitative binding assays have demonstrated high affinity specific binding to *Spodoptera* BBMVs for Cry1A, Cry1Bb, Cry1Ca, Cry1Ea, Cry1Fa and Vip3A toxins (Luo *et al.*, 1999; Rang *et al.*, 2004; Hernández-Rodríguez *et al.*, 2012; Chakroun and Ferré, 2014), but specific binding of Cry1Ia has only been demonstrated using qualitative assays (Bergamasco *et al.*, 2013). Data from Cry toxin binding assays do not correlate with susceptibility (Lu *et al.*, 2013), although low Cry1Ab toxicity was associated with high levels of non-specific Cry1Ab binding in midgut sections of *S. frugiperda* larvae (Aranda *et al.*, 1996).

As shown in Fig. 13.3, heterologous competition binding assays with *S. frugiperda* BBMVs using radiolabelled toxins have identified a shared population of Cry1A binding sites (site A) that is also recognized by Cry1Fa, Cry1Ja and Cry1A.105 (a chimeric protein with domains I and II from Cry1Ac and domain III almost identical to Cry1Fa) (Hernández and Ferré, 2005; Hernández-Rodríguez *et al.*, 2013). In these BBMVs, there is also evidence for the existence of an additional binding site (site B) recognized by the same toxins as site A, excepting Cry1Aa, as this toxin only displaces part of Cry1Ab or Cry1Ac binding. The existence of two Cry1A binding sites, one of them not recognized by Cry1Aa, has also been described in *S. exigua* BBMVs (Escriche *et al.*, 1997). Both Cry1Ca and Cry1Bb toxins share binding site C with Cry1Fa in *S. frugiperda* and *S. exigua* BBMVs (Luo *et al.*, 1999; Rang *et al.*, 2004), which does not have unique binding sites in these BBMVs. The displacement of Cry1Ca binding to *S. exigua* BBMVs by Cry9Ca suggests that Cry9Ca may recognize site C, but quantitative binding data has not been presented (Lambert *et al.*, 1996). Evidence for sites only recognized by Cry1Bb has also been provided, as neither Cry1Ca nor Cry1Fa competed all Cry1Bb binding (Luo *et al.*, 1999). In contrast, full displacement of Cry1Ba binding to *S. frugiperda* BBMVs was observed when using Cry1Fa as competitor (Rang *et al.*, 2004), suggesting that Cry1Bb may have binding sites in *Spodoptera* BBMVs not recognized by Cry1Ba (site D). The Cry1Ea toxin recognizes a unique population of binding sites in *S. frugiperda* (Rang *et al.*, 2004) and *S. littoralis* (Avisar *et al.*, 2004) BBMVs (site E). None of these sites interact with Cry2A toxins (Hernández-Rodríguez *et al.*, 2013), which bind to site F. Quantitative competition assays with radio-iodinated Vip3Aa and Cry1Ab have demonstrated that Vip3A toxins do not share binding sites with Cry1A toxins in *S. frugiperda* larvae (Sena *et al.*, 2009; Chakroun and Ferré, 2014), and a separate binding site (site G) is

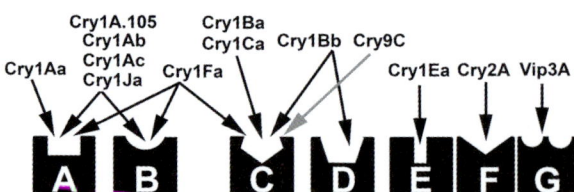

Fig. 13.3. Proposed binding site model with seven binding sites (A–G) for Cry and Vip toxins in brush border membrane vesicles from *Spodoptera* spp. The grey arrow indicates lack of reciprocal competition data (between labelled Cry9C and other toxins).

proposed for these. Binding assays with biotinylated Cry1Da toxin and *S. frugiperda* histological midgut sections revealed specific Cry1Da binding (Aranda *et al.*, 1996), though no information from heterologous competition binding studies is available to place this toxin in the current binding site model.

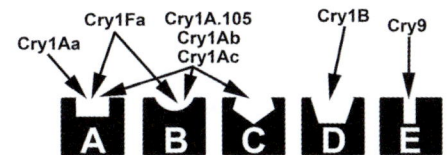

Fig. 13.4. Proposed binding site model with five binding sites (A–E) for Cry toxins in brush border membrane vesicles from *Ostrinia nubilalis*.

13.3.4 *Ostrinia nubilalis*

The European corn borer (*O. nubilalis*) is one of the most damaging pests of maize worldwide. A number of Cry toxins from Bt, namely Cry1Ab, Cry1Ac, Cry1A.105, Cry1B, Cry1Fa, Cry2Ab, Cry2Ae, Cry9C and Cry9E, have been reported to be effective against *O. nubilalis* larvae (van Frankenhuyzen, 2009; Hernández-Rodríguez *et al.*, 2013).

Binding experiments have shown an association between binding affinity and relative toxicity against *O. nubilalis* larvae for Cry1Ab and Cry1Ac (Denolf *et al.*, 1993b; Hua *et al.*, 2001). Part of radiolabelled Cry1Ab binding is competed by high Cry1Aa concentrations, supporting the existence of two binding sites for Cry1Ab (A and B, Fig. 13.4), one of them (site A) shared by all Cry1A toxins and the other (site B) by Cry1Ab and Cry1Ac (Hernández-Rodríguez *et al.*, 2013). Displacement of all labelled Cry1Fa binding by Cry1Ab or Cry1Ac demonstrates that these toxins share sites A and B (Hernández-Rodríguez *et al.*, 2013), yet the high Cry1Fa concentrations needed to compete for radiolabelled Cry1Ab toxin binding (Hua *et al.*, 2001; Hernández-Rodríguez *et al.*, 2013) suggest that these sites are of low affinity for Cry1Fa. As expected from the Cry1Ac and Cry1Fa toxin domains present in the chimeric Cry1A.105 toxin, this toxin also binds to sites A and B (Hernández-Rodríguez *et al.*, 2013). The observation that Cry1Fa could not compete for all Cry1A.105 or Cry1Ab binding suggests the potential existence of a third site (C in the model) not recognized by Cry1Fa (Hernández-Rodríguez *et al.*, 2013). In support of this hypothesis, resistance to Cry1Ab and high levels of cross-resistance to Cry1Ac and Cry1Aa were not associated

with cross-resistance to Cry1Fa in *O. nubilalis* strains (Siqueira *et al.*, 2004; Crespo *et al.*, 2011). Moreover, the reciprocal situation has been found for Cry1Fa-selected *O. nubilalis* strains (Pereira *et al.*, 2008). While reduced Cry1Aa binding was detected in BBMVs from Cry1Ab-resistant *O. nubilalis* larvae (Siqueira *et al.*, 2006; Crespo *et al.*, 2011), no altered binding was detected for Cry1Fa-resistant strains (Pereira *et al.*, 2010). Taken together, the resistance pattern and the binding results suggest the existence of a population of binding sites shared by Cry1A toxins that are not relevant to Cry1Fa toxicity. The reduction in Cry1Aa binding ability in the resistant strains might indicate a binding site alteration only detectable when using Cry1Aa, which would support the hypothesis of the occurrence of at least two different binding sites, one shared by the three Cry1A toxins (the one whose alteration confers multiple resistance) and the other only shared by Cry1Ab and Cry1Ac (Fig. 13.4). A third population of binding sites (site C) would only bind Cry1Ab, Cry1Ac and Cry1A.105 toxins. Unique sites D and E are proposed for Cry1B and Cry9 toxins, respectively (Denolf *et al.*, 1993b, Hua *et al.*, 2001, Lira *et al.*, 2013).

13.4 Binding Site Models in Coleoptera

13.4.1 *Diabrotica virgifera virgifera*

Research on Cry toxins and rootworms has mostly focused on the Western corn rootworm (*D. v. virgifera*) as target because larvae of this insect are the most economically damaging and difficult to control insect pest of maize. Data from bioassays and histo-pathological studies support Cry3Bb (Donovan *et al.*, 1992; Siegfried *et al.*, 2005), Cry6Aa (Li *et al.*, 2013) and the binary Cry34Ab and Cry35Ab toxins (expressed as Cry34/35Ab henceforth) (Ellis *et al.*, 2002) as active against larvae of *Diabrotica* spp. Modified Cry3Aa toxin (mCry3A) containing an additional chymotrypsin cleavage site, and a Cry3Aa-based hybrid toxin (eCry3.1Ab) containing domain III of Cry1Ab, have also been shown to be effective against rootworms (Walters *et al.*, 2008, 2010).

Binding experiments with radiolabelled Cry3Aa showed that this toxin binds specifically to BBMVs of *Diabrotica* spp., although it displays a low affinity, which correlates with its low activity against these larvae (Slaney *et al.*, 1992; Li *et al.*, 2013). In contrast, high-affinity specific binding to BBMVs from *D. v. virgifera* was shown for radiolabelled Cry34/35Ab, Cry6Aa and Cry8Ba (Li *et al.*, 2013). While Cry35Ab did bind to the BBMVs, its affinity was dramatically increased in the presence of Cry34Ab, thus supporting their action as binary toxins (Ellis *et al.*, 2002). Attempts to detect specific binding for Cry34Ab were unsuccessful owing to inactivation of the toxin during labelling (Li *et al.*, 2013). In heterologous competition binding assays with BBMVs from *D. v. virgifera*, Cry34/35 was shown not to share binding sites with Cry3Aa, Cry8Ba or Cry6Aa toxins (Li *et al.*, 2013), although the existence of shared binding sites between Cry3Aa, Cry8Ba and Cry6Aa was not tested. Binding assays with biotinylated mCry3A or eCry3.1Ab showed that these toxins bind specifically and do not share high-affinity binding sites in *D. v. virgifera* BBMVs (Walters *et al.*, 2008, 2010). However, based on eCry3A.1Ab containing domain II of Cry3Aa, it would be expected that eCry3A.1Ab shares some binding sites with Cry3Aa and m-Cry3Aa toxins (as denoted by the grey arrow in the model shown in Fig. 13.5). Remarkably, enzymatic cleavage dependent on the introduced chymotrypsin site in mCry3A was necessary for specific binding of the toxin to the BBMVs.

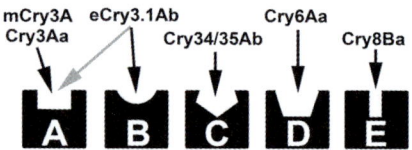

Fig. 13.5. Proposed binding site model with five binding sites (A–E) for Cry toxins in brush border membrane vesicles from *Diabrotica virgifera virgifera*. The grey arrow indicates expected, although not yet demonstrated, binding of eCry3A.1Ab toxin to Cry3Aa binding sites based on sequence identity in the domains involved in toxin binding.

Based on the available data, a model consisting of five binding sites is proposed for *D. v. virgifera* BBMVs (Fig. 13.5). Albeit not tested, and based on being identical in their toxin-binding domains, we hypothesize that Cry3Aa and mCry3A share binding site A. Binding sites B, C, D and E are recognized by eCry3.1Ab, Cry34/35Ab, Cry6Aa and Cry8Ba toxins, respectively. Future heterologous competition binding assays among Cry3Aa, eCry3.1Ab, Cry6Aa and Cry8Aa toxins would be needed to determine whether these toxins share any binding sites.

13.5 Conclusions

Our review of the available Cry toxin-binding data identifies a number of commonalities in the Lepidoptera. Most remarkable is the existence of a shared binding site for Cry1A toxins in all tested species. When tested, at least some of these binding sites are also shared with Cry1Fa and Cry1Ja toxins. The observations therefore suggest that these toxins should not be pyramided in transgenic crops owing to an increased risk of cross-resistance or, at least, they call for caution in those species where an additional binding site unique for Cry1Fa may be present. Another common feature in all the species studied is that Cry1B, Cry1C and Cry9C toxins do not share binding sites with Cry1A, Cry1F or Cry1J toxins. The only exception to this generalization is the sharing of binding sites between Cry1Ba and Cry1Ca in *Spodoptera* spp. BBMVs. Unique sites for Cry2A and Vip3A toxins have been found in all species tested. The available data for the Coleoptera supports some pyramid combinations, such as Cry3 and Cry34/35Ab toxins. Further work is necessary to expand current binding site models to include toxins with commercial interest that have not been tested for binding and to identify the specific proteins representing the proposed binding sites.

References

Aranda, E., Sánchez, J., Peferoen, M., Guereca, L. and Bravo, A. (1996) Interactions of *Bacillus thuringiensis* crystal proteins with the midgut epithelial cells of *Spodoptera frugiperda* (Lepidoptera: Noctuidae). *Journal of Invertebrate Pathology* 68, 203–212.

Avisar, D., Keller, M., Gazit, E., Prudovsky, E., Sneh, B. *et al.* (2004) The role of *Bacillus thuringiensis* Cry1C and Cry1E separate structural domains in the interaction with *Spodoptera littoralis* gut epithelial cells. *The Journal of Biological Chemistry* 279, 15779–15786.

Ballester, V., Escriche, B., Ménsua, J.L., Riethmacher, G.W. and Ferré, J. (1994) Lack of cross-resistance to other *Bacillus thuringiensis* crystal proteins in a population of *Plutella xylostella* highly resistant to Cry1A(b). *Biocontrol Science and Technology* 4, 437–443.

Ballester, V., Granero, F., Tabashnik, B.E., Malvar, T. and Ferré, J. (1999) Integrative model for binding of *Bacillus thuringiensis* toxins in susceptible and resistant larvae of the diamondback moth (*Plutella xylostella*). *Applied and Environmental Microbiology* 65, 1413–1419.

Bergamasco, V.B., Mendes, D.R.P., Fernandes, O.A., Desidério, J.A. and Lemos, M.V.F. (2013) *Bacillus thuringiensis* Cry1Ia10 and Vip3Aa protein interactions and their toxicity in *Spodoptera* spp. (Lepidoptera). *Journal of Invertebrate Pathology* 112, 152–158.

Bravo, A., Hendrickx, K., Jansens, S. and Peferoen, M. (1992) Immunocytochemical analysis of specific binding of *Bacillus thuringiensis* insecticidal crystal proteins to lepidopteran and coleopteran midgut membranes. *Journal of Invertebrate Pathology* 60, 247–253.

Caccia, S., Hernández-Rodríguez, C.S., Mahon,

R.J., Downes, S., James, W. *et al.* (2010) Binding site alteration is responsible for field-isolated resistance to *Bacillus thuringiensis* Cry2A insecticidal proteins in two *Helicoverpa* species. *PLoS One* 5(4): e9975.

Chakroun, M. and Ferré, J. (2014) *In vivo* and *in vitro* binding of Vip3Aa to *Spodoptera frugiperda* midgut and characterization of binding sites using ^{125}I-radiolabeling. *Applied and Environmental Microbiology* 80, 6258–6265.

Crespo, A.L., Rodrigo-Simón, A., Siqueira, H.A., Pereira, E.J., Ferré, J. *et al.* (2011) Cross-resistance and mechanism of resistance to Cry1Ab toxin from *Bacillus thuringiensis* in a field-derived strain of European corn borer, *Ostrinia nubilalis*. *Journal of Invertebrate Pathology* 107, 185–192.

Crickmore, N., Baum, J., Bravo, A., Lereclus, D., Narva, K. *et al.* (2014) *Bacillus thuringiensis* toxin nomenclature. Available at: www. btnomenclature.info/ (accessed 28 June 2014).

Cummings, C.E. and Ellar, D.J. (1994) Chemical modification of *Bacillus thuringiensis* activated δ-endotoxin and its effect on toxicity and binding to *Manduca sexta* midgut membranes. *Microbiology* 140, 2737–2747.

de Maagd, R.A., Bakker, P.L., Masson, L., Adang, M.J., Sangadala, S. *et al.* (1999) Domain III of the *Bacillus thuringiensis* delta-endotoxin Cry1Ac is involved in binding to *Manduca sexta* brush border membranes and to its purified aminopeptidase N. *Molecular Microbiology* 31, 463–471.

Denolf, P., Jansens, S., Van Houdt, S., Peferoen, M., Degheele, D. *et al.* (1993a) Biotinylation of *Bacillus thuringiensis* insecticidal crystal proteins. *Applied and Environmental Microbiology* 59, 1821–1827.

Denolf, P., Jansens, S., Peferoen, M., Degheele, D. and van Rie, J. (1993b) Two different *Bacillus thuringiensis* delta-endotoxin receptors in the midgut brush border membrane of the European corn borer, *Ostrinia nubilalis* (Hübner) (Lepidoptera: Pyralidae). *Applied and Environmental Microbiology* 59, 1828–1837.

Donovan, W.P., Rupar, M.J., Slaney, A.C., Malvar, T., Gawron-Burke, M.C. *et al.* (1992) Characterization of two genes encoding *Bacillus thuringiensis* insecticidal crystal proteins toxic to Coleoptera species. *Applied and Environmental Microbiology* 58, 3921–3927.

Ellis, R.T., Stockhoff, B.A., Stamp, L., Schnepf, H.E., Schwab, G.E. *et al.* (2002) Novel *Bacillus thuringiensis* binary insecticidal crystal proteins active on western corn rootworm, *Diabrotica virgifera virgifera* LeConte. *Applied and Environmental Microbiology* 68, 1137–1145.

Escriche, B., Ferré, J. and Silva, F.J. (1997) Occurrence of a common binding site in *Mamestra brassicae*, *Phthorimaea operculella*, and *Spodoptera exigua* for the insecticidal crystal proteins CryIA from *Bacillus thuringiensis*. *Insect Biochemistry and Molecular Biology* 27, 651–656.

Estela, A., Escriche, B. and Ferré, J. (2004) Interaction of *Bacillus thuringiensis* toxins with larval midgut binding sites of *Helicoverpa armigera* (Lepidoptera: Noctuidae). *Applied and Environmental Microbiology* 70, 1378-1384.

Ferré, J. and van Rie, J. (2002) Biochemistry and genetics of insect resistance to *Bacillus thuringiensis*. *Annual Review of Entomology* 47, 501–533.

Ferré, J., Real, M.D., van Rie, J., Jansens, S. and Peferoen, M. (1991) Resistance to the *Bacillus thuringiensis* bioinsecticide in a field population of *Plutella xylostella* is due to a change in a midgut membrane receptor. *Proceedings of the National Academy of Sciences of the United States of America* 88, 5119–5123.

Garczynski, S.F., Crim, J.W. and Adang, M.J. (1991) Identification of putative insect brush border membrane-binding molecules specific to *Bacillus thuringiensis* delta-endotoxin by protein blot analysis. *Applied and Environmental Microbiology* 57, 2816–2820.

Gómez, I., Pardo-López, L., Muñoz-Garay, C., Fernandez, L.E., Pérez, C. *et al.* (2007) Role of receptor interaction in the mode of action of insecticidal Cry and Cyt toxins produced by *Bacillus thuringiensis*. *Peptides* 28, 169–173.

Gouffon, C., van Vliet, A., van Rie, J., Jansens, S. and Jurat-Fuentes, J.L. (2011) Binding sites for *Bacillus thuringiensis* Cry2Ae toxin on heliothine brush border membrane vesicles are not shared with Cry1A, Cry1F, or Vip3A toxin. *Applied and Environmental Microbiology* 77, 3182–3188.

Granero, F., Ballester, V. and Ferré, J. (1996) *Bacillus thuringiensis* crystal proteins Cry1Ab and Cry1Fa share a high affinity binding site in *Plutella xylostella* (L.). *Biochemical and Biophysical Research Communications* 224, 779–783.

Hernández, C.S. and Ferré, J. (2005) Common receptor for *Bacillus thuringiensis* toxins Cry1Ac, Cry1Fa, and Cry1Ja in *Helicoverpa armigera*, *Helicoverpa zea*, and *Spodoptera exigua*. *Applied and Environmental Microbiology* 71, 5627–5629.

Hernández-Rodríguez, C.S., Van Vliet, A., Bautsoens, N., Van Rie, J. and Ferré, J. (2008) Specific binding of *Bacillus thuringiensis* Cry2A insecticidal proteins to a common site in the midgut of *Helicoverpa* species. *Applied and Environmental Microbiology* 74, 7654–7659.

Hernández-Rodríguez, C.S., Hernández-Martínez, P., van Rie, J., Escriche, B. and Ferré, J. (2012) Specific binding of radiolabeled Cry1Fa insecticidal protein from *Bacillus thuringiensis* to midgut sites in lepidopteran species. *Applied and Environmental Microbiology* 78, 4048–4050.

Hernández-Rodríguez, C.S., Hernández-Martínez, P., van Rie, J., Escriche, B. and Ferré, J. (2013) Shared midgut binding sites for Cry1A.105, Cry1Aa, Cry1Ab, Cry1Ac and Cry1Fa proteins from *Bacillus thuringiensis* in two important corn pests, *Ostrinia nubilalis* and *Spodoptera frugiperda*. *PLoS ONE* 8(7): e68164.

Herrero, S., González-Cabrera, J., Tabashnik, B.E. and Ferré, J. (2001) Shared binding sites in Lepidoptera for *Bacillus thuringiensis* Cry1Ja and Cry1A toxins. *Applied and Environmental Microbiology* 67, 5729–5734.

Higuchi, M., Haginoya, K., Yamazaki, T., Miyamoto, K., Katagiri, T. *et al.* (2007) Binding of *Bacillus thuringiensis* Cry1A toxins to brush border membrane vesicles of midgut from Cry1Ac susceptible and resistant *Plutella xylostella*. *Comparative Biochemistry and Physiology Part B: Biochemistry and Molecular Biology* 147, 716–724.

Hofmann, C., Lüthy, P., Hütter, R. and Pliska, V. (1988a) Binding of the delta endotoxin from *Bacillus thuringiensis* to brush-border membrane vesicles of the cabbage butterfly *(Pieris brassicae)*. *European Journal of Biochemistry* [now *The FEBS Journal*] 173, 85–91.

Hofmann, C., Vanderbruggen, H., Höfte, H., van Rie, J., Jansens, S. *et al.* (1988b) Specificity of *Bacillus thuringiensis* δ-endotoxins is correlated with the presence of high-affinity binding sites in the brush border membrane of target insect midguts. *Proceedings of the National Academy of Sciences of the United States of America* 85, 7844–7848.

Hua, G., Masson, L., Jurat-Fuentes, J.L., Schwab, G. and Adang, M.J. (2001) Binding analyses of *Bacillus thuringiensis* Cry δ-endotoxins using brush border membrane vesicles of *Ostrinia nubilalis*. *Applied and Environmental Microbiology* 67, 872–879.

Hua, G., Zhang, R., Abdullah, M.A. and Adang, M.J. (2008) *Anopheles gambiae* cadherin AgCad1 binds the Cry4Ba toxin of *Bacillus thuringiensis israelensis* and a fragment of AgCad1 synergizes toxicity. *Biochemistry* 47, 5101–5110.

Jurat-Fuentes, J.L. and Adang, M.J. (2001)

Importance of Cry1 δ-endotoxin domain II loops for binding specificity in *Heliothis virescens* (L.). *Applied and Environmental Microbiology* 67, 323–329.

Jurat-Fuentes, J.L. and Jackson, T.A. (2012) Bacterial entomopathogens. In: Vega, F.E. and Kaya, H.K. (eds) *Insect Pathology*, 2nd edn. Academic Press, San Diego, California, pp. 265–349.

Karim, S., Riazuddin, S. and Dean, D.H. (1999) Interaction of *Bacillus thuringiensis* δ-endotoxins with midgut brush border membrane vesicles of *Helicoverpa armigera*. *Journal of Asia-Pacific Entomology* 2, 153–162.

Kaur, R., Agrawal, N. and Bhatnagar, R. (2007) Purification and characterization of aminopeptidase N from *Spodoptera litura* expressed in Sf21 insect cells. *Protein Expression and Purification* 54, 267–274.

Lambert, B., Buysse, L., Decock, C., Jansens, S., Piens, C. *et al.* (1996) A *Bacillus thuringiensis* insecticidal crystal protein with a high activity against members of the family Noctuidae. *Applied and Environmental Microbiology* 62, 80–86.

Lee, M.K. and Dean, D.H. (1996) Inconsistencies in determining *Bacillus thuringiensis* toxin binding sites relationship by comparing competition assays with ligand blotting. *Biochemical and Biophysical Research Communications* 220, 575–580.

Lee, M.K., Miles, P. and Chen, J.S. (2006) Brush border membrane binding properties of *Bacillus thuringiensis* Vip3A toxin to *Heliothis virescens* and *Helicoverpa zea* midguts. *Biochemical and Biophysical Research Communications* 339, 1043–1047.

Li, H., González-Cabrera, J., Oppert, B., Ferré, J., Higgins, R.A. *et al.* (2004) Binding analyses of Cry1Ab and Cry1Ac with membrane vesicles from *Bacillus thuringiensis*-resistant and -susceptible *Ostrinia nubilalis*. *Biochemical and Biophysical Research Communications* 323, 52–57.

Li, H., Olson, M., Lin, G., Hey, T., Tan, S.Y. *et al.* (2013) *Bacillus thuringiensis* Cry34Ab1/Cry35Ab1 interactions with western corn rootworm midgut membrane binding sites. *PLoS ONE* 8(1): e53079.

Liang, Y., Patel, S.S. and Dean, D.H. (1995) Irreversible binding kinetics of *Bacillus thuringiensis* CryIA δ-endotoxins to gypsy moth brush border membrane vesicles is directly correlated to toxicity. *The Journal of Biological Chemistry* 270, 24719–24724.

Lira, J., Beringer, J., Burton, S., Griffin, S., Sheets,

J. *et al.* (2013) Insecticidal activity of *Bacillus thuringiensis* Cry1Bh1 against *Ostrinia nubilalis* (Hübner) (Lepidoptera: Crambidae) and other lepidopteran pests. *Applied and Environmental Microbiology* 79, 7590–7597.

Lu, Q., Cao, G.C., Zhang, L.L., Liang, G.M., Gao, X.W. *et al.* (2013) The binding characterization of Cry insecticidal proteins to the brush border membrane vesicles of *Helicoverpa armigera*, *Spodoptera exigua*, *Spodoptera litura* and *Agrotis ipsilon*. *Journal of Integrative Agriculture* 12, 1598–1605.

Luo, K. and Adang, M.J. (1994) Removal of adsorbed toxin fragments that modify *Bacillus thuringiensis* CryIC δ-endotoxin iodination and binding by sodium dodecyl sulfate treatment and renaturation. *Applied and Environmental Microbiology* 60, 2905–2910.

Luo, K., Sangadala, S., Masson, L., Mazza, A., Brousseau, R. *et al.* (1997) The *Heliothis virescens* 170 kDa aminopeptidase functions as "receptor A" by mediating specific *Bacillus thuringiensis* Cry1A δ-endotoxin binding and pore formation. *Insect Biochemistry and Molecular Biology* 27, 735–743.

Luo, K., Banks, D. and Adang, M.J. (1999) Toxicity, binding, and permeability analyses of four *Bacillus thuringiensis* Cry1 δ-endotoxins using brush border membrane vesicles of *Spodoptera exigua* and *Spodoptera frugiperda*. *Applied and Environmental Microbiology* 65, 457–464.

Luo, S., Wu, K., Tian, Y., Liang, G., Feng, X. *et al.* (2007) Cross-resistance studies of Cry1Ac-resistant strains of *Helicoverpa armigera* (Lepidoptera: Noctuidae) to Cry2Ab. *Journal of Economic Entomology* 100, 909–915.

Masson, L., Mazza, A. and Brousseau, R. (1994) Stable immobilization of lipid vesicles for kinetic studies using surface plasmon resonance. *Analytical Biochemistry* 218, 405–412.

Masson, L., Mazza, A., Brousseau, R. and Tabashnik, B. (1995) Kinetics of *Bacillus thuringiensis* toxin binding with brush border membrane vesicles from susceptible and resistant larvae of *Plutella xylostella*. *The Journal of Biological Chemistry* 270, 11887–11896.

Moar, W.J. and Anilkumar, K.J. (2007) Plant science: the power of the pyramid. *Science* 318, 1561–1562.

Oddou, P., Hartmann, H. and Geiser, M. (1991) Identification and characterization of *Heliothis virescens* midgut membrane proteins binding *Bacillus thuringiensis* δ-endotoxins. *European Journal of Biochemistry* [now *The FEBS Journal*] 201, 673–680.

Pereira, E.J., Storer, N.P. and Siegfried, B.D. (2008) Inheritance of Cry1F resistance in laboratory-selected European corn borer and its survival on transgenic corn expressing the Cry1F toxin. *Bulletin of Entomological Research* 98, 621–629.

Pereira, E.J., Siqueira, H.A., Zhuang, M., Storer, N.P. and Siegfried, B.D. (2010) Measurements of Cry1F binding and activity of luminal gut proteases in susceptible and Cry1F resistant *Ostrinia nubilalis* larvae (Lepidoptera: Crambidae). *Journal of Invertebrate Pathology* 103, 1–7.

Rang, C., Bergvingson, D., Bohorova, N., Hoisington, D. and Frutos, R. (2004) Competition of *Bacillus thuringiensis* Cry1 toxins for midgut binding sites: a basis for the development and management of transgenic tropical maize resistant to several stemborers. *Current Microbiology* 49, 22–27.

Raymond, B., Johnston, P.R., Nielsen-Leroux, C., Lereclus, D. and Crickmore, N. (2010) *Bacillus thuringiensis*: an impotent pathogen? *Trends in Microbiology* 18, 189–194.

Sanchis, V. (2011) From microbial sprays to insect-resistant transgenic plants: history of the biospesticide *Bacillus thuringiensis*. A review. *Agronomy for Sustainable Development* 31, 217–231.

Sanchis, V., Chaugaux, J. and Pauron, D. (1994) A comparison and analysis of the toxicity and receptor binding properties of *Bacillus thuringiensis* Cry1C δ-endotoxin on *Spodoptera littoralis* and *Bombyx mori*. *FEBS Letters* 353, 259–263.

Sayyed, A.H. and Wright, D.J. (2001) Cross-resistance and inheritance of resistance to *Bacillus thuringiensis* toxin Cry1Ac in diamondback moth (*Plutella xylostella* L) from lowland Malaysia. *Pest Management Science* 57, 413–421.

Sena, J.A., Hernández-Rodríguez, C.S. and Ferré, J. (2009) Interaction of *Bacillus thuringiensis* Cry1 and Vip3A proteins with *Spodoptera frugiperda* midgut binding sites. *Applied and Environmental Microbiology* 75, 2236–2237.

Sharma, P., Nain, V., Lakhanpaul, S. and Kumar, P.A. (2010) Synergistic activity between *Bacillus thuringiensis* Cry1Ab and Cry1Ac toxins against maize stem borer (*Chilo partellus* Swinhoe). *Letters in Applied Microbiology* 51, 42–47.

Siegfried, B.D., Vaughn, T.T. and Spencer, T. (2005) Baseline susceptibility of western corn rootworm (Coleoptera: Chrysomelidae) to Cry3Bb1 *Bacillus thuringiensis* toxin. *Journal of Economic Entomology* 98, 1320–1324.

Siqueira, H.A., Moellenbeck, D., Spencer, T. and Siegfried, B.D. (2004) Cross-resistance of Cry1Ab-selected *Ostrinia nubilalis* (Lepidoptera: Crambidae) to *Bacillus thuringiensis* δ-endotoxins. *Journal of Economic Entomology* 97, 1049–1057.

Siqueira, H.A., González-Cabrera, J., Ferré, J., Flannagan, R. and Siegfried, B.D. (2006) Analyses of Cry1Ab binding in resistant and susceptible strains of the European corn borer, *Ostrinia nubilalis* (Hubner) (Lepidoptera: Crambidae). *Applied and Environmental Microbiology* 72, 5318–5324.

Slaney, A.C., Robbins, H.L. and English, L. (1992) Mode of action of *Bacillus thuringiensis* toxin CryIIIA: an analysis of toxicity in *Leptinotarsa decemlineata* (Say) and *Diabrotica undecimpunctata howardi* Barber. *Insect Biochemistry and Molecular Biology* 22, 9–18.

Tabashnik, B.E., Malvar, T., Liu, Y.B., Finson, N., Borthakur, D. *et al.* (1996) Cross-resistance of the diamondback moth indicates altered interactions with domain II of *Bacillus thuringiensis* toxins. *Applied and Environmental Microbiology* 62, 2839–2844.

Tabashnik, B.E., Liu, Y.B., Finson, N., Masson, L. and Heckel, D.G. (1997a) One gene in diamondback moth confers resistance to four *Bacillus thuringiensis* toxins. *Proceedings of the National Academy of Sciences of the United States of America* 94, 1640–1644.

Tabashnik, B.E., Liu, Y.B., Malvar, T., Heckel, D.G., Masson, L. *et al.* (1997b) Global variation in the genetic and biochemical basis of diamondback moth resistance to *Bacillus thuringiensis*. *Proceedings of the National Academy of Sciences of the United States of America* 94, 12780–12785.

Vachon, V., Laprade, R. and Schwartz, J.L. (2012) Current models of the mode of action of *Bacillus thuringiensis* insecticidal crystal proteins: a critical review. *Journal of Invertebrate Pathology* 111, 1–12.

van Frankenhuyzen, K. (2009) Insecticidal activity of *Bacillus thuringiensis* crystal proteins. *Journal of Invertebrate Pathology* 101, 1–16.

van Rie, J., Jansens, S., Höfte, H., Degheele, D. and Van Mellaert, H. (1989) Specificity of *Bacillus thuringiensis* δ-endotoxins. Importance of specific receptors on the brush border membrane of the mid-gut of target insects. *European Journal of Biochemistry* [now *The FEBS Journal*] 186, 239–247.

van Rie, J., Jansens, S., Höfte, H., Degheele, D. and Van Mellaert, H. (1990) Receptors on the brush border membrane of the insect midgut as determinants of the specificity of *Bacillus thuringiensis* delta-endotoxins. *Applied and Environmental Microbiology* 56, 1378–1385.

Walters, F.S., Kulesza, C.A., Phillips, A.T. and English, L.H. (1994) A stable oligomer of *Bacillus thuringiensis* delta-endotoxin, CryIIIA. *Insect Biochemistry and Molecular Biology* 24, 963–968.

Walters, F.S., Stacy, C.M., Lee, M.K., Palekar, N. and Chen, J.S. (2008) An engineered chymotrypsin/cathepsin G site in domain I renders *Bacillus thuringiensis* Cry3A active against Western corn rootworm larvae. *Applied and Environmental Microbiology* 74, 367–374.

Walters, F.S., Defontes, C.M., Hart, H., Warren, G.W. and Chen, J.S. (2010) Lepidopteran-active variable-region sequence imparts coleopteran activity in eCry3.1Ab, an engineered *Bacillus thuringiensis* hybrid insecticidal protein. *Applied and Environmental Microbiology* 76, 3082–3088.

Wolfersberger, M.G. (1990) The toxicity of two *Bacillus thuringiensis* δ-endotoxins to gypsy moth larvae is inversely related to the affinity of binding sites on midgut brush border membranes for the toxins. *Experientia* 46, 475–477.

Wolfersberger, M., Luthy, P., Maurer, A., Parenti, P., Sacchi, V.F. *et al.* (1987) Preparation and partial characterization of amino acid transporting brush border membrane vesicles from the larval midgut of the cabbage butterfly (*Pieris brassicae*). *Comparative Biochemistry and Physiology Part A: Physiology* 86, 301–308.

14

Countering Pest Resistance with Genetically Modified Bt Toxins

Mario Soberón,[1]* Blanca Ines García-Gómez,[1] Sabino Pacheco,[1] Jorge Sánchez,[1] Bruce E. Tabashnik[2] and Alejandra Bravo[1]

[1]*Instituto de Biotecnología, Universidad Nacional Autónoma de México, Cuernavaca, Morelos, Mexico;* [2]*Department of Entomology, University of Arizona, Tucson, Arizona, USA*

Summary

Insecticidal crystalline (Cry) toxins from the bacterium *Bacillus thuringiensis* (Bt) used in sprays and transgenic crop plants have provided major benefits for pest control, including decreased reliance on broad-spectrum chemical insecticides. However, extensive use of Bt toxins has selected for resistance, thus reducing or eliminating these benefits against some populations of at least seven species of major crop pests. This chapter reviews efforts to counter pest resistance to native Bt toxins with genetically engineered toxins called Cry1AbMod and Cry1AcMod. We generated these modified toxins by trimming the genes encoding the native toxins Cry1Ab and Cry1Ac so they lack the nucleotides that code for a portion of the amino-terminal end of the protein, including helix α-1 and part of helix α-2. Consistent with the sequential binding model for the mode of action of the toxins, the Cry1AMod toxins formed oligomers without binding to cadherin, but the native toxins required cadherin for oligomer formation. The modified toxins were more potent than the corresponding native toxins in 13 of 19 pairwise comparisons with 12 resistant strains of nine species of Lepidoptera, including field-selected strains of *Plutella xylostella*, *Trichoplusia ni* and *Spodoptera frugiperda*. The potency of modified toxins relative to native toxins in these resistant strains did not depend on the resistance mechanism. Against susceptible strains, the modified toxins were less effective than their native counterparts in nearly all cases. Transgenic tobacco plants producing Cry1AbMod killed larvae of *Manduca sexta*, which represents progress towards commercial use of the modified toxins for pest management.

14.1 Introduction

Bacillus thuringiensis (Bt) produces crystalline (Cry) proteins that kill some important insect pests, but are not toxic to most other organisms (Mendelsohn *et al.*, 2003; Pardo-López *et al.*, 2013). Cry toxins have been used in sprays for decades to control insects (Sanahuja *et al.*, 2011). The genes encoding some Cry proteins have been incorporated into the genomes of maize and cotton since 1996, and more recently into soybean (James, 2013; Tabashnik *et al.*, 2013a). These transgenic plants, called Bt crops, produce Cry proteins that can suppress pests, reduce reliance on insecticide sprays and, in some cases, increase yield (Hutchison *et al.*, 2010; Tabashnik *et al.*, 2010; Edgerton *et al.*, 2012; Lu *et al.*, 2012).

* Corresponding author. E-mail address: mario@ibt.unam.mx

In 2013, farmers planted more than 75 million ha of Bt crops worldwide (James, 2013).

Extensive adoption of Bt crops selects for the evolution of insect resistance to Bt toxins; this entails a genetically based decrease in susceptibility to the toxins (Tabashnik et al., 2013a, 2014). Indeed, some degree of field-evolved resistance to Bt crops has been documented in nine major pest species (Tabashnik et al., 2008; Kruger et al., 2009; Liu et al., 2010; Storer et al., 2010; Dhurua and Gujar, 2011; Gassmann et al., 2011; Zhang et al., 2011; Tabashnik et al., 2013a). In five of these nine species, the resistance is severe enough in some populations to reduce the efficacy of the Bt crop with practical consequences for pest control (Tabashnik et al., 2013a, 2014; see also Tabashnik and Carrière, Chapter 1, Gao et al., Chapter 2 and Monnerat et al., Chapter 3). In addition, the evolution of resistance to Bt toxins used extensively in sprays has been documented in commercial agricultural settings for two major pest species (Tabashnik et al., 1990; Janmaat and Myers, 2003).

Approaches for dealing with resistance to Bt toxins can be classified into two main categories: sustaining the efficacy of the available toxins and expanding the set of available toxins. The widely adopted 'refuge' and 'pyramid' resistance management strategies exemplify the first category (Gould 1998; see also Tabashnik and Carrière, Chapter 1 and Huang, Chapter 16). The second category includes finding and deploying native Bt toxins that have not been used widely before, such as vegetative insecticidal proteins (Vips) (Estruch et al., 1996; see also Jakka et al., Chapter 13), and modifying native Bt toxins to enhance their toxicity (Bravo et al., 2013).

This chapter focuses on the Cry1AMod toxins (Cry1AbMod and Cry1AcMod) that we created by genetically engineering the native toxins Cry1Ab and Cry1Ac (Soberón et al., 2007). Unlike their native counterparts, Cry1AbMod and Cry1AcMod can form oligomers without binding to a particular midgut protein and can kill insects from strains of several lepidopteran pests that

have high levels of resistance to Cry1Ab and Cry1Ac (Soberón et al., 2007; Franklin et al., 2009; Muñoz-Garay et al., 2009; Tabashnik et al., 2011). In this chapter, we review the sequential binding model of the mode of action of Cry1A toxin, the role of cadherin in resistance to Cry1A toxins, the construction and characterization of Cry1AMod toxins, their toxicity to resistant and susceptible strains, and their potential use in the field.

14.2 Sequential Binding Model of Cry1A Mode of Action

Although models describing the mode of action of Cry toxins differ in some aspects (see Bravo et al., Chapter 6), they share the initial steps: Cry1A toxins produced as 130 kDa protoxins are solubilized in the lumen of the larval midgut, then converted by midgut proteases to a 60 kDa activated toxin composed of three domains (Li et al., 1991). Domain I is a seven helix bundle involved in toxin membrane insertion, oligomerization and pore formation. Domains II and III, which are composed mainly of β-sheets, are involved in binding to midgut proteins and are thus responsible for the specificity of these toxins (Fig. 14.1).

The sequential binding model (Fig. 14.2) proposes that the activated Cry1A toxins bind to abundant midgut proteins such as aminopeptidase N (APN) and alkaline phosphatase (ALP) that are anchored to the membrane by a glycolipid called glyco-sylphosphatidylinositol (GPI) (Pardo-López et al., 2013). This interaction concentrates activated toxin near the midgut epithelium, facilitating binding with a transmembrane cadherin protein that is less abundant than APN or ALP (Arenas et al., 2010). Binding to cadherin promotes oligomerization of the toxin (Gómez et al., 2002, 2014; Pacheco et al., 2009). The Cry1A oligomers bind with increased affinity to GPI-anchored APN and ALP, which facilitates insertion of the oligomer into the membrane. This insertion creates pores in the membrane that cause osmostic lysis of midgut cells and thereby kill larvae (Pacheco et al., 2009; Arenas et al., 2010). Resistance linked to

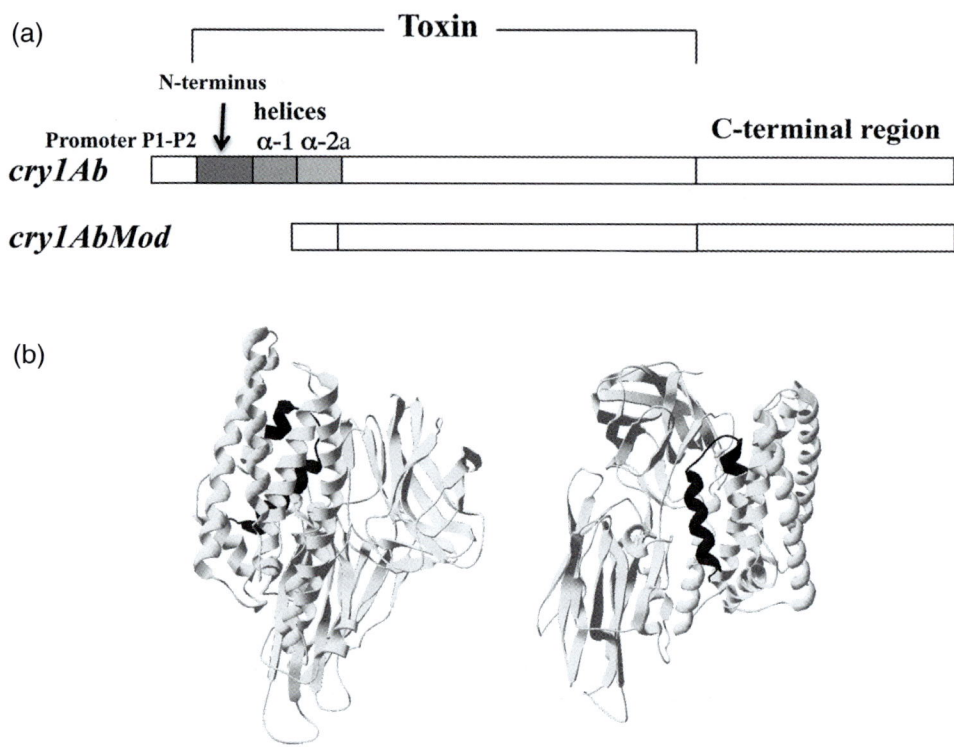

Fig. 14.1. The construction of Cry1AMod toxins. (a) The genes encoding wild type Cry1Ab and the modified toxin, Cry1AbMod. (b) Two views of the three-dimensional structure of the Cry1Ab toxin showing in black helix α-1 and the part of helix α-2 that are deleted in Cry1AbMod.

mutations in the ABC (ATP Binding Cassette) membrane transporter molecule ABCC2 in different lepidopteran insects indicates that this protein is involved in toxicity. The role of this protein in the mode of action remains unknown. It has been proposed that ABCC2 might be involved either in toxin binding or in facilitating oligomer insertion into the membrane (Heckel, 2012; Tanaka *et al.*, 2013; and see also Heckel, Chapter 9).

Recently, Gómez *et al.* (2014) found that the Cry1Ab protoxin also binds cadherin with a high affinity, resulting in the formation of Cry1Ab oligomeric structures that insert efficiently into insect membranes inducing pores. These authors also found that two distinct pre-pores are involved in pore formation and toxicity, depending on which Cry1A structure binds cadherin,

either monomeric activated toxin or protoxin (see Fig. 14.2; see also Bravo *et al.*, Chapter 6).

14.3 Cry1AMod Toxins: Construction and First Tests Against Insects with Altered Cadherin

As detailed by Fabrick and Wu in Chapter 7, the essential role of cadherin in the mode of action of Cry1A toxins is demonstrated by data indicating that reduced transcription or disruption of cadherin reduces susceptibility to these toxins. Binding to cadherin is thought to facilitate protease-mediated removal of the amino-terminal end of Cry1A toxins, including helix α-1 and part of helix α-2, thereby promoting oligomerization (Gomez *et al.*, 2002). Thus, we hypothesized

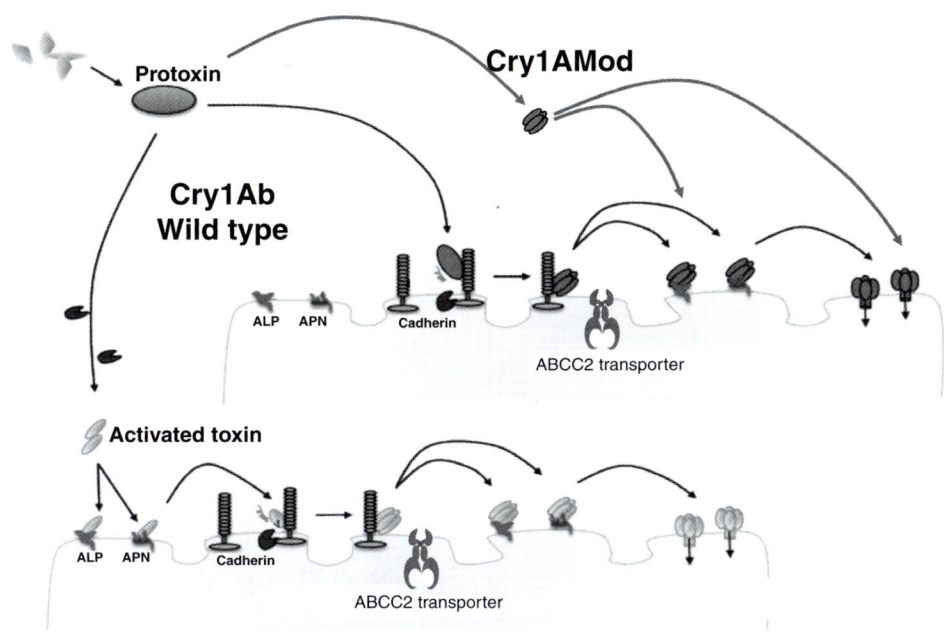

Fig. 14.2. The sequential binding model of the mode of action of wild type Cry1A and of the modified Cry1AMod toxins. The diagram shows sequential binding model of mode of action of Cry1A toxins, which includes the formation of two distinct pre-pores (Gómez *et al.*, 2014); it also shows the proposed mode of action of Cry1AMod toxins. The role of the ABCC2 membrane transporter protein in the mode of action of Cry1A still remains to be determined. The role of cadherin in facilitating oligomerization is supported by the toxicity of Cry1AMod toxins towards a strain of *Pectinophora gossypiella* in which resistance to Cry1A was linked to cadherin deletion mutations (Soberón *et al.*, 2007), but was not supported by its lack of activity towards *Helicoverpa armigera* and *Heliothis virescens* resistant populations that were also linked to cadherin mutations. Nevertheless, Cry1AMod has a significant loss of activity towards heliothine moths. The high toxicity of Cry1AMod against *Trichoplusia ni* and *Plutella xylostella* linked to mutations in the ABCC2 membrane transporter protein gene (Franklin *et al.*, 2009; Tabashnik *et al.*, 2011) could suggest that in some insects ABCC2 fulfils the role of cadherin in toxin oligomerization. ALP, alkaline phosphatase; APN, aminopeptidase N. See text for further details.

that: (i) Cry1A toxins genetically modified to lack the amino-terminal region would oligomerize in the absence of cadherin (see Fig. 14.2); and (ii) they would be toxic against insects with reduced susceptibility conferred partly or entirely by altered cadherin (Soberón *et al.*, 2007).

To test these hypotheses, we used genetic engineering of Cry1Ab and Cry1Ac to generate modified protoxins (Cry1AbMod and Cry1AcMod) lacking 56 amino acids at the amino-terminal end, including helix α-1 and part of helix α-2a (Soberón *et al.*, 2007) (see Fig. 14.1). Consistent with the hypothesis (i), *in vitro* experiments with trypsin-activated protoxins revealed that

the native toxins Cry1Ab and Cry1Ac required cadherin to form oligomers, but that Cry1AbMod and Cry1AcMod oligomerized without cadherin (Soberón *et al.*, 2007). We also found that Cry1AbMod was similar to Cry1Ab in binding to APN and in pore formation (Muñoz-Garay *et al.*, 2009). Consistent with hypothesis (ii), Cry1AbMod was much more effective than Cry1Ab against larvae of *Manduca sexta* in which cadherin transcription was reduced by RNA interference (RNAi; Soberón *et al.*, 2007). Furthermore, against the laboratory-selected AZP-R strain of *Pectinophora gossypiella*, in which resistance to Cry1Ac is genetically linked with cadherin deletion mutations, Cry1AbMod and Cry1AcMod were much more effective than Cry1Ab and Cry1Ac (Morin *et al.*, 2003; Soberón *et al.*, 2007).

To compare resistance levels between native and modified toxins, we used the resistance ratio, which is the concentration of toxin killing 50% of larvae (LC_{50}) for a resistant strain divided by the LC_{50} of a conspecific susceptible strain. We also calculated the potency of a modified toxin relative to its native counterpart, which is the LC_{50} of the native toxin, e.g. Cry1Ab, divided by the LC_{50} of the corresponding modified toxin, i.e. Cry1AbMod (Tabashnik *et al.*, 2011). Values >1 for the potency of a modified toxin relative to a native toxin indicate that the modified toxin is more potent than the native toxin and values <1 indicate the modified toxin is less potent than its native counterpart.

For the AZP-R strain of *P. gossypiella*, resistance ratios were >910 for Cry1Ab but only 2.8 for Cry1AbMod, and >3700 for Cry1Ac but only 0.41 for Cry1AcMod (Soberón *et al.*, 2007). Against this resistant strain, the potency of Cry1AcMod relative to Cry1Ac was more than 100 (Soberón *et al.*, 2007; Table 14.1). These results showed that Cry1AMod toxins could counter resistance linked with mutations in the cadherin gene.

14.4 Efficacy of Cry1AMod Toxins Against Lepidopteran Larvae with Diverse Mechanisms of Resistance

Inspired by the effectiveness of Cry1AMod toxins in the tests with *M. sexta* and *P. gossypiella*, we enlisted colleagues from eight laboratories on four continents to collaborate in testing the modified toxins and their native counterparts against resistant and susceptible strains of other insects. Overall, the Cry1AMod toxins and their native counterparts have been tested against 22 strains of ten species of Lepidoptera, including a susceptible strain of *M. sexta* with and without RNAi silencing of cadherin transcription (Soberón *et al.*, 2007), as well as 12 resistant strains and nine conspecific susceptible strains of eight other species

(Table 14.1). Because both Cry1AbMod and Cry1AcMod were compared with Cry1Ab and Cry1Ac against some strains, the results include 19 pairwise comparisons between modified and native toxins against 12 resistant strains. The modified toxins were more effective than their native counterparts in 13 of 19 (68%) comparisons, including at least one comparison for eight of the 12 (67%) resistant strains tested (Table 14.1).

We used these results to test hypothesis (ii) (see Section 14.3). We also tested a complementary hypothesis (iii), suggesting that against insects with resistance not conferred by altered cadherin, the modified toxins will not be more potent than their native counterparts (Tabashnik *et al.*, 2011).

Contrary to hypotheses (ii) and (iii), the proportion of cases in which potency was greater for the modified toxin than its native counterpart was not significantly associated with the resistance mechanism of altered cadherin (Fisher's exact test, $P = 1.0$, Table 14.1). The modified toxins were not consistently more potent than their native counterparts against insects with cadherin-based resistance, but they were more potent than their native counterparts against several strains in which resistance was not associated with altered cadherin. More specifically, in cases where resistance was caused partly or entirely by altered cadherin, the modified toxin was more potent than its native counterpart in six of nine comparisons (67%), but in cases where resistance was not associated with altered cadherin, the modified toxin was more potent than its native counterpart in five of eight comparisons (62%) (Table 14.1).

The strongest test of hypotheses (ii) and (iii) is based on the data from three resistant strains of *Heliothis virescens*: YFO had only cadherin-based resistance, YEE had resistance linked with the gene encoding the ABC protein ABCC2 and YHD3 had mutations affecting both cadherin and ABCC2 (Gahan *et al.*, 2010). Because all three strains had a similar genetic background and were tested in the same laboratory using the same methods, this is a particularly powerful set of comparisons

Table 14.1. Efficacy of modified toxins against insects with different mechanisms of resistance.[a]

Species	Strain	Selected[b]	Resistance ratio for native toxin[c]		Binding protein affected[d]	Potency of Mod-toxin relative to native toxin against resistant strain[e]		Reference/s[f]
			Cry1Ab	Cry1Ac		Cry1AbMod	Cry1AcMod	
Diatraea saccharalis	Bt-RR	Laboratory	22	8.2	Cadherin, APNs	2.8	0.15	1–3
Helicoverpa armigera	SCD-r1	Laboratory	39	2.2	Cadherin	0.53	–	1
Helicoverpa zea	AR1	Laboratory	–	120	ALP	<1	<1	4, 6
Heliothis virescens[g]	YEE	Laboratory	>2400	–	ABCC2	–	<0.35	1, 5
	YFO	Laboratory	850	–	Cadherin	–	<1	1, 5
	YHD3	Laboratory	>2400	–	Cadherin, ABCC2	–	>2.7	1, 5
Ostrinia nubilalis	KS	Laboratory	3100	–	APP	560	–	1, 15
Pectinophora gossypiella	BX-R[h]	Laboratory	270	310	Cadherin	17	43	7, 8
	AZP-R	Laboratory	>910	>3700	Cadherin	>100	>100	9, 10
Plutella xylostella	NO-QAGE	Field	>21,000	>110,000	ABCC2	>350	>540	1, 10
Spodoptera frugiperda	SfBt[i]	Field	4	2.5	Not known	8.5	5.7	11
Trichoplusia ni	GipBtR[j]	Greenhouse	580	1400	APN1, ABCC2	53	11	12–14

[a]Results based on diet bioassays with the protoxin form of the native and modified toxins.

[b]Where exposure to Bt toxins occurred: laboratory, field or greenhouse.

[c]Toxin concentration killing 50% of larvae (LC_{50}) for the resistant strain divided by the LC_{50} of a conspecific susceptible strain.

[d]ABCC2, ABC membrane transporter protein; ALP: alkaline phosphatase; APN: aminopeptidase N; APP: aminopeptidase P. Genetic linkage between resistance and the gene encoding the protein listed has been found except for the following four cases based on comparisons between resistant and susceptible strains: reduced transcription of cadherin and APNs for *D. saccharalis*, increased transcription of ALP for *H. zea*; reduced transcription and two predicted amino acid substitutions in APP for *O. nubilalis*; and reduced transcription of APN1 for *T. ni*; see references for details.

[e]For a given resistant strain, the LC_{50} of a native toxin divided by the LC_{50} of the corresponding modified toxin; values >1 indicate that the modified toxin was more effective than its native counterpart.

[f]References: 1, Tabashnik *et al.* (2011); 2, Yang *et al.* (2010); 3, Yang *et al.* (2011); 4, Caccia *et al.* (2012); 5, Gahan *et al.* (2010); 6, Fabrick and Wu, Chapter 7; 7, Tabashnik *et al.* (2013b); 8, Tabashnik *et al.* (2009) and unpublished data; 9, Soberón *et al.* (2007); 10, Morin *et al.* (2003); 11, Monnerat *et al.*,Chapter 3; 12, Franklin *et al.* (2009); 13, Tiewsiri and Wang (2011); 14, Baxter *et al.* (2011); 15, Khajuria *et al.* (2011).

[g]For *H. virescens*, resistance ratio and potency are based on the concentration of toxin causing 50% inhibition of larval growth (IC_{50}).

[h]In BX-R, resistance to Cry1Ac is associated with mutant cadherin (Tabashnik *et al.*, unpublished) and the mechanism of resistance to Cry2Ab is not known.

[i]SfBt resistance ratios were 33 for Cry1Aa and >10 for Cry1F (see Fabrick and Wu, Chapter 7).

[j]The mechanism of resistance was determined in the GLEN-Cry1Ac-BCS strain of *T. ni* (Baxter *et al.*, 2011; Tiewsiri and Wang, 2011) in which the resistance mutation reducing APN1 transcription is allelic with the resistance mutation in the GipBtR strain (P. Wang, personal communication, cited in Tabashnik *et al.*, 2011).

for evaluating the association between cadherin-based resistance and the efficacy of Cry1AcMod relative to Cry1Ac (Tabashnik *et al.*, 2011). Surprisingly, potency was lower for Cry1AcMod than Cry1Ac against YFO, which had only cadherin-based resistance, and Cry1AcMod was more potent than Cry1Ac against YHD3, which had resistance-conferring mutations affecting ABCC2 as well as cadherin (Table 14.1). Consistent with hypothesis (iii), Cry1AcMod was not more toxic than Cry1Ac against YEE, which had only ABCC2-based resistance (Table 14.1).

Although we did not see the expected association between efficacy of the modified toxins and cadherin-based resistance – either among the three resistant strains of *H. virescens* or across all of the resistant strains tested – we learned that greater efficacy of the modified toxins relative to their native counterparts was associated with increased resistance to the native toxins (Tabashnik *et al.*, 2011). The modified toxin was more potent than its native counterpart in ten of 11 cases (91%) when the resistance ratio was >200 for the native toxin, compared with only three of eight cases when the resistance ratio was <200 for the native toxin (38%) (Fisher's exact test, $P = 0.04$, Table 14.1). For example, among the three strains of *H. virescens*, YHD3 was much more resistant to Cry1Ac than YEE or YFO (Gahan *et al.*, 2010), and Cry1AcMod was more potent than Cry1Ac against YHD3, but not against YEE or YFO (Table 14.1).

The most spectacular effects of the modified toxins were against a strain of *Plutella xylostella* with >21,000-fold resistance to Cry1Ab and >110,000-fold resistance to Cry1Ac (Table 14.1). The resistance ratios were only 3.1 for Cry1AbMod and 4.8 for Cry1AcMod, representing reductions in the resistance ratio of >6900-fold and >23,000-fold, respectively (Tabashnik *et al.*, 2011). Against this strain, the modified toxins were more than 350 times more toxic than their native counterparts (Table 14.1). The resistance in this strain, which was derived from a field population in Hawaii selected with Bt sprays, was linked with a gene encoding ABCC2 and not with a cadherin

gene (Tabashnik *et al.*, 2000; Baxter *et al.*, 2005, 2011).

The modified toxins were also strikingly effective against a strain of *Trichoplusia ni* derived from a population that had evolved resistance to Bt sprays in a greenhouse in Canada (Table 14.1). The resistance ratios in this strain were 580 for Cry1Ab and 1400 for Cry1Ac, but only 5.5 for Cry1AbMod and 9.3 for Cry1AcMod (Franklin *et al.*, 2009). In this strain, resistance was also genetically linked with a mutation in the ABCC2 gene (Baxter *et al.*, 2011) and with reduction in APN1 transcript (Baxter *et al.*, 2011; Tiewsiri and Wang, 2011).

Cry1AbMod was also highly effective against a strain of *Ostrinia nubilalis* selected in the laboratory for 3100-fold resistance to Cry1Ab (Table 14.1). Cry1AbMod was 560 times more potent than Cry1Ab (Tabashnik *et al.*, 2011). This strain and another laboratory-selected strain had point mutations yielding two amino acid substitutions in an aminopeptidase P (APP) (Khajuria *et al.*, 2011).

Recent results show the efficacy of the modified toxins against the laboratory-selected BX-R strain of *P. gossypiella*, which had resistance ratios of 270 for Cry1Ab, 310 for Cry1Ac and 210 for Cry2Ab, but only 1.6 for Cry1AbMod and 2.1 for Cry1AcMod (Table 14.1; Tabashnik *et al.*, 2013b). The resistance to Cry1Ac in this strain is associated with a cadherin mutation (Tabashnik *et al.*, unpublished data). Its mechanism of resistance to Cry2Ab is not known, but almost certainly involves a gene other than cadherin, because cadherin-based resistance to Cry1Ac did not cause strong cross-resistance to Cry2Ab in this insect (Tabashnik *et al.*, 2002, 2009; Jurat-Fuentes *et al.*, 2003).

Against a strain of *Diatraea saccharalis* with resistance ratios of 28 for Cry1Ab and 8.2 for Cry1Ac, Cry1AbMod was 2.8 times more potent than Cry1Ab, but Cry1AcMod was less potent than Cry1Ac (Table 14.1; Tabashnik *et al.*, 2011). Resistance in this strain was associated with reduced transcript abundance for cadherin and three APNs (Yang *et al.*, 2010, 2011). Cry1AbMod and Cry1AcMod were less effective than their native counterparts against a laboratory-selected strain of *Helicoverpa zea* in which *c*.100-fold resistance to Cry1Ac was associated with increased levels of ALP in the midgut lumen (Table 14.1, Caccia *et al.*, 2012).

Cry1AbMod and Cry1AcMod were tested recently against a strain of *Spodoptera frugiperda* from Brazil that was isolated from Bt maize producing Cry1F and had a high level of field-evolved resistance to Cry1F, as well as cross-resistance to Cry1Aa, Cry1Ab and Cry1Ac (Monnerat *et al.*, Chapter 3). Against this strain, Cry1AbMod was 8.5 times more potent than Cry1Ab, and Cry1AcMod was 5.7 times more potent than Cry1Ac (Table 14.1). Furthermore, the efficacy of both Cry1AMod toxins against the Cry1F-resistant strain was similar to that of Cry1F against a susceptible laboratory strain (see Monnerat *et al.*, Chapter 3).

14.5 Lower Potency of Modified Toxins than Native Toxins Against Susceptible Insects

Whereas the potency of modified toxins relative to native toxins was greatest against the insect strains most resistant to native toxins, Cry1AMod toxins were almost uniformly less effective than the respective native toxins against susceptible strains. The potency of Cry1AbMod or Cry1AcMod relative to its native counterpart was <1 in 19 of 21 pairwise comparisons based on results from bioassays of the protoxin form of the proteins against susceptible strains of nine species of Lepidoptera (Soberón *et al.*, 2007; Franklin *et al.*, 2009; Muñoz-Garay *et al.*, 2009; Tabashnik *et al.*, 2011, 2013b; Caccia *et al.*, 2012; Gómez *et al.*, 2014; Portugal *et al.*, 2014; see also Monnerat *et al.*, Chapter 3). The two exceptions were with a susceptible strain of *S. frugiperda* where Cry1AbMod was 7.5 times more potent than Cry1Ab and Cry1AcMod was 13 times more potent than Cry1Ac (Monnerat *et al.*, Chapter 3). Excluding the results with *S. frugiperda*, for the 17 comparisons in which relative potency can be quantified precisely,

the median decrease in potency of the modified toxins relative to native toxins was tenfold, with a range from 2.8-fold for Cry1AbMod versus Cry1Ab against *D. saccharalis* to 370-fold for Cry1AcMod versus Cry1Ac against *H. virescens* (Tabashnik *et al.*, 2011).

We recently investigated the molecular basis of the reduced potency of Cry1AbMod relative to Cry1Ab in a susceptible strain of *M. sexta* (Gómez *et al.*, 2014), and found that the potency relative to Cry1Ab decreased by eightfold for Cry1AbMod activated toxin, but only twofold for Cry1AbMod protoxin (Gómez *et al.*, 2014). As explained above and in Chapter 5 (Janmaat *et al.*), two distinct pre-pores could be formed with Cry1Ab; one after activated toxin binding to cadherin, the other after protoxin binding to cadherin (Gómez *et al.*, 2014). Relative to Cry1Ab, oligomer formation and binding to cadherin were reduced for Cry1AbMod activated toxin, but not for Cry1AbMod protoxin (Gómez *et al.*, 2014). Therefore, we hypothesize that the reduced toxicity of Cry1AbMod relative to Cry1Ab against susceptible *M. sexta* is caused primarily by the reduced potency of activated Cry1AbMod toxin relative to activated Cry1Ab, reflecting reduced oligomer formation from Cry1AbMod activated toxin.

14.6 Potential Use of Cry1AMod Proteins to Counter Insect Resistance in the Field

Although the results summarized above are encouraging, we do not know yet whether the modified proteins will be useful in the field. None the less, Cry1AbMod produced in transgenic tobacco plants killed *M. sexta* larvae whose cadherin transcript was silenced by double-stranded RNA (dsRNA) as well as control larvae that were not treated with dsRNA (Porta *et al.*, 2011). These results suggest that expression of the modified toxins in other crop plants may be possible. We have patented the CryMod toxin technology (Pardo-López *et al.*, 2007) and the Universidad Nacional Autónoma de México has completed a licensing agreement to test and commercialize different transgenic crops that produce the modified toxins.

Combinations of native and modified toxins might be especially useful in transgenic crop pyramids (Moar and Anilkumar, 2007; Tabashnik *et al.*, 2013b). In diet bioassays with the resistant BX-R strain and a susceptible strain of *P. gossypiella*, antagonistic effects were not seen between Cry1AbMod and Cry1Ac, and the combination of Cry1AbMod plus Cry2Ab was mildly synergistic (Tabashnik *et al.*, 2013b). In principle, the native toxins would typically be most effective against susceptible insects, while the modified toxins could kill insects that are resistant to the native toxins. However, the usefulness of a toxin in transgenic plants depends on its ability to kill target pests when produced by the plants, not its potency relative to other toxins (Tabashnik *et al.*, 2013b). Thus, with sufficiently high expression, the modified toxins might be useful against susceptible as well as resistant insects. Moreover, in the exceptional case of *S. frugiperda*, the modified toxins were more potent than native toxins against a susceptible strain as well as against a strain with field-evolved resistance to Cry1F and cross-resistance to three native Cry1A toxins (see Monnerat *et al.*, Chapter 3).

We recently developed an improved expression system for CryMod toxins that uses a *cry3Aa* promoter and single cysteine mutations in the protoxin region (García-Gómez *et al.*, 2013). This system enables efficient production of Cry1AbMod that is toxic to *P. xylostella* and should facilitate further research with the modified toxins.

Recent results with the CF-1 lepidopteran cell line suggest that this experimental system can provide insights for better understanding of the mode of action of the modified toxins (Portugal *et al.*, 2014). Against CF-1, Cry1AbMod was 16-fold more potent than Cry1Ab and Cry1AcMod was 1.8-fold more potent than Cry1Ac (Portugal *et al.*, 2014). Both native and modified toxins formed oligomers after inserting into CF-1 cell membranes, but the ratio of oligomer to monomer was higher for the modified toxins.

The reduced potency of Cry1AMod toxins relative to their native counterparts against eight of nine susceptible strains of nine species of Lepidoptera that has been summarized above suggests that, if anything, the modified toxins may have a narrower spectrum of activity than native toxins. Nevertheless, bioassays of modified toxins against a wide variety of non-target organisms would be useful for guarding against unexpected harmful side effects.

Because the potency of modified toxins relative to native toxins varied widely among resistant strains (Table 14.1), the laboratory bioassay data suggest initial targets for the modified toxins in the field. For example, these toxins might be particularly useful against field populations of *P. xylostella* resistant to Cry1Ab and Cry1Ac and against *S. frugiperda* resistant to Cry1F (Table 14.1; see also Monnerat *et al.*, Chapter 3). Although several mechanisms of resistance to Bt toxins probably occur in *P. xylostella*, complementation tests indicate that the ABCC2 mutation in the NO-QAGE strain against which the modified toxins were so effective is widespread, occurring in insects from Hawaii, continental USA and the Philippines (Tabashnik *et al.*, 1997; Baxter *et al.*, 2005, 2011). Analogously, complementation tests indicate that the ABCC2 mutation in the GipBtR strain of *T. ni* tested with modified toxins also occurs in the GLEN-Cry1Ac-BCS strain of this pest (P. Wang, personal communication, cited in Tabashnik *et al.*, 2011). Thus, the high efficacy of the modified toxins (Table 14.1) might extend to greenhouse populations of *T. ni* in western Canada, where both resistant strains originated (Janmaat and Myers, 2003).

The efficacy of the modified toxins against two laboratory-selected strains of *P. gossypiella* with cadherin-based resistance to Cry1Ac (Table 14.1) suggests that the modified toxins might also be useful against populations of this pest in western India, where field-selected resistance to Cry1Ac is also associated with cadherin mutations

(Fabrick *et al.*, 2014). Future work will determine whether the CryMod toxins are useful against these and other pest populations in the field. Even though the modified toxins are certainly not a universal solution, they may prove to be an important addition to the suite of tools for pest management.

References

Arenas, I., Bravo, A., Soberón, M. and Gómez, I. (2010) Role of alkaline phosphatase from *Manduca sexta* in the mechanism of action of *Bacillus thuringiensis* Cry1Ab toxin. *The Journal of Biological Chemistry* 285, 12497–12503.

Baxter, S.W., Zhao, J.-Z., Gahan, L.J., Shelton, A.M., Tabashnik, B.E. and Heckel, D.G. (2005) Novel genetic basis of field-evolved resistance to Bt toxins in *Plutella xylostella*. *Insect Molecular Biology* 14, 327–334.

Baxter, S.W., Badenes-Pérez, F.R., Morrison, A., Vogel, H., Crickmore N. *et al.* (2011) Parallel evolution of Bt toxin resistance in Lepidoptera. *Genetics* 189, 675–679.

Bravo, A., Gómez, I., Porta, H., García-Gómez, B.I., Rodriguez-Almazan, C. *et al.* (2013) Evolution of *Bacillus thuringiensis* Cry toxins insecticidal activity. *Microbial Biotechnology* 6, 17–26.

Caccia, S., Moar, W.J., Chandrashekhar, J., Oppert, C., Anilkumar. K.J. *et al.* (2012) Association of Cry1Ac toxin resistance in *Helicoverpa zea* (Boddie) with increased alkaline phosphatase levels in the midgut lumen. *Applied and Environmental Microbiology* 78, 5690–5698.

Dhurua, S. and Gujar, G.T. (2011) Field-evolved resistance to Bt toxin Cry1Ac in the pink bollworm, *Pectinophora gossypiella* (Saunders) (Lepidoptera: Gelechiidae), from India. *Pest Management Science* 67, 898–903.

Edgerton, M.D., Fridgen, J., Anderson, J.R. Jr, Ahigrim, J., Criswell, M. *et al.* (2012) Transgenic insect resistance traits increase corn yield and yield stability. *Nature Biotechnology* 30, 493–496.

Estruch, J., Warren, G., Mullins, M., Nye, G. and Craig, J.K. (1996) Vip3A, a novel *Bacillus thuringiensis* vegetative insecticidal protein with a wide spectrum of activities against lepidopteran insects. *Proceedings of the National Academy of Sciences of the United States of America* 93, 5389–5394.

Fabrick, J.A., Ponnuraj, J., Singh, A., Tanwar, R.K., Unnithan, G.C. *et al.* (2014) Alternative splicing and highly variable cadherin transcripts associated with field-evolved resistance of pink bollworm to Bt cotton in India. *PLoS One* 9(5): e97900.

Franklin, M.T., Nieman, C.L., Janmaat, A.F., Soberón, M., Bravo, A. *et al.* (2009) Modified *Bacillus thuringiensis* toxins and a hybrid *B. thuringiensis* strain counter greenhouse-selected resistance in *Trichoplusia ni*. *Applied and Environmental Microbiology* 75, 5739–5741.

Gahan, L.J., Pauchet, Y., Vogel, H. and Heckel, D.G. (2010) An ABC transporter mutation is correlated with insect resistance to *Bacillus thuringiensis* Cry1Ac toxin. *PLoS Genetics* 6(12): e1001248.

García-Gómez, B.I., Sánchez, J., Martínez de Castro, D.L., Ibarra, J.E., Bravo, A. *et al.* (2013) Efficient production of *Bacillus thuringiensis* Cry1AMod toxins under regulation of *cry3Aa* promoter and single cysteine mutations in the protoxin region. *Applied and Environmental Microbiology* 79, 6969–6973.

Gassmann, A.J., Petzold-Maxwell, J.L., Keweshan, R.S. and Dunbar, M.W. (2011) Field-evolved resistance to Bt maize by western corn rootworm. *PLoS ONE* 6(7): e22629.

Gómez, I., Sánchez, J., Miranda, R., Bravo, A. and Soberón, M. (2002) Cadherin-like receptor binding facilitates proteolytic cleavage of helix α-1 in domain I and oligomer pre-pore formation of *Bacillus thuringiensis* Cry1Ab toxin. *FEBS Letters* 513, 242–246.

Gómez, I., Sanchez, J., Muñoz-Garay, C., Matus, V., Gill, S.S. *et al.* (2014) *Bacillus thuringiensis* Cry1A toxins are versatile-proteins with multiple modes of action: two distinct pre-pores are involved in toxicity. *Biochemical Journal* 459, 383–396.

Gould, F. (1998) Sustainability of transgenic insecticidal cultivars: integrating pest genetics and ecology. *Annual Review of Entomology* 43, 701–726.

Heckel, D.G. (2012) Learning the ABCs of Bt: ABC transporters and insect resistance to *Bacillus thuringiensis* provide clues to a crucial step in toxin mode of action. *Pesticide Biochemistry and Physiology* 104, 103–110.

Hutchison, W.D., Burkness, E.C., Mitchell, P.D., Moon, R.D., Leslie, T.W. *et al.* (2010) Areawide suppression of European corn borer with Bt maize reaps savings to non-Bt maize growers. *Science* 330, 222–225.

James, C. (2013) *Global Status of Commercialized Biotech/GM Crops: 2013*. ISAAA Brief 46, International Service for the Acquisition of Agri-Biotech Applications, Ithaca, New York.

Janmaat, A.F. and Myers J.H. (2003) Rapid evolution and the cost of resistance to *Bacillus thuringiensis* in greenhouse populations of cabbage loopers, *Trichoplusia ni*. *Proceedings of the Royal Society B: Biological Sciences* 270, 2263–2270.

Jurat-Fuentes, J.L., Gould, F.L. and Adang, M.J. (2003) Dual resistance to *Bacillus thuringiensis* Cry1Ac and Cry2Aa toxins in *Heliothis virescens* suggests multiple mechanisms of resistance. *Applied and Environmental Microbiology* 69, 5898–5906.

Khajuria, C., Buschman, L.L., Chen, M.-S., Siegfried, B.D. and Zhu K.Y. (2011) Identification of a novel aminopeptidase P-like gene (*OnAPP*) possibly involved in Bt toxicity and resistance in a major corn pest (*Ostrinia nubilalis*). *PloS ONE* 6(8): e23983.

Kruger, M.J., van Rensburg, J.B.J. and Van den Berg, J. (2009) Perspective on the development of stem borer resistance to Bt maize and refuge compliance at the Vaalharts irrigation scheme in South Africa. *Crop Protection* 28, 684–689.

Li, J., Carrol, J. and Ellar, D.J. (1991) Crystal structure of insecticidal δ-endotoxin from *Bacillus thuringiensis* at 2.5 Å resolution. *Nature* 353, 815–821.

Liu, F., Xu, Z., Zhu, Y.C., Huang, F., Wang, Y. *et al.* (2010) Evidence of field-evolved resistance to Cry1Ac-expressing Bt cotton in *Helicoverpa armigera* (Lepidoptera:Noctuidae) in northern China. *Pest Management Sciences* 66, 155–161.

Lu, Y., Wu, K., Jiang, Y., Guo, Y. and Desneux, N. (2012) Widespread adoption of Bt cotton and insecticide decrease promotes biocontrol services. *Nature* 487, 362–365

Mendelsohn, M., Kough, J., Vaituzis, Z. and Matthews, K. (2003) Are Bt crops safe? *Nature Biotechnology* 21, 1003–1009.

Moar, W.J. and Anilkumar, K.J. (2007) The power of the pyramid. *Science* 318, 1561–1562.

Morin, S., Biggs, R.W., Shriver, L., Ellers-Kirk, C., Higginson, D. *et al.* (2003) Three cadherin alleles associated with resistance to *Bacillus thuringiensis* in pink bollworm. *Proceedings of the National Academy of Sciences of the United States of America* 100, 5004–5009.

Muñóz-Garay [Muñoz-Garay], C., Portugal, L., Pardo-López, L., Jiménez-Juárez, N., Arenas, I. *et al.* (2009) Characterization of the mechanism of action of the genetically modified Cry1AbMod toxin that is active against Cry1Ab-resistant insects. *Biochimica et Biophysica Acta (BBA) Biomembranes* 1788, 2229–2237.

Pacheco, S., Gómez, I., Arenas, I., Saab-Rincon, G., Rodríguez-Almazán, C. *et al.* (2009) Domain II loop 3 of *Bacillus thuringiensis* Cry1Ab toxin is involved in a "ping pong" binding mechanism with *Manduca sexta* aminopeptidase-N and cadherin receptors. *The Journal of Biological Chemistry* 284, 32750–32757.

Pardo-López, L., Tabashnik, B.E., Soberón M. and Bravo A. (2007) Suppression of resistance in insects to *Bacillus thuringiensis* Cry toxins, using toxins that do not require the cadherin receptor. US Patent Application 20100186123 A1. US Patent and Trademark Office, Alexandria, Virginia. Available at: http://appft1. uspto.gov/netacgi/nph-Parser?Sect1=PTO1&S ect2=HITOFF&d=PG01&p=1&u=/netahtml/ PTO/srchnum.html&r=1&f=G&l=50&s1=20100 186123.PGNR. (accessed 27 October 2014).

Pardo-López, L., Soberón, M. and Bravo, A. (2013) *Bacillus thuringiensis* insecticidal 3-domain Cry toxins: mode of action, insect resistance and consequences for crop protection. *FEMS Microbiology Reviews* 37, 3–22.

Porta, H., Jiménez, G., Cordoba, E., León, P., Soberón, M. *et al.* (2011) Tobacco plants expressing the Cry1AbMod toxin suppress tolerance to Cry1Ab toxin of *Manduca sexta* cadherin-silenced larvae. *Insect Biochemistry and Molecular Biology* 41, 513–519.

Portugal, L., Gringonten, J.L., Caputo, G.F., Soberón, M., Muñoz-Garay, C. *et al.* (2014) Toxicity and mode of action of insecticidal Cry1A proteins from *Bacillus thuringiensis* in an insect cell line, CF-1. *Peptides* 53, 292–299.

Sanahuja, G., Banakar, R., Twyman, R., Capell, T. and Christou, P. (2011) *Bacillus thuringiensis*: a century of research, development and commercial applications. *Plant Biotechnology Journal* 9, 283–300.

Soberón, M., Pardo-López, L., López, I., Gómez, I., Tabashnik, B. *et al.* (2007) Engineering modified Bt toxins to counter insect resistance. *Science* 318, 1640–1642.

Storer, N.P., Babcock, J.M., Schlenz, M., Meade, T., Thompson, G.D. *et al.* (2010) Discovery and characterization of field resistance to Bt maize: *Spodoptera frugiperda* (Lepidoptera: Noctuidae) in Puerto Rico. *Journal of Economical Entomology* 103, 1031–1038.

Tabashnik, B.E., Cushing, N.L., Finson, N. and Johnson M.W. (1990) Field development of resistance to *Bacillus thuringiensis* in diamondback moth (Lepidoptera: Plutellidae). *Journal of Economic Entomology* 83, 1671–1676.

Tabashnik, B.E., Liu, Y.-B., Malvar, T., Heckel, D.G., Masson, L. *et al.* (1997) Global variation in the genetic and biochemical basis of diamondback moth resistance to *Bacillus thuringiensis*.

Proceedings of the National Academy of Sciences of the United States of America 94, 12780–12785.

Tabashnik, B.E., Johnson, K.W., Engleman, J.T. and Baum, J.A. (2000) Cross-resistance to *Bacillus thuringiensis* toxin Cry1Ja in a strain of diamondback moth adapted to artificial diet. *Journal of Invertebrate Pathology* 76, 81–83.

Tabashnik, B.E., Dennehy, T.J., Sims, M.A., Larkin, K., Head, G.P. *et al.* (2002) Control of resistant pink bollworm by transgenic cotton with *Bacillus thuringiensis* toxin Cry2Ab. *Applied and Environmental Microbiology* 68, 3790–3794.

Tabashnik, B.E., Gassmann, A.J., Crowder, D.W. and Carrière, Y. (2008) Insect resistance to Bt crops: evidence versus theory. *Nature Biotechnology* 26, 199–202.

Tabashnik, B.E., Unnithan, G.C., Masson, L., Crowder, D.W., Li, X. *et al.* (2009) Asymmetrical cross-resistance between *Bacillus thuringiensis* toxins Cry1Ac and Cry2Ab in pink bollworm. *Proceedings of the National Academy of Sciences of the United States of America* 105, 11889–11894.

Tabashnik, B.E., Sisterson, M.S., Ellsworth, P.C., Dennehy, T.J., Antilla, N. *et al.* (2010) Suppressing resistance to Bt cotton with sterile insect releases. *Nature Biotechnology* 28, 1304–1307.

Tabashnik, B.E., Huang, F., Ghimire, M.N., Leonard, B.R., Siegfried, B.D. *et al.* (2011) Efficacy of genetically modified Bt toxins against insects with different mechanisms of resistance. *Nature Biotechnology* 29, 1128–1131.

Tabashnik, B.E., Brévault, T. and Carrière, Y. (2013a) Insect resistance to Bt crops: lessons from the first billion acres. *Nature Biotechnology* 31, 510–521.

Tabashnik, B.E., Fabrick, J.A., Unnithan, G.C.,

Yelich, A.J., Masson, L. *et al.* (2013b) Efficacy of genetically modified Bt toxins alone or in combinations against pink bollworm resistant to Cry1Ac and Cry2Ab. *PloS ONE* 8(11): e80496.

Tabashnik, B.E., Mota-Sanchez, D., Whalon, M.E., Hollingworth, R.M., and Carrière, Y. (2014) Defining terms for proactive management of resistance to Bt crops and pesticides. *Journal of Economic Entomology* 107, 496–507.

Tanaka S., Miyamoto, K., Noda, H., Jurat-Fuentes, J.L., Yoshizaw, Y. *et al.* (2013) The ATP-binding cassette transporter subfamily C member 2 in *Bombyx mori* larvae is a functional receptor for Cry toxins from *Bacillus thuringiensis*. *The FEBS Journal* 280, 1782–1794.

Tiewsiri, K. and Wang, P. (2011) Differential alteration of two aminopeptidases N associated with resistance to *Bacillus thuringiensis* toxin Cry1Ac in cabbage looper. *Proceedings of the National Academy of Sciences of the United States of America* 108, 14037–14042.

Yang, Y., Zhu, Y.C., Ottea, J., Hussenede, C., Leonard B.R. *et al.* (2010) Molecular characterization and RNA interference of three midgut aminopeptidase N isozymes from *Bacillus thuringiensis*-susceptible and -resistant strains of sugarcane borer, *Diatraea saccharalis*. *Insect Biochemistry and Molecular Biology* 40, 592–603.

Yang, Y., Zhu, Y.C., Ottea, J., Husseneder, C., Leonard, B.R. *et al.* (2011) Down regulation of a gene for cadherin, but not alkaline phosphatase, associated with Cry1Ab resistance in sugarcane borer *Diatraea saccharalis*. *PloS ONE* 6(10): e25783.

Zhang, H., Yin, W., Zhao, J., Jin, L., Yang, Y. *et al.* (2011) Early warning of cotton bollworm resistance associated with intensive planting of Bt cotton in China. *PloS ONE* 6(8): e22874.

15 RNA Interference Strategy for Crop Protection Against Insect Pests

Sneha Yogindran and Manchikatla V. Rajam*

Department of Genetics, University of Delhi South Campus, New Delhi, India

Summary

RNA interference (RNAi) has significantly accelerated functional genomic research on insects. Since its discovery, numerous reports have reported efforts to apply RNAi approaches to insect species lacking characterized genomes. The technique also has substantial potential for use in the control of insect pests that cause severe loss of crop yield and quality. Several approaches have been exploited to control these pests, including the application of different agrochemicals and the development of resistant crops by breeding and transgenic approaches. However, there are certain limitations with these strategies, and therefore novel alternative strategies are required for the development of stress-tolerant plants. RNAi strategies have proven to be such a novel and potential alternative for disease and pest control. This technique essentially involves the production of hairpin double-stranded RNA (dsRNA) in transgenic plants that targets pest genes that are essential for their growth and development. This chapter provides an insight into the practical aspects of this gene silencing methodology for the development of tools for pest management.

15.1 Introduction

Insects are real threat to plant growth and development. Crop plants especially are greatly affected by insect attack, and billions of dollars are spent every year worldwide to manage the crop losses caused by insects. The continuous use of insecticides can lead to many health issues and also poses the risk of developing resistance in insects. To reduce dependence on the use of insecticides, genetically engineered plants have been developed that express insecticidal proteins encoded by genes from the soil bacterium *Bacillus thuringiensis* (Bt) (Sanahuja *et al.*, 2011). Bt produces a diverse group of insecticidal toxins: crystal (Cry) proteins, cytolytic (Cyt) proteins and vegetative insecticidal proteins (Vips). These toxins interact with specific receptors in the insect midgut and form a pore-like structure in the host cell membrane. The formation of this pore leads to ion leakage from the cells and finally kills the insects (Bravo *et al.*, 2007). Bt toxins have been used as pesticides for over 40 years and, in comparison with synthetic pesticides, are non-toxic to vertebrates. Similarly, the use of transgenic Bt crops has led to reduced use of insecticides, increased yields and higher profits to farmers (Carpenter, 2010; Edgerton *et al.*, 2012; Lu *et al.*, 2012).

Although most pests are susceptible to Bt, field-evolved resistance towards Bt crops has been reported for some pest populations (Tabashnik *et al.*, 2013), and transgenic plants expressing multiple Bt toxins have been developed to combat this resistance. For instance, the development of transgenic

* Corresponding author. E-mail address: rajam.mv@gmail.com or venkat.rajam@south.du.ac.in

cotton expressing Cry1Ac and Cry2Ab has conferred resistance to the cotton bollworm, *Helicoverpa zea*, the fall army-worm, *Spodoptera frugiperda*, and the beet armyworm, *S. exigua* (Zhao *et al.*, 2003). Instances of the field resistance of insects to Bt crops has been observed and reported since 2005. For example, *H. zea*, and the pink bollworm, *Pectinophora gossypiella*, have been reported as showing resistance against Cry1Ac in cotton across a number of continents (Moar *et al.*, 2008; Tabashnik *et al.*, 2008a,b, 2013). Also, in Puerto Rico, *S. frugiperda*, showed resistance to Cry1F in maize, leading to the discontinuation of this Bt crop (Matten *et al.*, 2008). Hence, there is an urgent need to find a cost-effective and long-lasting pest control strategy. RNA interference (RNAi) technology provides a rapid method for the functional analysis of genes by loss-of-function strategies in both *in vitro* and *in vivo* conditions, especially in organisms in which stable transgenic lines are not available (Turner *et al.*, 2006; Tian *et al.*, 2009; Trivedi, 2010). The technique is now emerging, as a potential tool for crop protection against insect pests, and probably one with fewer biosafety issues as no transgene protein is expressed in the transgenic lines (Price and Gatehouse, 2008; Huvenne and Smagghe, 2010; Rajam 2011, 2012a,b).

15.2 Mechanism of RNAi

The discovery of RNAi in the nematode worm *Caenorhabditis elegans*, which is used as a model organism, brought about a revolution in molecular biology in which double-stranded RNA (dsRNA) was shown to precisely downregulate gene expression by cleaving its mRNA counterpart (Fire *et al.*, 1998). The mechanism of this action relies on the specificity of the sequence of one strand of the dsRNA and that of the corresponding complementary transcript. The pathway mainly comprises small-interfering RNAs (siRNAs) and micro RNAs (miRNAs), both of which bind to their complementary bases on the target mRNA and lead to their silencing. RNAi is, therefore, a post-transcriptional gene silencing mechanism, which leads to the degradation of cognate mRNA.

dsRNA is cleaved by the RNase III Dicer enzyme into approximately 21 bp fragments known as siRNAs with a 2 bp 3′ overhang. Passenger strands (sense strands) of these 21 bp duplexes are eliminated and the guide strands (antisense strands) are retained. The guide strand is then incorporated into the RNA-induced silencing complex (RISC), a multi-protein complex. Argonaute protein, the catalytic component of RISC, cleaves single-stranded RNA (ssRNA) molecules that have sequence complementarity with the guide strand of the duplex (Shabalina and Koonin, 2008). An RNA-dependent RNA polymerase (RdRP) interacts with RISC and generates a fresh lot of dsRNA based on the partially digested target template (Filipowicz, 2005). The antisense siRNAs with their 3′ OH terminal ends annealed to the ssRNA (target), are essential to elongating the dsRNA by the action of RdRP (Fig. 15.1). The amplification of the signal for efficient RNAi occurs at three different steps in *C. elegans*. The first level is cleavage of dsRNA into siRNAs by Dicer. The second level results from the recycling of siRNAs after degradation of the target mRNA. The third step involves the activity of RdRPs for the production of secondary siRNAs (Sijen *et al.*, 2001). The gene encoding RdRP is found in plants and fungi, but its homologue in insects and mammals has not yet been reported. There may, thus, be a possibility that these organisms may not require RdRP for RNAi signal amplification (Stein *et al.*, 2003).

15.2.1 Types of RNAi

RNAi responses can be categorized into three classes, namely cell autonomous, environmental and systemic. The last two can be designated as non-cell autonomous responses.

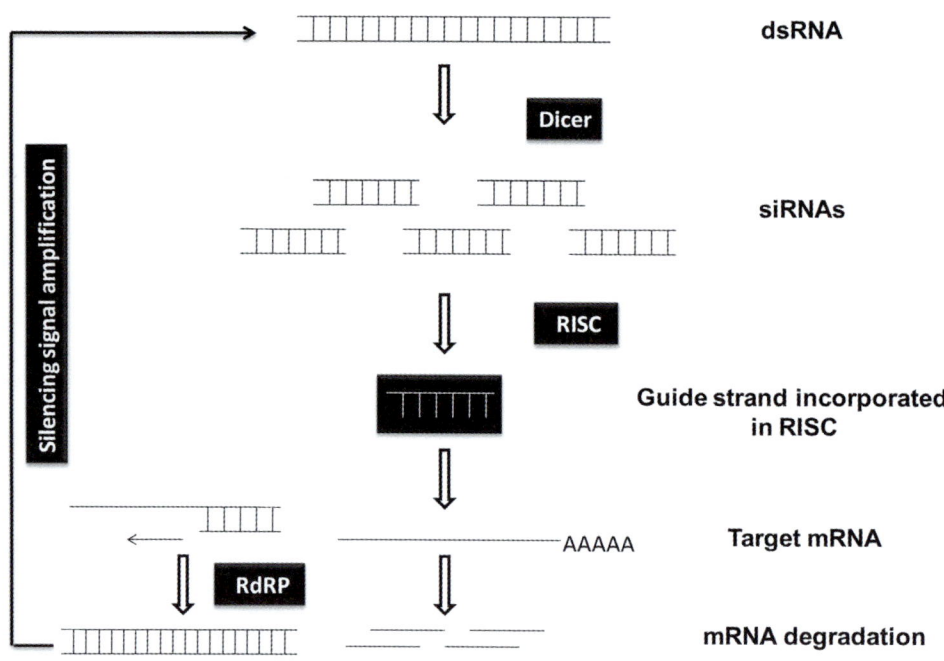

Fig. 15.1. An overview of the RNA interference (RNAi) mechanism. Key: 'AAAAA', the polyadenylate tail of the target mRNA; Dicer, ribonuclease (RNase III) Dicer enzyme; dsRNA, double-stranded RNA; mRNA, messenger RNA; RdRP, RNA-dependent RNA polymerase; RISC, the RNA-induced silencing complex; siRNAs, small-interfering RNAs.

Cell autonomous RNAi

As the name suggests, gene silencing by this type of RNAi occurs within the cells where dsRNA is constitutively expressed or introduced externally. In this case, the dsRNA is cleaved by an RNase III (Dicer) into 21–25 bp long siRNAs, which are then incorporated into RISC, leading to mRNA degradation. Hence, it mainly works on conserved genetic machinery, and is similar in a wide range of organisms (Whangbo and Hunter, 2008; Siomi and Siomi, 2009).

Non-cell autonomous RNAi

The gene silencing effect takes place in tissues/cells in a location that is different from the location of application or production of the dsRNA. It was first observed in C. elegans after the injection of dsRNA into the body through head or tail region, when gene silencing was demon-

strated throughout the injected animal as well as in its progeny (Fire *et al.*, 1998). Non-cell autonomous RNAi effects are of two types: environmental RNAi and systemic RNAi. While environmental and systemic RNAi share some overlapping machinery, the two processes are different, as environmental RNAi has also been observed in unicellular organisms such as protozoans.

ENVIRONMENTAL RNAi. Environmental RNAi has been observed in a wide range of species, but not in vertebrates. It includes the process in which dsRNA is taken up from the surrounding environment by the cell. Studies in C. elegans have shown how dsRNA from the environment enters into the organisms to trigger RNAi. Environmental RNAi in C. elegans involves the following steps: (i) dsRNA uptake by the intestinal cell; (ii) export of either dsRNAs or dsRNA-derived siRNAs from the intestinal cell; (iii) import of silencing signals into other tissues, such as muscle, epidermal and germline cells; and (iv) gene silencing via cell autonomous RNAi (Whangbo and Hunter, 2008). Environmental RNAi has also been shown in *Helicoverpa armigera* cell lines by soaking cells in dsRNA solution, which provided a link between the Bt toxin Cry1Ac and aminopeptidase N (HaAPN1) in the larval gut. The soaking experiment with dsRNA reduced the levels of *Haapn1* mRNA compared with those in controls, which ultimately decreased the sensitivity of the insect cells to the Cry1Ac protein (Sivakumar *et al.*, 2007).

SYSTEMIC RNAi. Systemic RNAi is well known in multicellular organisms as the signal for silencing travels from one cell to another. It is well documented in *C. elegans*, where the silencing signal is amplified and spread through a multi-step process (Sijen *et al.*, 2001; May and Plasterk, 2005), but it has not been reported in insects. If a similar system were present in insect pests, it would make target selection easier as targets other than those that are gut specific could be selected. Also, the presence of the RNAi amplification step would eradicate the need for a continuous supply of dsRNA, and so would compensate for the loss of dsRNA due to instability in the insect gut.

Mutation analysis in *C. elegans* came up with the gene involved in the systemic RNAi response in that organism. The gene, *systemic RNA interference deficient-1 (sid-1)*, was found to be essential for the systemic RNAi effect in *C. elegans*. When *sid-1* was expressed in *Drosophila* S2 cell lines, which reportedly lack a long-lasting systemic RNAi, the cells showed enhanced dsRNA uptake at suboptimal dsRNA concentrations (Feinberg and Hunter, 2003). Further analysis enabled the identification of another *C. elegans* dsRNA uptake mutant, *sid-2*. The functional characterization of *sid-2* was also carried out in a related nematode, *C. briggsae*, which is defective in the uptake of dsRNA from the gut lumen. The transformation of *C. briggsae* with *C. elegans sid-2* restored its systemic RNAi response (Winston *et al.*, 2007). Recently, SID-5, a *C. elegans* endosome-

associated protein, has been reported as required for an efficient systemic RNAi response to both ingested and expressed dsRNA (Hinas *et al.*, 2012).

Tribolium castaneum, the flour beetle, has been used as a model system to study systemic RNAi in insects, as this phenomenon is not reported in the usual model, *Drosophila*. The injection of dsRNA against the bristle-forming gene *Tc-achaete-scute (Tc-ASH)* into larvae at a single site resulted in the loss-of-bristle phenotype over the entire epidermal region of insects. This result suggested the presence of systemic RNAi in *T. castaneum* (Tomoyasu and Denell, 2004). However, completion of the sequencing of the *Tribolium* genome has come up with an interesting observation; *Tribolium* lacks the gene for RdRP, which is required for amplification of the silencing signal, and so systemic RNAi in *Tribolium* must either be based on a different gene with similar activity or on a different mechanism altogether (Price and Gatehouse, 2008).

15.3 Success of RNAi in Insect Pest Management

The accomplishment of RNAi depends mainly upon the mode of dsRNA delivery into the target cell. This is, therefore, a major limiting factor and has to be properly considered for a successful RNAi strategy in insect studies. There are three main delivery methods that are used regularly in entomological research. The first, micro-injection, is the most commonly used method for dsRNA delivery in insects (Arakane *et al.*, 2004; Suzuki *et al.*, 2008). The second method comprises the oral feeding of dsRNA, which can either be mixed into the insect diet, or fed to the insects by making them feed on transgenic plants expressing dsRNA (Baum *et al.*, 2007). The third method, soaking and spraying, is also applicable to the induction of RNAi, but this is usually utilized in cell cultures (Clemens *et al.*, 2000; Whyard *et al.*, 2009). The following section briefly describes these methods and their use in pest control.

15.3.1 Microinjection

Microinjection is widely and efficiently used as a tool to introduce dsRNA into organisms *in vivo*, in nematodes as well as arthropods. The dsRNA can be synthesized *in vitro* based on the target gene of the insect concerned, and is injected into the haemocoel (Dzitoyeva *et al.*, 2001). Microinjections in insects were first carried out in *D. melanogaster*, in which the *frizzled* and *frizzled 2* genes were downregulated (Kennerdell and Carthew, 1998).

In Lepidoptera, dsRNA delivery by injection has been shown to be successful, although the technique is difficult in these insects compared with its use in other classes of insects. A large variation is observed in the success rate of these experiments, which could be due to the factors such as life stage of the insect, concentration and amount of dsRNA used and the tissue where the target gene is expressed. Members of the Saturniidae family seem to be highly sensitive to RNAi administered by microinjection (Terenius *et al.*, 2011). Other successful reports citing injection as a successful mode of dsRNA delivery have been from the honeybee, *Apis mellifera* (Farooqui *et al.*, 2003; Aronstein and Saldivar, 2005), the pea aphid, *Acyrthosiphon pisum* (Jaubert-Possamai *et al.*, 2007), the cockroach, *Blattella germanica* (Martín *et al.*, 2006; Bellés, 2010; Huang and Lee, 2011) and the cricket, *Gryllus bimaculatus* (Moriyama *et al.*, 2008; Nakamura *et al.*, 2008).

The advantage of injecting dsRNA into insects is that it allows a direct delivery of dsRNA to the tissue of choice or into the haemolymph for an efficient inhibition of gene expression. It also allows an accurate interpretation of the results as the exact amount of dsRNA injected is known. However, there are some shortcomings to this method, which prevent it from being commonly used: the technique requires expertise and is time-consuming; additionally, it depends upon factors such as the choice of optimal volume, needle and site of injection. Furthermore, cuticle damage caused during injection may stimulate the insect's immune system, which can affect the data analysis (Han *et al.*, 1999; Yu *et al.*, 2013).

15.3.2 Oral delivery

The feeding or oral delivery of dsRNA has become an attractive alternative to microinjection because of its convenience and ease of manipulation. It is a simpler method for introducing dsRNA into an insect body and also causes less damage to the insect than injection (Chen *et al.*, 2010). dsRNA feeding experiments can be carried out by two methods: (i) the expression of dsRNA in bacteria; (ii) the *in vitro* synthesis of dsRNA. These products can either be mixed with the artificial diet or supplied in droplets.

C. elegans fed on dsRNA-expressing *Escherichia coli* bacteria showed a loss-of-function phenotype similar to that of mutants (Timmons and Fire, 1998; Timmons *et al.*, 2001). When *E. coli* HT115 (DE3) expressing dsRNAs of the chitin synthase A gene of *S. exigua* (*SeCHSA*) was fed to larvae of *S. exigua* by mixing with their artificial diet, the larvae showed disturbed development and higher mortality when compared with controls (Tian *et al.*, 2009). This also suggested the presence of systemic RNAi in *S. exigua*, as *SeCHSA* is a non-midgut gene expressed in the cuticle and trachea.

Oral delivery of *in vitro* synthesized dsRNA specific to the E-subunit of the vacuolar ATPase (V-ATPase) gene (*v-ATPase*) of *T. castaneum*, *A. pisum* and the tobacco hornworm, *Manduca sexta*, gave 50–75% mortality in all three insect species (Whyard *et al.*, 2009). The ingestion of dsRNA of the gut carboxylesterase gene (*EposCXE1*) by larvae of the light brown apple moth, *Epiphyas postvittana*, reduced the transcript level of the gene to less than half within 2 days of feeding compared with controls (Turner *et al.*, 2006). Thereafter, the feeding of dsRNA and siRNA solutions for the knock-down of target pest genes has been successfully demonstrated in *H. armigera* (Kumar *et al.*, 2009), the brown planthopper, *Nilaparvata lugens* (Chen *et al.*, 2010), the whitefly, *Bemisia tabaci* (Upadhyay *et al.*, 2011) and the mirid bug *Apolygus lucorum* (Zhou *et al.*, 2014), where it led to a strong decline in the expression of the target gene and could be used to explore gene functions. Chemically synthesized siRNA molecules also were applied on to the surface of tobacco leaves and fed to the larvae of *H. armigera* to silence the target gene for acetylcholinesterase (*AChE*) (Kumar *et al.*, 2009).

The development of plants expressing the Bt toxin emerged as a real boon for the control of lepidopteran pests (Naranjo, 2011), though pesticides are still the major method used to control phloem sap-sucking insects such as aphids, whiteflies, plant hoppers and plant bugs, as there is no Bt toxin known with potential insecticidal effects on these types of pests (Gatehouse and Price, 2011). Hence, there is an urgent need to develop techniques for the control of these insect pests. The feeding of dsRNA to control these insects by using transgenic plants expressing the desired dsRNA of vital insect genes is emerging as a potential alternative. Transgenic plants expressing dsRNAs against genes of Lepidoptera, Coleoptera and Hemiptera are common (Mao *et al.*, 2007). The advantage of using genetically engineered plants expressing dsRNAs is their continuous and stable expression of those dsRNAs (Rajam, 2011, 2012a,b).

Baum *et al.* (2007) fed Western corn rootworm larvae (WCR, *Diabrotica virgifera*) 290 different dsRNAs and observed that 14 of them caused significant mortalities at doses ≤ 5.2 ng cm^{-2}. They transformed maize to produce dsRNA specific to the gene encoding the A subunit of the V-ATPase proton pump and demonstrated that these plants showed a significant reduction in WCR feeding damage. In another study, Mao *et al.* (2007) targeted in cotton bollworm the gut-specific cytochrome P450 gene *CYP6AE14*, which confers resistance to gossypol, a polyphenol from cotton plants. They first fed cotton bollworm larvae on transgenic tobacco and *Arabidopsis* plants that expressed the *CYP6AE14*-specific dsRNA, and showed that the insects were subsequently sensitive to gossypol in

artificial diets. Further, Mao *et al.* (2011) demonstrated that the introduced dsRNA of *CYP6AE14* was stably expressed not only in the T_1 but also in the T_2 generation, and that the enhanced resistance to bollworms was recorded in both generations. These three reports (Baum *et al.*, 2007; Mao *et al.*, 2007, 2011) indicate that the RNAi pathway could be exploited to control insect pests via *in planta* expression of dsRNA against well-chosen target genes of insects. The *v-ATPaseA* gene has also been shown to be a potent target for controlling the whitefly population. Plant-mediated pest resistance was achieved against whiteflies by the genetic transformation of tobacco, which generated siRNAs against the whitefly *v-ATPaseA* gene. The transcript levels of *v-ATPaseA* in whiteflies were reduced up to 62% after feeding on the transgenic plants, leading to mortality of the insects (Thakur *et al.*, 2014).

A number of later reports have also suggested the success of this technology for the control of insect pests. When transgenic tobacco plants expressing dsRNA against *EcR-USP*, the gene for the ecdysone receptor-ultraspiracle particle (Zhu *et al.*, 2012), *AChE*, the gene for acetylcholinesterase (Kumar, 2011) and *HR3*, a gene involved in the regulation of moulting and development in *H. armigera* (Xiong *et al.*, 2013) were fed to larvae, all of them resulted in larval developmental deformities and lethality. The gene *HMGR* for the enzyme 3-hydroxy-3-methylglutaryl coenzyme A reductase (HMG-CoA reductase), a key enzyme in the mevalonate pathway in insects, has also been shown to be a potential target for insect control using RNAi (Wang *et al.*, 2013). Recently, Di Lelio *et al.* (2014) showed that oral delivery to the African cotton leafworm, *S. littoralis*, of dsRNA molecules against a gene highly similar to *P102* of the tobacco budworm, *Heliothis virescens*, strongly suppressed the encapsulation and melanization response, while haemocoel injections did not result in evident phenotypic variations. This suggested that the protein is functionally conserved and plays a role in insect immunity. Hence, based on immunosuppression, dsRNA oral delivery

may be exploited to develop novel technologies of pest control.

The oral delivery of dsRNA is
advantageous because it is labour saving,
cost-effective and easy to perform (Tian *et
al.*, 2009). It is also less invasive than
microinjection and can be used for high-
throughput gene screening, even in very
small insects such as aphids and first and
second instar larval stages (Araujo *et al.*,
2006). None the less, it may not be suitable
for all species and the efficiency of RNAi by
ingestion also varies between different
species, possibly as a result of different gut
environments. Another limitation is that
the amount of dsRNA taken in by the insect
is hard to determine (Surakasi *et al.*, 2011),
which can affect data interpretation.

15.3.3 Soaking and spraying

The soaking of organisms in a dsRNA
solution was first demonstrated in *C. elegans*
as triggering an RNAi response (Tabara *et
al.*, 1998). In insects, soaking experiments
have been reported mostly from cell lines, as
the insect cuticle poses a barrier when
soaking is performed with whole insect.
Only certain insect cells and tissues, as well
as insects of specific developmental stages,
are suitable for soaking that readily absorbs
dsRNA from solution, and so this method is
used rarely. In insects, soaking was first
reported in *Drosophila melanogaster* S2 cells
derived from embryos (Clemens *et al.*,
2000). Recently, solutions of *cycE* and *ago*
dsRNA have been shown to effectively
inhibit the expression of these two cell cycle
genes, thereby elevating the levels of protein
synthesis in S2 cells of *D. melanogaster*
(March and Bentley, 2007).

If dsRNA can penetrate the insect body
and enter into the body cavity, then it can be
applied in the field in a manner similar to
that of traditional insecticides. Despite the
difficulties in uptake, direct spraying of
dsRNA for the DS10 gene (for a
chymotrypsin-like serine proteinase from
the Asian corn borer, *Ostrinia furnalalis*) and
the DS28 gene (a gene of unknown function
from *H. armigera*) on to newly hatched *O.*

furnalalis larvae gave 40–50% mortality of
newly hatched larvae (Wang *et al.*, 2011).
Spraying will be a feasible method only if
dsRNA can be mass produced cheaply and
can reduce the pest population faster than
conventional insecticides. However, dsRNA
spraying is a useful technique for both
functional studies and pest control.

15.4 Conclusion and Prospects of RNAi for Insect Control

RNAi holds a great potential for crop
improvement against insect attack. It takes
the advantage of the pathway involved in
gene regulation and works with high
specificity in gene targeting, making it a safe
method for insect pest management. The
commercial success of RNAi will depend
upon the choice of a target gene that can
effectively kill the pests. Targeting multiple
genes would further strengthen the
application of this technology (Rajam, 2011).

The spraying of dsRNA ensures its use at
a particular stage of plant life that is
vulnerable to insect damage. A crude
preparation of dsRNA has been shown to be
effective in silencing plant viral genes when
sprayed on to the leaves and also exists
stably for several days (Tenllado *et al.*, 2004).
As a potential bioinsecticide, the use of
dsRNA needs to consider its cost, stability
and ecological issues. Alternatively, bacteria
expressing dsRNA specific to the target
insect gene can be used as an insect control
spray (Rajam, 2011). At this point, it seems
as if the use of RNAi transgenic plants as
feed might be a better way to go as this
technique will produce dsRNA inside the
plants without any contact with the
environment. The use of transgenic plants
expressing dsRNA would also be cost-
effective as they will produce dsRNA during
the entire life of the plant concerned.

Another potential application of RNAi
involves overcoming the problem of
pesticide resistance and its use for pest
management. Specific gene silencing will
increase the sensitivity of insects and also
pose little harm to other organisms.
Different approaches are now being studied

thoroughly to avoid off-target effects, which are considered to a major limiting factor. A number of target genes for RNAi have been tested in insects. The genes for the production of glutathione S-transferases, α-tubulin, β-tubulin, cytochrome P450, V-ATPase and acetylcholinesterase are some examples. Next-generation sequencing (NGS) and RNAi target sequencing (RIT-seq) have provided a platform for the precise selection of target genes (Haas and Zody, 2010; Alsford et al., 2011). Genome sequencing of D. melanogaster, A. mellifera, T. castaneum and the silkworm, Bombyx mori, has opened the way for testing new targets in these insects as well as in related organisms.

Although RNAi technology for insect control looks promising, several aspects have to be studied before its practical applicability, such as finding specific target genes, the stabilization of dsRNAs during and after delivery, and the development of methods that are cheap for large-scale use. Gene families can be targeted using a single sequence if the selected target sequence is conserved, which can lead to higher silencing and also reduce the chances of evolving resistance in insects. Also, the target gene can be selected in a manner that can provide efficient insect mortality while maintaining less selective pressure. Mutations are major cause of insects gaining resistance against transgenic crops. As long as dsRNA is being produced by RNAi plants, which eventually generate different siRNAs, mutation in the target gene may not affect the effectiveness of siRNAs, and too many mutations in the target gene would be detrimental to the insect itself. Moreover, siRNAs can tolerate some mismatches within the target if they are not in the 9–11 bp region (Du et al., 2005). Furthermore, mutations in the genes that are involved in the RNAi machinery of the target insect is another possible mechanism for resistance as it avoids the effect of insect target gene-specific siRNAs, though this may also prove to be harmful to the insect itself.

Successful experiments have shown the potential of RNAi for pest management, but the long-term effectiveness will be achieved only after field trials. To date, Bt crops constitute a major insect biocontrol approach, but the pace with which RNAi is being used for developing pest-resistant plants shows its likely significant role in the near future. RNAi-mediated crop protection has been considered to be a potential strategy for insect control that can complement existing pest control methods. However, there is a need for new and effective resistance management strategies to minimize the evolution of resistance in insects for the continuing benefits of the use of RNAi.

Acknowledgements

MVR is grateful to the Department of Biotechnology and the Department of Science and Technology (DST), New Delhi, for generously funding RNAi programmes in the laboratory. We thank the University Grants Commission for a Special Assistance Program, and DST for the FIST and DU-DST PURSE program. A research fellowship to SY by the INSPIRE-DST is acknowledged.

References

Alsford, S., Turner, D.J., Obado, S.O., Sanchez-Flores, A., Glover, L. et al. (2011) High-throughput phenotyping using parallel sequencing of RNA interference targets in the African trypanosome. Genome Research 21, 915–924.

Arakane, Y., Hogenkamp, D.G., Zhu, Y.C., Kramer, K.J., Specht, C.A. et al. (2004) Characterization of two chitin synthase genes of the red flour beetle, Tribolium castaneum, and alternate exon usage in one of the genes during development. Insect Biochemistry and Molecular Biology 34, 291–304.

Araujo, R.N., Santos, A., Pinto, F.S., Gontijo, N.F., Lehane, M.J. et al. (2006) RNA interference of the salivary gland nitrophorin 2 in the triatomine bug Rhodnius prolixus (Hemiptera: Reduviidae) by dsRNA ingestion or injection. Insect Biochemistry and Molecular Biology 36, 683–693.

Aronstein, K. and Saldivar, E. (2005) Characterization of a honeybee Toll related receptor gene Am18w and its potential involvement in antimicrobial immune defense. Apidologie 36, 3–14.

Baum, J.A., Bogaert, T., Clinton, W., Heck, G.R., Feldmann, P. *et al.* (2007) Control of coleopteran insect pests through RNA interference. *Nature Biotechnology* 25, 1322–1326.

Bellés, X. (2010) Beyond *Drosophila*: RNAi *in vivo* and functional genomics in insects. *Annual Review of Entomology* 55, 111–128.

Bravo, A., Gill, S. and Soberón, M. (2007) Mode of action of *Bacillus thuringiensis* Cry and Cyt toxins and their potential for insect control. *Toxicon* 49, 423–435.

Carpenter, J.E. (2010) Peer-reviewed surveys indicate positive impact of commercialized GM crops. *Nature Biotechnology* 28, 319–321.

Chen, J., Zhang, D., Yao, Q., Zhang, J., Dong, X. *et al.* (2010) Feeding-based RNA interference of a trehalose phosphate synthase gene in the brown planthopper, *Nilaparvata lugens. Insect Molecular Biology* 19, 777–786.

Clemens, J.C., Worby, C.A., Simonson-Leff, N., Muda, M., Maehama, T. *et al.* (2000) Use of double-stranded RNA *Drosophila* cell lines to dissect signal transduction pathways. *Proceedings of the National Academy of Sciences of the United States of America* 97, 6499–6503.

Di Lelio, I., Varricchio, P., Di Prisco, G., Marinelli, A., Lasco, V. *et al.* (2014) Functional analysis of an immune gene of *Spodoptera littoralis* by RNAi. *Journal of Insect Physiology* 64, 90–97.

Du, Q., Thonberg, H., Wang, J., Wahlestedt, C. and Liang, Z. (2005) A systematic analysis of the silencing effects of an active siRNA at all single-nucleotide mismatched target sites. *Nucleic Acids Research* 33, 1671–1677.

Dzitoveya, S., Dimitrijevic, N. and Manev, H. (2001) Intra-abdominal injection of double-stranded RNA into anesthetized adult *Drosophila* triggers RNA interference in the central nervous system. *Molecular Psychiatry* 6, 665–670.

Edgerton, M.D., Fridgen, J., Anderson, J.R. Jr, Ahlgrim, J., Criswell, M. *et al.* (2012) Transgenic insect resistance traits increase corn yield and yield stability. *Nature Biotechnology* 30, 493–496.

Farooqui, T., Robinson, K., Vaessin, H. and Smith, B.H. (2003) Modulation of early olfactory processing by an octopaminergic reinforcement pathway in the honeybee. *Journal of Neuroscience* 23, 5370–5380.

Feinberg, E.H. and Hunter, C.P. (2003) Transport of dsRNA into cells by the transmembrane protein SID-1. *Science* 301, 1545–1517.

Filipowicz, W. (2005) RNAi: the nuts and bolts of the RISC machine. *Cell* 122, 17–20.

Fire, A., Xu, S.Q., Montgomery, M.K., Kostas, S.A., Driver, S.E. *et al.* (1998) Potent and specific genetic interference by double-stranded RNA in *Caenorhabditis elegans. Nature* 391, 806–811.

Gatehouse, J.A. and Price, D.R.G. (2011) Protection of crops against insect pests using RNA interference. *Insect Biotechnology* 2, 145–168.

Haas, B.J. and Zody, M.C. (2010) Advancing RNA-Seq analysis. *Nature Biotechnology* 28, 421–423.

Han, Y.S., Chun, J., Schwartz, A., Nelson, S. and Paskewitz, S.M. (1999) Induction of mosquito hemolymph proteins in response to immune challenge and wounding. *Development and Comparative Immunology* 23, 553–562.

Hinas, A., Wright, A.J. and Hunter, C.P. (2012) SID-5 is an endosome-associated protein required for efficient systemic RNAi in *C. elegans. Current Biology* 22, 1938–1943.

Huang, J.H. and Lee, H.J. (2011) RNA interference unveils functions of the hypertrehalosemic hormone on cyclic fluctuation of hemolymph trehalose and oviposition in the virgin female *Blatella germanica. Journal of Insect Physiology* 57, 858–864.

Huvenne, H. and Smagghe, G. (2010) Mechanisms of dsRNA uptake in insects and potential of RNAi for pest control: a review. *Journal of Insect Physiology* 56, 227–235.

Jaubert-Possamai, S., Le Trionnaire, G., Bonhomme, J., Christophides, G.K., Rispe, C. *et al.* (2007) Gene knockdown by RNAi in the pea aphid *Acyrthosiphon pisum. BMC Biotechnology* 7:63.

Kennerdell, J.R. and Carthew, R.W. (1998) Use of dsRNA mediated genetic interference to demonstrate that *frizzled* and *frizzled 2* act in the wingless pathway. *Cell* 95, 1017–1026.

Kumar, M. (2011) RNAi-mediated targeting of acetylcholinesterase gene of *Helicoverpa armigera* for insect resistance in transgenic tomato and tobacco. PhD thesis, Department of Genetics, University of Delhi, New Delhi.

Kumar, M., Gupta, G.P. and Rajam, M.V. (2009) Silencing of acetylcholinesterase gene of *Helicoverpa armigera* by siRNA affects larval growth and its life cycle. *Journal of Insect Physiology* 55, 273–278

Lu, Y., Wu, K., Jiang, Y., Guo, Y. and Desneux, N. (2012) Widespread adoption of Bt cotton and insecticide decrease promotes biocontrol services. *Nature* 487, 362–365.

Mao, Y.-B., Cai, W.-J., Wang, J.-W., Hong, G.-J., Tao, X.-Y. *et al.* (2007) Silencing a cotton bollworm P450 monooxygenase gene by plant-mediated RNAi impairs larval tolerance of gossypol. *Nature Biotechnology* 25, 1307–1313.

Mao, Y.-B., Tao, X.-Y., Xue, X.-Y., Wang, L.-J. and

Chen, X.-Y. (2011) Cotton plants expressing *CYP6AE14* double-stranded RNA show enhanced resistance to bollworms. *Transgenic Research* 20, 665–673.

March, J.C. and Bentley, W.E. (2007) RNAi-based tuning of cell cycling in *Drosophila* S2 cells: effects on recombinant protein yield. *Applied Microbiology and Biotechnology* 73, 1128–1135.

Martín, D., Maestro, O., Cruz, J., Mané-Padrós, D. and Bellés, X. (2006) RNAi studies reveal a conserved role for RXR in molting in the cockroach *Blattella germanica*. *Journal of Insect Physiology* 52, 410–416.

Matten, S., Head, G. and Quemada, H. (2008) How governmental regulation can help or hinder the integration of Bt crops into IPM programs. In: Romeis J., Shelton, A.M. and Kennedy, G.G. (eds) *Integration of Insect-resistant Genetically Modified Crops within IPM Programs. Progress in Biological Control, Volume 5*. Springer, Dordrecht, The Netherlands, pp. 27–39.

May, R.C. and Plasterk, R.H. (2005) RNA interference spreading in *C. elegans*. *Methods in Enzymology* 392, 308–315.

Moar, W., Roush, R., Shelton, A., Ferré, J., MacIntosh, S. *et al.* (2008) Field-evolved resistance to Bt toxins. *Nature Biotechnology* 26, 1072–1076.

Moriyama, Y., Sakamoto, T., Karpova, S.G., Matsumoto, A., Noji, S. *et al.* (2008) RNA interference of the clock gene period disrupts circadian rhythms in the cricket *Gryllus bimaculatus*. *Journal of Biological Rhythms* 23, 308–318.

Nakamura, T., Mito, T., Miyawaki, K., Ohuchi, H. and Noji, S. (2008) EGFR signaling is required for re-establishing the proximodistal axis during distal leg regeneration in the cricket *Gryllus bimaculatus* nymph. *Developmental Biology* 319, 46–55.

Naranjo, S.E. (2011) Impacts of Bt transgenic cotton on integrated pest management. *Journal of Agricultural and Food Chemistry* 59, 5842–5851.

Price, D.R.G. and Gatehouse, J.A. (2008) RNAi-mediated crop protection against insects. *Trends in Biotechnology* 26, 393–400.

Rajam, M.V. (2011) RNA interference: a new approach for the control of fungal pathogens and insects. In: Muralidharan, K. and Siddiq, E.A. (eds) *Proceedings of the National Symposium on 'Genomics and Crop Improvement: Relevance and Reservations', Held at the Acharya N.G. Ranga Agricultural University during February 25–27, 2010*, pp. 220–229.

Rajam, M.V. (2012a) Micro RNA interference: a new platform for crop protection. *Cell and Developmental Biology* 1:e115.

Rajam, M.V. (2012b) Host induced silencing of fungal pathogen genes: an emerging strategy for disease control in crop plants. *Cell and Developmental Biology* 1: e118.

Sanahuja, G., Banakar, R., Twyman, R., Capell, T. and Christou, P. (2011) *Bacillus thuringiensis*: a century of research, development and commercial applications. *Plant Biotechnology Journal* 9, 283–300.

Shabalina, S.A. and Koonin, E.V. (2008) Origins and evolution of eukaryotic RNA interference. *Trends in Ecology and Evolution* 23, 578–587.

Sijen, T., Fleenor, J., Simmer, F., Thijssen, K.L., Parrish, S. *et al.* (2001) On the role of RNA amplification in dsRNA-triggered gene silencing. *Cell* 107, 465–476.

Siomi, H. and Siomi, M.C. (2009) On the road to reading the RNA-interference code. *Nature* 457, 396–404.

Sivakumar, S., Rajagopal, R., Venkatesh, G.R., Srivastava, A. and Bhatnagar, R.K. (2007) Knockdown of aminopeptidase-N from *Helicoverpa armigera* larvae and in transfected Sf21 cells by RNA interference reveals its functional interaction with *Bacillus thuringiensis* insecticidal protein Cry1Ac. *The Journal of Biological Chemistry* 282, 7312–7319.

Stein, P., Svoboda, P., Anger, M. and Schultz, R.M. (2003) RNAi: mammalian oocytes do it without RNA-dependent RNA polymerase. *RNA* 9, 187–192.

Surakasi, V.P., Mohamed, A.A.M. and Kim, Y. (2011) RNA interference of β1 integrin subunit impairs development and immune responses of the beet armyworm, *Spodoptera exigua*. *Journal of Insect Physiology* 57, 1537–1544.

Suzuki, Y., Truman, J.W. and Riddiford, L.M. (2008) The role of *broad* in the development of *Tribolium castaneum*: implications for the evolution of the holometabolous insect pupa. *Development* 135, 569–577.

Tabara, H., Grishok, A. and Mello, C.C. (1998) RNAi in *C. elegans*: soaking in the genome sequence. *Science* 282, 430–431.

Tabashnik, B.E., Gassmann, A.J., Crowder, D.W. and Carrière, Y. (2008a) Insect resistance to Bt crops: evidence versus theory. *Nature Biotechnology* 26, 199–202.

Tabashnik, B.E., Gassmann, A.J., Crowder, D.W. and Carrière, Y. (2008b) Field-evolved resistance to Bt toxins. [Reply to Moar *et al.*, 2008]. *Nature Biotechnology* 26, 1074–1076.

Tabashnik, B.E., Brévault, T. and Carrière, Y. (2013) Insect resistance to Bt crops: lessons from the first billion acres. *Nature Biotechnology* 31, 510–521.

Tenllado, F., Llave, C. and Diaz-Ruiz, J.R. (2004) RNA interference as a new biotechnological tool for the control of virus diseases in plants. *Virus Research* 102, 85–96.

Terenius, O., Papanicolaou, A., Garbutt, J.S., Eleftherianos, I., Huvenne, H. *et al.* (2011) RNA interference in Lepidoptera: an overview of successful and unsuccessful studies and implications for experimental design. *Journal of Insect Physiology* 57, 231–245.

Thakur, N., Upadhyay, S.K., Verma, P.C., Chandrashekar, K., Tuli, R. *et al.* (2014) Enhanced whitefly resistance in transgenic tobacco plants expressing double stranded RNA of *v-ATPase A* gene. *PLoS ONE* 9(3): e87235.

Tian, H., Peng, H., Yao, Q., Chen, H., Xie, Q. *et al.* (2009) Developmental control of a lepidopteran pest *Spodoptera exigua* by ingestion of bacterial expressing dsRNA of a non-midgut gene. *PLoS One* 4(7): e6225.

Timmons, L. and Fire, A. (1998) Specific interference by ingested dsRNA. *Nature* 395, 854.

Timmons, L., Court, D.L. and Fire, A. (2001) Ingestion of bacterially expressed dsRNAs can produce specific and potent genetic interference in *Caenorhabditis elegans*. *Gene* 263,103–112.

Tomoyasu, Y. and Denell, R.E. (2004) Larval RNAi in *Tribolium* (Coleoptera) for analyzing adult development. *Development Genes and Evolution* 214, 575–578.

Trivedi, B. (2010) Bug silencing: the next generation of pesticides. *New Scientist* 205, 34–37.

Turner, C.T., Davy, M.W., MacDiarmid, R.M., Plummer, K.M., Birch, N.P. *et al.* (2006) RNA interference in the light brown apple moth, *Epiphyas postvittana* (Walker) induced by double-stranded RNA feeding. *Insect Molecular Biology* 15, 383–391.

Upadhyay, S.K., Chandrashekar, K., Thakur, N., Verma, P.C. and Borgio, J.F. (2011) RNA interference for the control of whiteflies (*Bemisia tabaci*) by oral route. *Journal of Bioscience* 36, 153–161.

Wang, Y., Zhang, H., Li, H. and Miao, X. (2011) Second-generation sequencing supply an effective way to screen RNAi targets in large scale for potential application in pest insect control. *PloS One* 6(4): e18644.

Wang, Z., Dong, Y., Desneux, N. and Niu, C. (2013) RNAi silencing of the HaHMG-CoA reductase gene inhibits oviposition in the *Helicoverpa armigera* cotton bollworm. *PLoS One* 8(7): e67732.

Whangbo, J.S. and Hunter, C.P. (2008) Environmental RNA interference. *Trends in Genetics* 24, 297–305.

Whyard, S., Singh, A.D. and Wong, S. (2009) Ingested double stranded RNAs can act as species-specific insecticides. *Insect Biochemistry and Molecular Biology* 39, 824–832.

Winston, W.M., Sutherlin, M., Wright A.J., Feinberg E.H. and Hunter C.P. (2007) *Caenorhabditis elegans* SID-2 is required for environmental RNA interference. *Proceedings of the National Academy of Sciences of the United States of America* 104, 10565–10570.

Xiong, Y., Zeng, H., Zhang, Y., Xu, D. and Qiu, D. (2013) Silencing the *HaHR3* gene by transgenic plant-mediated RNAi to disrupt *Helicoverpa armigera* development. *International Journal of Biological Sciences* 9, 370–381.

Yu, N., Christiaens, O., Liu, J., Niu, J., Cappelle, K. *et al.* (2013) Delivery of dsRNA for RNAi in insects: an overview and future directions. *Insect Science* 20, 4–14.

Zhao, J., Cao, J., Li, Y., Collins, H., Roush, R. *et al.* (2003) Transgenic plants expressing two *Bacillus thuringiensis* toxins delay insect resistance evolution. *Nature Biotechnology* 21, 1493–1497.

Zhou, Y.-L., Zhu, X.-Q., Gu, S.-H., Cui, H.-H., Guo, Y.-Y. *et al.* (2014) Silencing in *Apolygus lucorum* of the olfactory coreceptor Orco gene by RNA interference induces EAG response declining to two putative semiochemicals. *Journal of Insect Physiology*, 60, 31–39.

Zhu, J.-Q., Liu, S., Ma, Y., Zhang, J.-Q., Qi, H.-S. *et al.* (2012) Improvement of pest resistance in transgenic tobacco plants expressing dsRNA of an insect-associated gene *EcR*. *PLoS One* 7(6): e38572.

16 Resistance Management for Bt Maize and Above-ground Lepidopteran Targets in the USA: From Single Gene to Pyramided Traits

Fangneng Huang*

Department of Entomology, Louisiana State University Agricultural Center, Baton Rouge, Louisiana, USA

Summary

Since first being commercialized in 1996, transgenic maize expressing *Bacillus thuringiensis* (Bt) proteins has gained widespread acceptance in the world. In 2013, nearly 50 Mha of Bt maize were planted in 15 countries. In the same year, growers in the USA alone planted *c.*30 Mha of Bt maize, which accounted for 76% of the total Bt maize area of the country. Up to now, Bt maize technology can be classified into two generations. The first generation of Bt maize contains only a single Bt gene for a target. In 2010, the second generation of Bt maize became commercially available and this expresses two or more pyramided Bt proteins. Currently, the pyramided products are predominant in the USA. The major lepidopteran targets of Bt maize in the USA are corn borers (Crambidae), the corn earworm, *Helicoverpa zea*, and the fall armyworm, *Spodoptera frugiperda*. To counter the threat of insect resistance, two resistance management strategies for Bt maize, 'high dose/refuge' and gene pyramiding, have been implemented. The long-term use of Bt maize against the major agricultural pests in North America provides a good opportunity to analyse the effectiveness of the adopted insect resistance management

(IRM) plans. Analysis of the available data shows that all corn borer species remain susceptible to Bt proteins and that no field resistance has occurred after nearly two decades of intensive use of Bt maize in the continent. Pyramided Bt maize is effective in controlling corn earworm and fall armyworm, although recent studies indicate that field resistance to single-gene Cry1F maize in the fall armyworm has occurred in the south-east coastal areas of the US mainland. Knowledge of the resistance management gained from the USA should be useful for other countries in their sustainable use of Bt crop technology.

16.1 Introduction

Maize, *Zea mays*, is one of the most widely planted field crops in the world, with a total production of 853 million t in 2012 (National Corn Growers Association, 2013). In the USA, it is the most widely planted field crop, and in 2012, 273.8 million t was produced with a crop value of US$79.8 billion. Many arthropods are associated with maize production, from sowing to harvesting. Chemical and biological control, and host plant resistance, have been common methods used in the management of maize

* Corresponding author. E-mail address: fhuang@agcenter.lsu.edu

pests. Advances in molecular engineering have allowed scientists to transfer foreign genes into species that are not related to one another. One of the most successful bio-engineering achievements is the trans-formation of crop plants with insecticidal genes from the soil bacterium, *Bacillus thuringiensis* (Bt). The resulting transgenic Bt crops produce insecticidal Bt proteins within the plant's tissues that directly kill insects when those plant tissues are consumed. Since their first commercializa-tion in 1996, Bt crops – predominantly maize and cotton – have been rapidly adopted for insect pest management worldwide (James, 2013). In 2013, over 74 Mha of Bt crops were planted in 23 countries, including 49.6 Mha of Bt maize, 22.6 Mha of Bt cotton and 2.2 Mha of Bt soybean (James, 2013) (Table 16.1). The USA has being the leading country in the planting of Bt crops.

The rapid acceptance of Bt crops creates a threat to the long-term durability of the technology. Intensive planting of Bt crops over a wide region can place a high selection pressure on the pest populations and thus accelerate resistance development. To date, field resistance that results in reduced efficacy or failure of pest control (Huang *et al.*, 2011) has been documented in at least four cases (Storer *et al.*, 2010; van Rensburg, 2007; Dhurua and Gujar, 2011; Gassmann *et al.*, 2011); see also Chapters 1 (Tabashnik and Carrière), and 3 (Monnerat *et al.*). In the USA, in order to conserve the susceptibility of pests to Bt crops, insect resistance management (IRM) plans are now man-datory for the planting of these crops (Matten *et al.*, 2012).

Because of the recent occurrence of field resistance in the western corn rootworm, *Diabrotica virgifera virgifera*, to Cry3Bb1

Table 16.1. Global adoption of Bt maize and Bt cotton in 2013.

Country[a]	Bt maize		Bt cotton		Total Bt crop area (Mha)
	Area (Mha)	% total maize	Area (Mha)	% total cotton	
USA	29.94	76	3.03	75	32.97
Brazil	12.35	69	0.33	33	12.68
India	–	–	11.02	95	11.02
China	–	–	4.14	90	4.14
Argentina	3.11	75	0.36	79	3.47
Pakistan	–	–	2.80	86	2.80
South Africa	1.93	71	0.01	95	1.94
Canada	1.25	77	–	–	1.25
Philippines	0.71	57	–	–	0.71
Australia	–	–	0.42	94	0.42
Myanmar	–	–	0.31	85	0.31
Spain	0.13	–[b]	–	–	0.08
Mexico	–	–	0.10	–[b]	0.10
Colombia	0.077	–[b]	0.024	–[b]	0.101
Sudan	–	–[b]	0.062	–[b]	0.062
Chile	0.024	–[b]	–	–	0.024
Honduras	0.018	–[b]	–	–	0.018
Others	0.014 (5 countries)	–[b]	<0.01 (1 country)	–[b]	0.024
Total	49.55		22.62		72.17

[a]Data for the USA are from NASS (2013) and for Canada from Dunlop (2013). All other data were calculated based on James (2013). In addition to Bt maize and Bt cotton, Brazil planted 2.2 Mha of Bt soybean (James, 2013). Other countries not listed above that planted Bt crops with an area of <0.01 Mha in 2013 included Portugal, Cuba, the Czech Republic, Costa Rica, Romania and Slovakia.
[b]A very small percentage.

maize in the USA, intensive studies have been performed on the Bt resistance of these coleopteran pests (Gassmann *et al.*, 2011). In this chapter, a review is presented of the implementation of IRM for Bt maize targeting above-ground lepidopteran pests in the USA over nearly 20 years. The review focuses on the European corn borer (ECB; *Ostrinia nubilalis*), southwestern corn borer (SWCB; *Diatraea grandiosella*), sugarcane borer (SCB; *Diatraea saccharalis*), corn earworm (CEW; *Helicoverpa zea*) and fall armyworm (FAW; *Spodoptera frugiperda*) because there is substantial information on these five targets. The long-term use of Bt maize in managing these major agricultural pests in the USA also provides a good opportunity to analyse the effectiveness of the IRM plans that have been adopted.

*c.*5% of the nation's total maize planting area (Fig. 16.1). Since then, both the percent and area of Bt maize have increased every year except for 2000 and 2001, which were caused by the StarLink issue (Taylor and Tick, 2000). In 2013, the USA planted nearly 30 Mha of Bt maize, which accounted for 76% of the country's total maize area and 60% of the world's total Bt maize area. The rapid adoption of Bt maize in the USA is revealed with an annual increase rate of 18.8% in the area basis during the past 18 years. Up to date, the commercial Bt maize products that have planted in the U.S. can be classified into two generations. Products of the 1st generation Bt maize produce only a single Bt protein for a target pest, while the second generation expresses ≥2 Bt proteins that act on a same insect.

16.2 Adoption of Bt Maize in the USA

The year 1996 marked the first year that Bt maize was commercially planted in the USA over an area of 1.6 Mha, which accounted for

16.2.1 First-generation Bt maize

Before 2010, all Bt maize products planted in the USA expressed only a single Bt protein for a target pest, and this usually

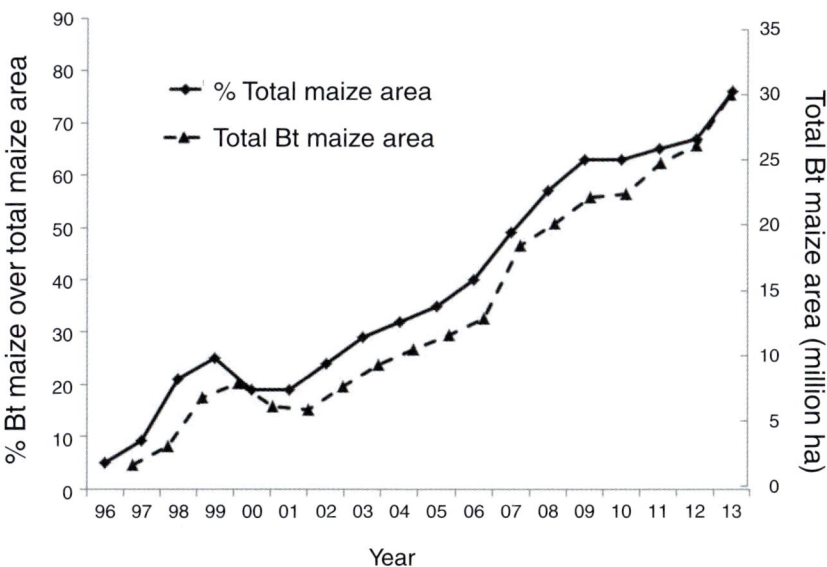

Fig. 16.1. Percentage and total area of maize transformed with genes from *Bacillus thuringiensis* (Bt maize) planted in the USA from 1996 to 2013. Most of the data are based on the annual reports of crop acreage from the USDA National Agricultural Statistics Service (NASS, 1996–2013). Percentage of Bt maize in a year was calculated based on the Bt maize area divided by the total maize area in that year.

Table 16.2. Examples of first-generation Bt maize products.

Bt maize event[a]	Bt gene/s	Year of release	Examples of Bt maize products	Major target insects[b]
MON810	Cry1Ab	1996	YieldGard	Corn borers
Bt11	Cry1Ab	1996	YieldGard (later also called Agrisure Corn Borer)	Corn borers
TC1507	Cry1F	2001	Herculex® I	Corn borers, fall armyworm
MON810 + MON863	Cry1Ab + Cry3Bb1	2003	YieldGard® Plus corn	Cry1Ab for corn borers and Cry3Bb1 for corn rootworms
TC1507 + DAS-59122-7	Cry1F + Cry34/35Ab1	2005	Hexculex® XTRA corn	Cry1F for corn borers and fall armyworm, and Cry34/35Ab1 for corn rootworms
MON810 + MON863	Cry1Ab + Cry3Bb1	2007	YieldGard® VT Triple	Cry1Ab for corn borers and Cry3Bb1 for corn rootworms

[a]An event is a specific genetic modification in a specific species.
[b]Corn borer species may include the European corn borer (*Ostrinia nubilalis*), southwestern corn borer (*Diatraea grandiosella*), sugarcane borer (*Diatraea saccharalis*), southern corn stalk borer (*Diatraea crambidoides*) and lesser corn stalk borer (*Elasmopalpus lignosellus*). Corn rootworm species may include the western corn rootworm (*Diabrotica virgifera virgifera*), northern corn rootworm (*Diabrotica barberi*) and Mexican rootworm (*Diabrotica virgifera zeae*). The fall armyworm is *Spodoptera frugiperda*.

has a relatively narrow insecticidal spectrum (see Table 16.2). For example, YieldGard corn, the most widely used single-gene Bt corn, expresses only the Cry1Ab protein and is effective against some lepidopteran corn borers, such as ECB and SWCB. Some of the first-generation Bt maizes did contain two Bt genes, but these were for different targets. For example, YieldGard Plus corn, a commonly planted first-generation Bt maize, contains the genes for both Cry1Ab and Cry3Bb1, but the Cry1Ab is for controlling lepidopteran corn borers, while the Cry3Bb1 is for managing the coleopteran corn rootworm. The Cry1Ab protein does not have any activity against Coleoptera and the Cry3Ab1 protein is not toxic to Lepidoptera. In the literature, the transfer of more than one foreign gene into a same plant, each gene for different targets/purposes, is often called 'gene stacking' (Huang et al., 2011). An exception to this is the Bt maize Herculex®RW *Rootworm Protection,* which contains two Bt genes, those for the Cry34Ab1 and Cry35Ab1 proteins, which are both active against corn rootworms. However, the two are nearly identical in gene structure and mode of action, so they function like a single gene.

The main above-ground lepidopteran targets of first-generation Bt maize in the USA are stalk borers, including ECB in the entire US Corn Belt (Ostlie et al., 1997), SWCB in the central and southern regions (Trisyono and Chippendale, 2002), SCB in the mid-south (Castro et al., 2004), the southern corn stalk borer, *Diatraea crambidoides*, in the south-east (Reisig and Roberson, 2011) and the lesser corn stalk borer, *Elasmopalpus lignosellus*, in the south (Vilella et al., 2002). FAW is also listed as a main target of the single-gene Cry1F maize (e.g. Herculex®I). In general, the field performance of the first-generation Bt maize has been outstanding for controlling the target pests (Huang et al., 1999; Buschman et al., 2001), and growers in the USA have recognized the great economic and environmental benefits offered by Bt maize (Hutchison et al., 2010). First-generation Cry1Ab maize also can suppress some secondary lepidopteran pests, such as CEW and FAW (Buntin et al., 2004; Chilcutt et al., 2007), but the suppression levels are usually not high (US EPA, 2001). For this reason, neither CEW nor FAW are listed as targets of the first-generation Bt maize, except that FAW is a target of the single-gene Cry1F maize (US EPA, 2005).

16.2.2 Second-generation Bt maize

Second-generation transgenic maize that expresses more than one dissimilar Bt protein targeting above-ground Lepidoptera was first commercially planted in the USA and Canada in 2010. An example is one of the most widely used second-generation products, Genuity® SmartStax™, which contains three dissimilar Cry genes, those

for the proteins Cry1A.105, Cry2Ab2 and Cry1F, which all target Lepidoptera. The technique of transferring more than one different Bt gene into the same plant for same target pest/s is usually called 'gene pyramiding' in the literature (Huang *et al.*, 2011). Since the first release of second-generation maize in 2010, many pyramided products have then been commercialized in the USA (see Table 16.3). Most of these products contain multiple Bt genes for managing underground rootworms.

Besides its high efficacy for corn borer control, pyramided maize is also very effective against some noctuid species

Table 16.3. Currently available major Bt maize products targeting above-ground lepidopteran pests in the USA.

Event[a]	Bt genes	Major target insects[b]	Examples of Bt maize products
Singe Bt gene products			
Bt 11	Cry1Ab	CB	Agrisure CB/LL, Agrisure GT/CB/LL, Agrisure 300GT, Agrisure CB/LL/RW, Agrisure Artesian 3011A
MON810	Cy1Ab	CB	YieldGard VT Triple
TC1507	Cry1F	CB, FAW	Herculex® I, Herculex® XTRA; Optimum® AcreMax®1, Optimum® TRIsect
Products containing two pyramided Bt genes			
MON89034	Cry1A.105, Cry2Ab2	CB, CEW, FAW	Genuity® VT Double PRO®, Genuity®VT Triple PRO®
Bt11 and MIR162	Cry1Ab, Vip3A	CB, CEW, FAW	Agrisure Viptera® 3110, Agrisure Viptera® 3111
Bt11 and TC1507	Cry1Ab, Cry1F	CB, CEW, FAW	Agrisure 3122, Agrisure Duracade ™5122
MON810 and TC1507	Cry1Ab, Cry1F	CB, CEW, FAW	Optimum®AcreMax®, Optimum®AcreMax® Xtra, Optimum®AcreMax® Xtreme, Optimum® Intrasect™, Optimum® Intrasect™ Xtra, Optimum® Intrasect™ Xtreme
Products containing three pyramided Bt genes			
MON89034 and TC1507	Cry1A.105, Cry2Ab2, Cry1F	CB, CEW, FAW	Genuity® SmartStax®, SmartStax®
Bt11, TC1507 and MIR162	Cry1Ab, Cry1F, Vip3A	CB, CEW, FAW	Agrisure Duracada™5222

[a]An event is a specific genetic modification in a specific species.
[b]CB, corn stalk borer species, which may include the European corn borer (*Ostrinia nubilalis*), southwestern corn borer (*Diatraea grandiosella*), sugarcane borer (*Diatraea saccharalis*), southern corn stalk borer (*Diatraea crambidoides*) and lesser corn stalk borer (*Elasmopalpus lignosellus*); CEW, corn earworm (*Helicoverpa zea*); FAW, fall armyworm (*Spodoptera frugiperda*).

(Burkness *et al.*, 2010; Siebert *et al.*, 2012; Reay-Jones and Reisig, 2014; Rule *et al.*, 2014) and so CEW and FAW are listed as targets for all pyramided products that control above-ground Lepidoptera (Table 16.3) (DiFonzo and Cullen, 2013). However, there are arguments about the economic benefits of using of Bt maize for controlling CEW, as some studies have shown that there are no significant relationships between maize yield and CEW injury (Reay-Jones and Reisig, 2014).

Since pyramided Bt maize became available in 2010, areas planted with single-gene products in the USA have been reduced rapidly. Although there are no official reports of the actual area planted with pyramided products, it is obvious that predominant type of Bt maize currently planted in the USA contains pyramided traits. Because of the increasing challenge of insect resistance, it is expected that single-gene Bt maize will be completely replaced by pyramided products in the near future.

16.3 Resistance Management for Bt Maize Targeting Above-ground Lepidopteran Pests

Resistance development in target pest populations is a great threat to the sustainable use of Bt maize technology (Tabashnik *et al.*, 2013). To delay the development of resistance, IRM plans are required for the planting of Bt maize in the USA. Basically, two different strategies are currently used for Bt maize IRM: (i) a 'high dose/refuge' strategy; and (ii) a gene pyramiding method.

16.3.1 High dose/refuge strategy

Since Bt maize was first planted in 1996, a high dose/refuge IRM strategy has been used in the USA (Matten *et al.*, 2012). This involves planting/sowing a portion of the maize to high-dose Bt maize that can kill insects carrying only one copy of the resistant gene/s (heterozygotes, RS) (US

EPA, 2001). The remaining maize crop is established with non-Bt maize that serves as a refuge for susceptible insects (SS). The relatively large (homozygote) SS populations from refuge plants should then mate with the rare surviving individuals carrying two copies of the resistant gene (homozygotes, RR). Thus, most of their offspring that carry resistance alleles will be RS. Because RS individuals will be killed by high-dose Bt plants, the frequencies of resistance alleles in pest populations should be maintained at low levels for a long period of time (Huang *et al.*, 2011).

The success of the high dose/refuge strategy depends on three key assumptions, that: (i) Bt maize hybrids express a 'high dose'; (ii) the initial frequency of resistance alleles is low (e.g. <0.001); and (iii) a sufficient SS refuge population is available in the nearby environment. A high-dose Bt maize means that the Bt protein concentrations within it are sufficiently high to kill ≥95% of RS individuals (US EPA, 2001). In other words, Bt resistance is functionally recessive. However, the high-dose qualification has not been directly evaluated for most target pests because resistance traits have not been found in those insects. A US Environmental Protection Agency (EPA) Scientific Advisory Panel defined a high dose as 25 times the protein concentration needed to kill SS individuals and recognized five methods to demonstrate that a transgenic crop expresses a high dose of Bt proteins (US EPA, 2001); a Bt plant must meet at least two of these five criteria to qualify as a high-dose type. This definition has been used to evaluate the high-dose qualification of Bt crops against corn borers and other target species.

Currently, the required non-Bt refuge maize is planted in two ways: (i) by structured planting; and (ii) by using a seed mixture. Before 2010, non-Bt refuge maize could only be planted structurally with the Bt maize in an area. For single-gene Bt maize, the high dose/refuge IRM plan requires growers to plant ≥20% structured non-Bt maize as a refuge in the Corn Belt, but they need to plant a minimum of 50% non-Bt maize in the southern region where cotton is planted. The structured non-Bt refuge should be planted within 800 m of the Bt maize field on every farm (Ostlie et al., 1997). Growers' compliance with the structured refuge requirements has been an issue. During the early years, a relatively high rate of compliance (e.g. 86–92%) was reported, but this has declined considerably in recent years (Smith et al., 2012).

Because of the compliance issue with the use of the 'structured refuge', in 2010 the US EPA approved a seed mixture refuge strategy (also called 'refuge-in-the-bag' or RIB) as an alternative for planting pyramid Bt maize in the US Corn Belt (Matten et al., 2012). This strategy has not been approved in the southern region. For RIB, a defined percentage of non-Bt maize seeds is mixed with Bt maize seeds in each bag by seed companies before it is sold to farmers. Hence, compliance will no longer be an issue. The currently used RIB is at a rate of 95% Bt mixed with 5% non-Bt seeds. Mathematical models show that RIB could be an effective strategy (Carroll et al., 2012), though scientific data to support the strategy are still very limited.

With the structured refuge strategy, the dispersal behaviour of adults is important, but with RIB, the major concern is larval movement among Bt and non-Bt plants. For example, movement of susceptible larvae from non-Bt refuge plants to Bt plants in an RIB could cause greater mortality to SS populations than in a structured refuge planting, and so result in a lower refuge population. More importantly, maize is a cross-pollinating crop in which most pollination results from pollen dispersed by wind and gravity (Burkness and Hutchison, 2012). Pollen movement from the surrounded Bt plants to the non-Bt refuge plants in an RIB planting could result in Bt expression in refuge kernels and thereby could directly kill susceptible refuges, especially for kernel feeders such as CEW. Furthermore, pollen movement could also create sublethal exposure and promote selection pressure for resistance by increas-ing the survival of RS individuals or of individuals carrying minor resistance alleles. All of these essential parameters are largely unknown. The lack of such crucial scientific data is a major reason that has prevented approval of the use of RIB in the southern USA.

16.3.2 Gene pyramiding

The second method currently used for Bt maize IRM is gene pyramiding. In order to be effective, the genes pyramided should lack cross-resistance so that resistance to one Bt protein is not also resistance to others (Zhao et al., 2003; Moar and Anilkumar, 2007). Currently, most pyramided Bt maize products are planted mixed with 5% non-Bt maize in the US Corn Belt (Matten et al., 2012). Compared with the use of single-gene Bt maize, modellings show that resistance development to pyramided Bt maize can be delayed considerably (Carroll et al., 2013).

Published empirical data support the use of gene pyramiding. Zhao et al. (2003) conducted a greenhouse study using an artificial population of diamondback moth, Plutella xylostella, carrying genes for resistance to Cry1Ac and Cry1C. After 24 generations of selection, resistance to pyramided two-gene plants was significantly delayed compared with the single-gene plants that were deployed in mosaics. Several recent studies have also shown that pyramided Bt maize products could be used for managing insect populations that have developed high levels of resistance to single Bt proteins. For example, SCB populations that are resistant to single-gene Cry1Ab maize are highly susceptible to the pyramided maize products Genuity® VT Double Pro™ or SmartStax™ (Wangila et al., 2012). Similarly, pyramided Bt maize is also effective for controlling Cry1F-resistant FAW (Niu et al., 2014). These results showcase gene pyramiding as a possibly useful tool for delaying resistance development.

16.4 Current Status of Bt Susceptibility of Above-ground Lepidopteran Targets in the USA

As mentioned above, there are several lepidopteran pests targeted by Bt maize in the USA (Tables 16.2 and 16.3). The available data show that corn borer populations in the USA are still susceptible to Bt maize. Pyramided Bt maize is also effective for CEW and FAW. Huang *et al.* (2011) analysed the possible reasons that have contributed to the long-term success of Bt maize for managing corn borers in the North America. They found that the fundamental assumptions of the 'high-dose/refuge' IRM strategy have been met for both ECB and SWCB. However, recent studies have shown that field resistance to single-gene Cry1F maize has occurred in FAW in south-eastern coastal areas of the USA. Below, an analysis is given of the status of Bt susceptibility/ resistance in five lepidopteran targets of Bt maize. Other above-ground lepidopteran targets are minor pests on which there is not much information available.

16.4.1 European corn borer

ECB is the most economically important corn borer and the primary target of Bt maize in the USA, and before the use of Bt maize, annual losses caused by ECB were estimated at US$1 billion (Ostlie *et al.*, 1997). Bt maize was first planted to control ECB in 1996, and its field performance has been excellent (Huang *et al.*, 1999), with no field control failures reported. Monitoring over 10 years (1995–2005) showed that all field populations evaluated have remained susceptible to Cry1Ab across the major maize production areas of the USA (Siegfried *et al.*, 2007). In the past several years, field populations of ECB have been maintained very low in both Bt and non-Bt maize fields (Bohnenblust *et al.*, 2014). A study by Hutchison *et al.* (2010) has shown that the nationwide suppression of ECB is associated with the use of Bt maize, which has offered considerable economic benefits to the US maize growers. In addition, >800 isofemale

lines and 131 ECB males collected nation-wide have been examined for Cry1Ab resistance using F_2 screens (Andow *et al.*, 1998, 2000; Bourguet *et al.*, 2003; Stodola *et al.*, 2006). No major resistance alleles were found in these populations. Based on the F_2 screens, resistance allele frequency to Cry1Ab in ECB is estimated to be <0.00086 with 95% probability in the USA. The results confirm that the frequency of the Bt resistance allele in ECB is very low in the USA and meets the rare resistance criterion required for the high dose/refuge strategy.

16.4.2 Southwestern corn borer

SWCB is a major corn borer in the central and southern USA (Trisyono and Chippendale, 2002; Castro *et al.*, 2004). Bt maize hybrids are also effective against SWCB (Huang *et al.*, 1999; Buschman *et al.*, 2001). The Bt susceptibility status of SWCB is similar to that of ECB in that no major Bt resistance genes have been found. All Bt maize products targeting Lepidoptera remain effective in controlling SWCB, although compared with ECB, the research associated with SWCB has been limited. An early study using dose-response bioassays indicated that the Cry1Ab susceptibility of the populations collected from several US states did not increase from 1998 to 2000 (Trisyono and Chippendale, 2002). A field population collected in Louisiana in 2005 also remained susceptible to both Cry1Ab protein and Cry1Ab maize plants (Huang *et al.*, 2006). In addition, >400 SWCB individuals collected in Louisiana in 2005 were examined for Cry1Ab resistance using an F_2 screen, but no major resistance was detected (Huang *et al.*, 2007). The results also indicate that Bt resistance allele frequency in SWCB is low and should meet the rare resistance requirement of the high dose/refuge strategy.

16.4.3 Sugarcane borer

SCB is a major corn borer in the mid-southern USA (Castro *et al.*, 2004). It is also a pest of sugarcane, rice and grain sorghum. Compared with ECB and SWCB, SCB is less susceptible to Cry1Ab protein (Huang et al., 2006), but Bt maize hybrids do still remain effective against SCB and field resistance has not occurred (Huang et al., 2012).

Since 2004, >3000 SCB individuals collected from Louisiana, Texas and Mississippi have been examined for Cry1Ab resistance using F_1/F_2 screens. The results showed that Cry1Ab resistance allele frequencies in SCB were low in the populations in Louisiana during 2004–2008 (0.0011), in Texas in 2007 (<0.0016) and in Mississippi in 2009 (<0.061) (Huang et al., 2009, 2012). Laboratory bioassays also showed that 18 field populations collected from Louisiana and Texas during 2004–2006 remained susceptible to Cry1Ab (Huang et al., 2008). However, the resistance frequency in the Louisiana populations increased significantly in 2009, reaching 0.0176 (Huang et al., 2012). Since 2010, SCB populations in maize fields as well as on sorghum and rice in the mid-southern USA have decreased considerably. The reason for this decline is not certain, but it may be related to the area-wide adoption of Bt maize as was reported for the decline of ECB in the North Corn Belt (Hutchison et al., 2010). If the decrease is due to the use of Bt maize, a conservative estimate of the net benefit from planting Bt maize to Louisiana maize growers alone is >US$20 million annually (F. Huang, unpublished data).

16.4.4 Corn earworm

CEW is considered to be the most costly crop pest in North America. The pest has >200 host plants, many of which are economically important crops, such as maize, cotton, grain sorghum and soybean. Damage to maize is mainly caused by larvae feeding on the ear kernels. In the US South, CEW is known to overwinter in the pupal stage, but it cannot usually overwinter in most areas of the Corn Belt. It is a highly mobile insect, capable of long-distance migration (US EPA, 2001). In the South, CEW moves to other hosts such as cotton, grain sorghum and

soybean for two to three additional generations after maize has matured. Because CEW is also a major target of Bt cotton in the South, it presents a significant challenge for IRM, as there is the potential for multiple exposures to Bt maize and Bt cotton across generations.

Because the first-generation Bt maize offers only partial control of CEW, this insect was not listed as a target of the single-gene Bt maize and so most studies on Bt resistance in CEW have been associated with Bt cotton (see Chapter 1, Tabashnik and Carrière; Chapter 2, Gao et al.; and Chapter 4, Van den Berg and Campagne; and other published articles). Because of the improved efficacy of the second-generation Bt maize, CEW is currently listed as a target for all pyramided Bt maize products (DiFonzo and Cullen, 2013). To date, all pyramided Bt maize remains effective for the control of CEW and no field resistance has been reported in the USA. None the less, there is a major concern in implementing RIB because of the cross-pollinating properties of maize, which can cause Bt proteins to be present in refuge maize kernels in seed mixtures. A recent study showed that cross-pollination in a 95:5% RIB planting of SmartStax™ caused >90% of refuge kernels to express more than one Bt protein (Yang et al., 2014). The expression of Bt proteins in the refuge ears reduced CEW survivorship to only 4.6%, a reduction of 88.1% relative to larval feeding on the ears of pure non-Bt maize plantings. The results demonstrate that at the currently implemented rate, the RIB approach will not be effective in providing refuge for ear-feeding pests such as CEW.

16.4.5 Fall armyworm

FAW is distributed throughout most of the USA and it is believed to be able to overwinter successfully only in south Florida and south Texas in the US mainland. It is also a well-known long-distance migratory insect. This pest also has a wide range of host plants. Similar to CEW, the FAW is also a target of pyramided Bt cotton in the USA.

Most single-gene Bt maize products are not very effective against FAW and so FAW is excluded from the target list for the 1st generation Bt maize except for the Cry1F maize. Unfortunately, after only a few years of intensive use of Cry1F maize in Puerto Rico, field resistance in FAW occurred in the territory (Storer et al., 2010). As a result, Cry1F maize was withdrawn from commercial use in Puerto Rico shortly after the resistance was identified. The pyramided Bt maize is more effective (Burkness et al., 2010; Siebert et al., 2012) and so FAW is listed as a target for this product (DiFonzo and Cullen, 2013).

One study reported that field populations of FAW collected in 2010–2011 in the US mainland, including Florida, were susceptible to Cry1F (Storer et al., 2012). In contrast, unexpected survival of FAW on Cry1F maize plants has been reported in recent years on several occasions in the south-eastern USA and Brazil (see also Chapter 3, Monnerat et al.). However, scientific documentation of field resistance has not been reported from anywhere except Puerto Rico (Storer et al., 2010). Since 2011, F_2 screens, diet-incorporated bioassays, greenhouse tests and field studies have been conducted in four south-eastern US states to determine whether this unexpected survival is due to resistance. F_2 screens showed that Cry1F resistance allele frequency in a FAW population collected in 2011 in south Florida reached 0.293 (Huang et al., 2014). Diet-incorporated bioassays showed that populations collected in 2012–2013 from Florida and North Carolina exhibited a significant level of Cry1F resistance. Reduced efficacy and control failure of Cry1F maize for FAW were also documented in field trials in south Florida. Another independent study also found that a population collected from Palm Beach, Florida, in 2011 showed a high Cry1F resistance allele frequency of 0.247 (Vélez et al., 2013). These results documented that field resistance to Cry1F maize has occurred in FAW in the south-east coastal areas of the US mainland.

It is necessary to develop mitigation strategies before this resistance becomes widespread. A study with leaf tissue bioassays showed that the Cry1F-resistant FAW also exhibited a significant level of cross-resistance to three of four pyramided Bt maize products tested (Niu et al., 2013), although the levels of cross-resistance observed in these bioassays were not sufficient enough to allow the resistant FAW to survive on whole plants of pyramided Bt maize (Niu et al., 2014). The results suggest that pyramided Bt maize can be used for managing the Cry1F-resistant FAW. Timely switching from single-gene to pyramided products in the USA should be helpful in maintaining the continued success of Bt maize for managing FAW, at least temporarily.

16.5 Prospects

Since 1996, >275 Mha of Bt maize have been planted in the USA and crop growers have gained considerable economic benefits from planting the crop, which will definitely encourage more countries and more crop growers to adopt the technology. In addition, transgenic RNA interference (RNAi) crops that express different insecticidal toxins are expected to become available soon. The rapid increase in global demand for food/feed and fuel energy makes the need for such technologies even more urgent. Therefore, it is widely expected that the adoption of Bt crops will continue to increase in the future. Thus, IRM will continue to be a great challenge for the sustainable use of those crops.

As in industrialized countries, the use of Bt crops has also delivered substantial benefits in developing countries (James, 2013). During 2013, >40 Mha of Bt crops were planted in 17 developing countries, which accounted for >50% of the world's total Bt crop area (Table 16.1), and as more Bt crops are planted in these countries, the implementation of effective IRM plans will be critical for the sustainability of the technology. The knowledge and experience that has been gained with Bt crops in the USA should be helpful for the implementation of Bt crop IRM in developing countries.

Note

This paper reports research results only. Mention of a proprietary product name does not constitute an endorsement for its use by Louisiana State University Agricultural Center. The article is published with the approval of the Director of the Louisiana Agricultural Experiment Station as manuscript number No. 2014-234-15702.

References

Andow, D.A., Alstad, D.N., Pang, Y.H., Bolin, P.C. and Hutchison, W.D. (1998) Using an F_2 screen to search for resistance alleles to *Bacillus thuringiensis* toxin in European corn borer (Lepidoptera: Crambidae). *Journal of Economic Entomology* 91, 579–584.

Andow, D.A., Olson, D.M., Hellmich, R.L., Alstad, D.N. and Hutchison, W.D. (2000) Frequency of resistance to *Bacillus thuringiensis* toxin Cry1Ab in an Iowa population of European corn borer (Lepidoptera: Crambidae). *Journal of Economic Entomology* 93, 26–30.

Bohnenblust, E.W., Breining, J.A., Shaffer, J.A., Fleischer, S.J., Roth, G.W. *et al.* (2014) Current European corn borer, *Ostrinia nubilalis*, injury levels in the northeastern US and the value of Bt field corn. *Pest Management Science* 70, 1711–1719.

Bourguet, D., Chaufux, J., Séguin, M., Buisson, C., Hinton, J.L. *et al.* (2003) Frequency of alleles conferring resistance to Bt maize in French and US corn belt populations of the European corn borer, *Ostrinia nubilalis*. *Theoretical and Applied Genetics* 106, 1225–1233.

Buntin, G.D., All, J.N., Lee, R.D. and Wilson, D.M. (2004) Plant-incorporated *Bacillus thuringiensis* resistance for control of fall armyworm and corn earworm (Lepidoptera: Noctuidae) in corn. *Journal of Economic Entomology* 97, 1603–1611.

Burkness, E.C. and Hutchison, W.D. (2012) Bt pollen dispersal and Bt kernel mosaics: integrity of non-Bt refugia for lepidopteran resistance management in maize. *Journal of Economical Entomology* 105, 1477–1870.

Burkness, E.C., Dively, G., Patton, T., Morey, A.C. and Hutchison, W.D. (2010) Novel Vip3A *Bacillus thuringiensis* (Bt) maize approaches high-dose efficacy against *Helicoverpa zea* (Lepidoptera: Noctuidae) under field conditions. *GM Crops and Food* 1, 1–7.

Buschman L., Sloderbeck, P. and Witt, M. (2001) Efficacy of Cry1F corn for the control of southwestern corn borer and corn earworm. In: *2001 Southwest Research-Extension Center Field Day, Kansas State University. Report of Progress* 877. Garden City, Kansas, pp. 67–70.

Carroll, M.W., Head, G. and Caprio, M. (2012) When and where a seed mix refuge makes sense for managing insect resistance to plants. *Crop Protection* 38, 74–79.

Carroll, M.W., Head, G., Caprio, M. and Stork, L. (2013) Theoretical and empirical assessment of a seed mix refuge in corn for southwestern corn borer. *Crop Protection* 49, 58–65.

Castro, B.A., Riley, T.J., Leonard, B.R. and Baldwin, J. (2004) Borers galore: emerging pest in Louisiana corn, grain sorghum and rice. *Louisiana Agriculture* 47, 4–6.

Chilcutt, C.F., Odvody, G.N., Correa, J.C. and Remmers, J. (2007) Effects of *Bacillus thuringiensis* transgenic corn on corn earworm and fall armyworm (Lepidoptera: Noctuidae) densities. *Journal of Economic Entomology* 100, 327–334.

Dhurua, S. and Gujar, G.T. (2011) Field-evolved resistance to Bt toxin Cry1Ac in the pink bollworm, *Pectinophora gossypiella* (Saunders) (Lepidoptera: Gelechiidae), from India. *Pest Management Science* 67, 898–903.

DiFonzo, C. and E. Cullen, E. (2013) Handy Bt Trait Table. Department of Entomology, University of Wisconsin–Madison, Madison, Wisconsin. Available at: http://corn.agronomy.wisc.edu/Management/pdfs/Handy_Bt_Trait_Table.pdf (accessed 29 October 2014).

Dunlop, G. (2013) *Bt Corn IRM Compliance in Canada. 2013 Canadian Corn Pest Coalition Report 13.* Available at: http://www.cornpest.com/index.cfm/news-archive/2013-bt-corn-irm-compliance-study-ccpc-final-report/ (accessed 29 October 2014).

Gassmann, A.J., Petzold-Maxwell, J.L., Keweshan, R.S. and Dunbar, M.W. (2011) Field-evolved resistance to Bt maize by western corn rootworm. *PLoS One* 6(7): e22629.

Huang, F., Higgins, R.A. and Buschman, L.L. (1999) Transgenic Bt plants: successes, challenges, and strategies. *Pestology* 23, 2–29.

Huang, F., Leonard, B.R. and Gable, R.H. (2006) Comparative susceptibility of European corn borer, southwestern corn borer, and sugarcane borer (Lepidoptera: Crambidae) to Cry1Ab protein in a commercial Bt-corn hybrid. *Journal of Economic Entomology* 99, 194–202.

Huang, F., Leonard, B.R., Cook, D.R., Lee, D.R., Andow, D.A. *et al.* (2007) Frequency of alleles conferring resistance to *Bacillus thuringiensis* maize in Louisiana populations of southwestern corn borer (Lepidoptera: Crambidae). *Entomologia Experimentalis et Applicata* 122, 53–58.

Huang, F., Leonard, B.R., Moore, S.H., Yue, B., Parker, R. *et al.* (2008) Geographical susceptibility of Louisiana and Texas populations of sugarcane borer, *Diatraea saccharalis* (F.) (Lepidoptera: Crambidae) to *Bacillus thuringiensis* Cry1Ab protein. *Crop Protection* 27, 799–806.

Huang, F., Parker, R., Leonard, B.R., Yang, Y. and Liu, J. (2009) Frequency of resistance alleles to *Bacillus thuringiensis*-corn in Texas populations of the sugarcane borer, *Diatraea saccharalis* (F.) (Lepidoptera: Crambidae). *Crop Protection* 28, 174–180.

Huang, F., Andow, D.A. and Buschman, L.L. (2011) Success of the high dose/refuge resistance management strategy after 15 years of Bt crop use in North America. *Entomologia Experimentalis et Applicata* 140, 1–16.

Huang, F., Ghimire, M.N., Leonard, B.R., Daves, C.D., Levy, R. *et al.* (2012) Extended monitoring of resistance to *Bacillus thuringiensis* Cry1Ab maize in *Diatraea saccharalis* (Lepidoptera: Crambidae). *GM Crops and Food* 3, 245–254.

Huang, F., Qureshi, J.A., Meagher, R.L. Jr, Reisig, D.D., Head, G.P. *et al.* (2014) Cry1F resistance in fall armyworm *Spodoptera frugiperda*: single gene versus pyramided Bt maize. *PLos ONE* 9(11): e112958.

Hutchison, W.D., Burkness, E.C., Mitchell, P.D., Moon, R.D., Leslie, T.W. *et al.* (2010) Areawide suppression of European corn borer with Bt maize reaps savings to non-Bt maize growers. *Science* 330, 222–225.

James, C. (2013) *Global Status of Commercialized Biotech/GM Crops: 2013*. ISAAA Brief 46, International Service for the Acquisition of Agri-Biotech Applications, Ithaca, New York.

Matten, S.R., Frederick, R.J. and Reynolds, A.H. (2012) United States Environmental Protection Agency insect resistance management programs for plant-incorporated protectants and use of simulation modeling. In: Wozniak, C.A. and McHughen, A. (eds) *Regulation of Agricultural Biotechnology: The United States and Canada*. Springer, Dordrecht, The Netherlands, pp. 175–267.

Moar, W.J. and Anilkumar, K.J. (2007) The power of the pyramid. *Science* 318, 1561–1562.

NASS (1996–2013) Annual reports of crop acreage. US Department of Agriculture National Agricultural Statistics Service, Washington, DC. Available at: http://www.nass.usda.gov/Statistics_by_Subject/index.php?sector=CROPS (accessed 29 October 2014).

NASS (2013) *Acreage*, June 2013. Released June 28, 2013, by the National Agricultural Statistics Service (NASS), Agricultural Statistics Board, United States Department of Agriculture, Washington, DC. Available at: http://usda.mann lib.cornell.edu/usda/nass/Acre//2010s/2013/ Acre-06-28-2013.pdf (accessed 4 November 2014).

National Corn Growers Association (2013) *The World of Corn: Unlimited Possibilities*. Washington, DC.

Niu, Y., Meagher R.L. Jr, Yang, F. and F. Huang, F. (2013) Susceptibility of field populations of the fall armyworm (Lepidoptera: Noctuidae) from Florida and Puerto Rico to purified Cry1F protein and corn leaf tissue containing single and pyramided Bt genes. *Florida Entomologist* 96, 701–713.

Niu, Y., Yang, F., Dangal, V. and Huang, F. (2014) Larval survival and plant injury of Cry1F-susceptible, -resistant, and -heterozygous fall armyworm (Lepidoptera: Noctuidae) on non-Bt and Bt corn containing single or pyramided genes. *Crop Protection* 59, 22–28.

Ostlie, K.R., Hutchinson, W.D. and Hellmich, R.L. (eds) (1997) *Bt Corn and European Corn Borer: Long-term Success through Resistance Management*. North Central Regional Extension Publication NCR 602, University of Minnesota Extension, East Lansing, Michigan. Updated version available at: http://www.extension. umn.edu/agriculture/corn/pest-management/ bt-corn-and-european-corn-borer/ (accessed 29 October 2014).

Reay-Jones, F.P.F. and Reisig, D.D. (2014) Impact of corn earworm injury on yield of transgenic corn producing Bt toxins in the Carolinas. *Journal of Economic Entomology* 107, 1101–1109.

Reisig, D. and Roberson, S. (2011) Transgenic Bt corn. Department of Entomology, College of Agriculture and Life Sciences, North Carolina State University, Raleigh, North Carolina. Available at: http://www.ces.ncsu.edu/plymouth/ ent/btcorn.html (accessed 29 October 2014).

Rule, D.M., Nolting, S.P, Prasifka, P.L., Storer, N.P., Hopkins, B.W. *et al.* (2014) Efficacy of pyramided Bt proteins Cry1F, Cry1A.105, and Cry2Ab2 expressed in SmartStax corn Hybrids against lepidopteran insect pests in the northern United States. *Journal of Economic Entomology* 107, 403–409.

Siebert, M.W., Nolting, S.P., Hendrix, W., Dhavala, S., Craig, C. *et al.* (2012) Evaluation of corn hybrids expressing Cry1F, Cry1A.105, Cry2Ab2, Cry34Ab1/Cry35Ab1, and Cry3Bb1 against southern United States insect pests. *Journal of Economic Entomology* 105, 1825–1834.

Siegfried, B.D., Spencer, T., Crespo, A.L., Storer,

N.P., Head, G.P. *et al.* (2007) Ten years of Bt resistance monitoring in the European corn borer: what we know, what we don't know, and what we can do better. *American Entomologist* 53, 208–214.

Smith, M.J., Storer, N.P., Garden, J. and Guyer, D. (2012) *2011 Insect Resistance Management (IRM) Compliance Assurance Program Report for Corn Borer-protected Bt Corn, Corn Rootworm-protected Bt Corn, Corn Borer/Corn Rootworm-protected Stacked and Pyramided Bt Corn.* US Environmental Protection Agency, Washington, DC.

Stodola, T.J., Andow, D.A., Hyden, A.R., Hinton, J.L., Roark, J.J. *et al.* (2006) Frequency of resistance to *Bacillus thuringiensis* toxin Cry1Ab in Southern US Corn Belt populations of European corn borer (Lepidoptera: Crambidae). *Journal of Economic Entomology* 99, 502–507.

Storer, N.P., Babcock, J.M., Schlenz, M., Meade, T., Thompson, G.D. *et al.* (2010) Discovery and characterization of field resistance to Bt maize: *Spodoptera frugiperda* (Lepidoptera: Noctuidae) in Puerto Rico. *Journal of Economic Entomology* 103, 1031–1038.

Storer, N.P., Kubiszak, M.E., King, J.E., Thompson, G.D. and Santos, A.C. (2012) Status of resistance to Bt maize in *Spodoptera frugiperda*: lessons from Puerto Rico. *Journal of Invertebrate Pathology* 110, 294–300.

Tabashnik, B.E., Brévault, T. and Carrière, Y. (2013) Insect resistance to Bt crops: lessons from the first billion acres. *Nature Biotechnology* 31, 510–521.

Taylor, M.R. and Tick, J. (2001) *The StarLink Case: Issues for the Future.* Resources for the Future/Pew Initiative on Food and Biotechnology, Washington, DC. Available at: http://rff.org/RFF/Documents/RFF-RPT-StarLink.pdf (accessed 29 October 2014).

Trisyono, Y.A. and Chippendale, G.M. (2002) Susceptibility of field-collected populations of the southwestern corn borer, *Diatraea grandiosella*, to *Bacillus thuringiensis*. *Pest Management Science* 58, 1022–1028.

US EPA (2001) *Biopesticides Registration Action Document – Bacillus thuringiensis Plant-incorporated Protectants.* Biopesticides and Pollution Prevention Division, US Environmental Protection Agency, Washington, DC. Available at: http://www.epa.gov/oppbppd1/biopesticides/pips/bt_brad.htm (accessed 29 October 2014).

US EPA (2005) *Biopesticide Registration Action Document – Bacillus thuringiensis Cry1F Corn, August 2005, Updated August 2005.* US Environmental Protection Agency, Washington, DC. Available at: bch.cbd.int/database/attachment/?id=10711 (accessed 29 October 2014).

van Rensburg, J.B.J. (2007) First report of field resistance by the stem borer, *Busseola fusca* (Fuller) to Bt-transgenic maize. *South African Journal of Plant and Soil* 24, 147–151.

Vélez, A.M., Spencer, T.A., Alves, A.P., Moellenbeck, D., Meagher, R.L. *et al.* (2013) Inheritance of Cry1F resistance, cross-resistance and frequency of resistant alleles in *Spodoptera frugiperda* (Lepidoptera: Noctuidae). *Bulletin of Entomological Research* 103, 700–713.

Vilella, F.M.F., Waquil, J.M., Vilella, E.F., Viana, P.A., Lynch, R.E. *et al.* (2002) Resistance of Bt transgenic maize to lesser cornstalk borer (Lepidoptera: Pyralidae). *Florida Entomologist* 85, 652–653.

Wangila, D.S., Leonard, B.R., Bai, Y., Head, G.P. and Huang, F. (2012) Larval survival and plant injury of Cry1Ab-susceptible, -resistant, and -heterozygous genotypes of the sugarcane borer on transgenic corn containing single or pyramided Bt genes. *Crop Protection* 42, 108–115.

Yang, F., Kerns, D.L., Head, G.P., Leonard, B.R., Levy, R. *et al.* (2014) A challenge for the seed mixture refuge strategy in Bt maize: impact of cross-pollination on an ear-feeding pest, corn earworm. *PLos ONE* 9(11): e112962.

Zhao, J.Z., Cao, J., Li, Y., Collins, H.L., Roush, R.T. *et al.* (2003) Transgenic plants expressing two *Bacillus thuringiensis* toxins delay insect resistance evolution. *Nature Biotechnology* 21, 1493–1497.

17 Insect Resistance Management and Integrated Pest Management for Bt Crops: Prospects for an Area-wide View

William D. Hutchison*

Department of Entomology, University of Minnesota, St Paul, Minnesota, USA

Summary

Throughout this book, several authors have reviewed the pest resistance challenges within the context of the use of genetically modified (GM) crops, the solutions that are necessary to mitigate the evolution of insect pest resistance and the continued need for effective insect resistance management (IRM). Clearly, the current selection pressure has resulted from the extensive adoption of GM crops by millions of farmers worldwide due, in part, to their real or perceived benefits. Many of the benefits of GM maize and cotton have been well documented. They include increased yields, reduced yield variability, increased economic returns to farmers, reductions in insecticide use, reductions in pesticide exposure to farm workers, the subsequent conservation of beneficial insects, and the environmental benefits resulting from less tillage. These benefits, however, are not universal for all GM crops, and several important insect pests have exhibited 'field-evolved resistance' to crops engineered to express toxins from *Bacillus thuringiensis* (Bt). In several cases, 'practical resistance' and yield losses have been confirmed. From these case studies, we can review the factors that often contribute to resistance and thereby develop more proactive IRM programmes that are not only compatible with integrated pest management (IPM), but are fully integrated with IPM. Several compatible IPM tactics, such as biological control and cultural controls (e.g. crop rotation) can be quite effective with Bt crops and should further reduce selection pressure. Within the context of IRM and IPM, a primary question remains: what can we learn from the case studies of Bt resistance versus longer term success in order to better design future IRM plans? In this chapter, the benefits of Bt crops are reviewed as a basis for understanding grower decision making and the rationale for Bt crop use. I then summarize the key IRM elements that are necessary to facilitate the sustainability of Bt crop use within an IPM context, and subsequently review several reasons why the goals of IRM and IPM may be best understood and implemented from a landscape, or area-wide, management perspective.

17.1 Introduction

Ever since Melander observed the resistance of San José scale to lime sulfur (Melander, 1914), the 100 year history of insect resistance to insecticides has taken a fascinating, if not frustrating, path to seek

* Corresponding author. E-mail address: hutch002@umn.edu

new ways to delay resistance, and to develop alternative and more ecologically based pest management approaches, particularly for agricultural pests and for pests that transmit human disease (Perkins, 1982; Onstad, 2014). The early response to excessive use of insecticides was also met with concerns from the public, and culminated in publication of the seminal book *Silent Spring* (Carson, 1962). By the 1950s, the problems of resistance and loss of biological control in agricultural crops prompted entomologists in California to publish the well-known treatise on 'The integrated control concept' (Stern *et al.*, 1959). The concepts proposed in this paper became a precursor to the foundations of integrated pest management (IPM) that are still relevant today. Indeed, various manifestations of IPM have been developed and implemented worldwide over the past 40 years (Pedigo and Rice, 2006; Radcliffe *et al.*, 2009; Gray, 2011; Peshin and Pimentel, 2014). Likewise, and by necessity as insecticide resistance issues continue to be documented, particularly for arthropod pests, the need for proactive insect resistance management (IRM) has never been more compelling (Onstad, 2014; Tabashnik *et al.*, 2014).

Throughout this book, several authors have reviewed the pest resistance challenges within the context of the use of genetically modified (GM) crops, solutions that are necessary for mitigating the evolution of insect pest resistance and the continued need for effective IRM. Clearly, the foundation for continued selection pressure for several important insect pests has resulted from the wide-scale adoption of GM crops, due in part to the extensive list of their benefits as experienced by millions of farmers worldwide (James, 2013; Barfoot and Brookes, 2014). Many of the benefits, for GM maize and cotton for example, have been well documented. They include: increased yields, reduced yield variability, increased economic returns to farmers, reductions in insecticide use, reductions in farm worker exposure to pesticides, the subsequent conservation of beneficial insects and natural enemies, and other environmental benefits resulting from less

use of tillage (Shelton *et al.*, 2002; Qaim and Zilberman, 2003; Naranjo *et al.*, 2008; Romeis *et al.*, 2008; Naranjo, 2011; Lu *et al.*, 2012; Qaim and Kouser, 2013; Shi *et al.*, 2013; Barfoot and Brookes, 2014). In a comprehensive study of the environmental benefits of all GM crops for the period 1996–2012, Barfoot and Brookes (2014) also reported that total use of pesticide sprays had been reduced by 503 million kg (−8.8%) and that, as a result, there is a decreased environmental impact associated with herbicide and insecticide use on GM crops of 18.7%, as measured by the Environmental Impact Quotient (EIQ) indicator. The technology has also facilitated a significant reduction in the release of greenhouse gas emissions from GM cropping areas. Since 1996, the benefits experienced at farm and regional levels, have resulted in continued demand and an increase in grower adoption of these crops. As of 2013, a record 18 million farmers planted ~175 million ha of GM crops, with 54% of the global GM production now occurring in developing countries (James, 2013). GM crop use in developing countries first exceeded that of industrial nations in 2012 (James, 2013).

Despite the many benefits of GM crops, concerns about insect pest resistance to plants engineered to express toxins produced by the bacterium *Bacillus thuringiensis* (Bt), continue to dominate the future outlook (Gassmann and Hutchison, 2012; Siegfried and Hellmich, 2012; Tabashnik *et al.*, 2013; 2014; Onstad, 2014). Concerns continue because expanding Bt crop use rates create high selection pressure against the targeted pest species, and not all Bt crops necessarily meet the preferred criteria for successful IRM programmes, which include a 'high dose' of Bt toxin for all insect pests. Conversely, these concerns are juxtaposed against the desire of growers, crop advisors and industry to achieve long-term sustainability of Bt crops, while avoiding the fate of many conventional insecticides over the past 60 years (Fleischer *et al.*, 2014; Tabashnik *et al.*, 2014).

To avoid or delay the evolution of insect resistance to Bt crops, and for long-term sustainability, the technology should not be viewed as the next 'silver bullet' for insect control, but be deployed as one element of a broader IPM scheme, within which multiple IPM tactics are truly integrated with other complementary tactics (Kennedy, 2008; Radcliffe et al., 2009; Gray, 2011; Naranjo, 2011; Fleischer et al., 2014). Many complementary tactics are indeed available for Bt crop–pest systems, including: biological control (Romeis et al., 2008; Liu et al., 2014; White et al., 2014) and cultural controls such as crop rotation and the manipulation of planting dates to avoid pest pressure (Pedigo and Rice, 2006; Gassmann et al., 2012). Most recently, in the southwestern USA, a novel area-wide sterile-insect release programme was integrated successfully with Bt cotton for the control of pink bollworm (Pectinophora gossypiella) (Tabashnik et al., 2012). Whether intended or not, growers who integrate Bt crops with other compatible pest management tactics (and unique modes of action), are generally more successful in extending the efficacy of Bt traits (e.g. Romeis et al., 2008). In some cases, reduced insecticide use has also enhanced IPM options for non-target or secondary pests, as with Bt cotton in the south-western USA (Naranjo et al., 2008; Naranjo, 2011).

Within the context of IRM and IPM, there is a primary question: what can we learn from case studies of Bt resistance, versus longer term success, to enable the better design of future IRM plans? In this chapter, I summarize the benefits of Bt crops, as this provides insights on grower decision making and conveys the rationale for their use. I then summarize key IRM elements necessary for the sustainability of Bt crops within an IPM context and subsequently review several reasons why the goals of IRM and IPM may be best understood from a landscape, or area-wide, management perspective.

17.2 Benefits of Bt Crops and Influence on Grower Decision Making

During the early commercialization of Bt maize in the USA, two important farmer surveys were conducted to measure the benefits of the technology, or the perceived benefits anticipated by growers, during early adoption. During the first 3 years of Bt maize use in the Midwestern USA, Pilcher et al. (2002) found that most growers rapidly reduced insecticide use for the primary targeted pest, the European corn borer (Ostrinia nubilalis), with the interesting finding that most growers had under-estimated the yield losses due to this pest. For example, insecticide use for the pest declined (over five states) from 25% spraying at least once a year in 1996 to 14% using insecticide by 1998, just 3 years after introduction of the first Bt maize cultivars in this area. Importantly, although Bt maize was commercialized in the USA in 1996, there was limited seed available in that initial year. For the first time, season-long control of the corn borer had provided yield protection not previously experienced with conventional scouting and insecticide use. Given the history of yield losses in several states, most growers were willing to plant Bt hybrids without knowing the degree of pest pressure in the next growing season, and the mean adoption rate of Bt maize increased from 16% in 1996 to 40% by 1998. A subsequent farmer survey in 2001 (Wilson et al., 2005) for the same Midwest region found that the two main advantages of Bt maize, from a farmer viewpoint, included less farmer exposure to insecticide (69.9%), and less insecticide in the environment (68.5%). In addition, across all states, the survey indicated that approximately 21% of the farmers indicated they had experienced yield increases of ~40 bu ha^{-1}, given that (for example) average yields during this period were ~375–400 bu ha^{-1} (Wilson et al., 2005).

In subsequent surveys for both Bt maize and Bt cotton, many of the benefits experienced by the early adopters have continued, including increased yields, improved consistency of yields, reduced foliar insecticide use and reduced need to handle conventional insecticides. Subsequent surveys based on actual Bt maize and Bt cotton use conducted over the past 15–17 years, confirmed many of these anticipated benefits by farmers (e.g. Romeis et al., 2008; Naranjo, 2011; Barfoot and Brookes, 2014). Given the high adoption rates of Bt maize and Bt cotton in both developed and developing countries, it would be informative to conduct additional research in order to better understand current farmer attitudes to GM crop technology, and how these views influence farmer decision making to plant or not to plant GM crops. This information can be quite valuable in designing IRM programmes that also account for the logistical aspects that farmers must deal with to implement IRM and IPM at the farm level (Mitchell and Hutchison, 2009; Onstad, 2014).

Some of the benefits of GM field crops, particularly less insecticide use, are also substantial for selected GM vegetable crops (Shelton et al., 2008), such as Bt sweetcorn and Bt brinjal, two crops that historically receive high levels of insecticide use (e.g. Shelton et al., 2008; Burkness et al., 2011). As noted by Shelton et al. (2013), and as of 2010, global insecticide sales data indicate that ~45% of all insecticide purchased is used on fruit and vegetable crops, whereas insecticide use is much lower on cotton (14.1%) and maize (7.6%). Going forward, it seems logical that additional Bt vegetable crops would benefit by further reductions in conventional insecticide use, and thereby further contribute to food security and global nutritional needs (Park et al., 2011; James, 2013; Qaim and Kouser, 2013).

17.3 Insect Resistance to Bt Crops

Before we can fully discuss the concerns about Bt resistance, it is necessary to determine how resistance itself should be defined, particularly within the context of insect resistance to Bt crops. Clear definitions are not only necessary to objectively measure a population change in pest susceptibility to Bt crops, they are also necessary to generate a consistent response from regulatory agencies. For example, the US Environmental Protection Agency (EPA) requires resistance monitoring methodology and resistance criteria to know how to respond to specific pest resistance incidents

(Matten et al., 2008; Onstad, 2014). Developing clear, accurate and operational definitions for resistance to pesticides is not a trivial matter; the topic has been the source of much debate in IRM circles for at least the past 30 years (e.g. Brent, 1986; Tabashnik et al., 2014). Based on the historical context of resistance definitions highlighted by Brent (1986), and more recent discussions about Bt crop case studies, Tabashnik et al. (2014) recommended definitions for 50 terms related to resistance and IRM. The definitions are equally applicable to arthropod resistance to conventional pesticides and to insect resistance to Bt crops. Moreover, the definitions that follow are not limited to pest species, but are also applicable to levels of resistance in certain strains of beneficial predators and parasitoids.

The primary definitions most germane to the discussion of IRM for Bt crops include:

1. Resistance: a 'genetically based decrease in susceptibility to a pesticide'.
2. Laboratory-selected resistance: 'a genetically based decrease in susceptibility to a pesticide in a population caused by exposure of the population to the pesticide in the laboratory'.
3. Field-evolved resistance: 'a genetically based decrease in susceptibility to a pesticide in a population caused by exposure of the population to the pesticide in the field'.
4. Practical resistance: 'field-evolved resistance that reduces pesticide efficacy and has practical consequences for pest control'.

Tabashnik et al. (2014) further defined the worst case scenario of practical resistance, in which >50% of individuals in a population show resistance (e.g. via a discriminatory dose), as well the requirement that reduced efficacy in the field has been reported. Given this criterion, since the inception of Bt crops, there have been five species within which some populations can be defined as having practical resistance concerns as of 2014. These are: the maize stalk borer, *Busseola fusca* (Cry1Ab, maize) in South Africa (Kruger et al., 2011); western corn rootworm, *Diabrotica virgifera* in the USA (Gassmann et al., 2011, 2012); the corn earworm, *Helicoverpa zea* (Cry1Ac, cotton) in the USA; (Tabashnik *et al.*, 2014); pink bollworm (Cry1Ac, cotton) in India (Dennehy *et al.*, 2010); and fall armyworm, *Spodoptera frugiperda* in Puerto Rico (Storer *et al.*, 2012a) and the USA (see Tabashnik *et al.*, 2014, and references therein). A new report of Bt resistance in *S. frugiperda* in Brazil has also recently been published (Farias *et al.*, 2014).

Conversely, several targeted insect pests continue to remain susceptible to Bt toxins, in both Bt maize and Bt cotton, with no evidence to date of sustained field-evolved resistance in any of these species. Insect species that have remained susceptible to Bt traits for several years following commercialization, include: the European corn borer in the USA (Hutchison *et al.*, 2010; Siegfried and Hellmich, 2012); the Asian corn borer (*Ostrinia furnacalis*) in the Philippines (Alcantara *et al.*, 2011); the spotted stem borer (*Chilo partellus*) in South Africa (J. Van den Berg, personal communication); pink bollworm in the southwest USA (Tabashnik *et al.*, 2011, 2012); and the tobacco budworm (*Heliothis virescens*) in the southern USA (Blanco 2012).

17.4 Insect Resistance Management (IRM) for Bt Crops

By far the most common IRM strategy developed for Bt crops has been the HDR strategy (e.g. Romeis *et al.*, 2008; Tabashnik *et al.*, 2013), which essentially consists of the following requirements: (i) Bt plants are engineered to produce high-dose concentrations of one or more Bt insecticidal toxins; and (ii) farmers are expected to maintain a certain portion of non-Bt plants, usually on each farm, to produce non-Bt-selected individuals and, thereby, over time, produce a high ratio of non-Bt-selected adults that can then mate with any potential Bt-selected adult survivors that may emerge from the Bt portion of a given field or farm (Matten *et al.*, 2008; Onstad, 2014; Sappington, 2014). For these requirements, critical assumptions of the HDR strategy

include the following: (i) resistant allele(s) are very rare (e.g. $<10^{-4}$); (ii) random mating among non-Bt-selected and Bt-selected adults, implying the need for knowledge about the adult dispersal range of the targeted pest species relative to Bt crop field and farm size; (iii) if resistance genes are present in a given pest population, the inheritance of those genes is functionally recessive; and (iv) the 'high dose' is a true high dose for the key, targeted insect pests (Tabashnik *et al.*, 2012, 2013). As noted elsewhere in this volume, and in recent review articles, the HDR strategy has been quite successful thus far for most Bt crop–pest systems where these critical assumptions are met and, in some cases, assisted by the fact that there may be several additional non-Bt host crops for a given insect pest, or weed hosts, that also serve as 'natural' refuge areas (Siegfried and Hellmich 2012; Storer *et al.*, 2012b; Tabashnik *et al.*, 2013). In some cases, resistance has also been delayed because resistant alleles also imparted fitness costs. However, where resistant alleles appeared to not be rare, as judged by multiple and geographically dispersed resistance events, resistance has evolved quickly in both the laboratory and the field.

Tabashnik *et al.* (2013, 2014) reviewed the primary factors most likely to be responsible for the evolution of Bt resistance in selected target species. These include: (i) lack of a true high dose among one or more Bt toxins against the selected pest (e.g. Cry3Bb1 and western corn rootworm); (ii) lack of sufficient 'structured non-Bt crop refuge' (and/or lack of natural non-Bt crop plants), as often required by regulatory authorities (e.g. the US EPA, Matten *et al.*, 2008); (iii) lack of random mating between selected and non-selected adult populations, resulting in more rapid evolution of resistance; and (iv) that the genetics of resistance is not functionally recessive.

It is important to note that the consistency of the first factor listed above, a high-dose strategy, is dependent upon the technology providers (industry or university scientists) and the degree to which Bt (or alternative) toxins are developed to achieve and maintain a high dose under field conditions. In the field, a high dose generally assumes that there is ≥99% control of the targeted pest, even when the population harbours a resistance gene in the heterozygote state (see Burkness et al., 2011). For regulatory purposes, as applied to Bt maize and Bt cotton, the high dose was initially defined by the US EPA as the dose expressed by a transgenic plant that is at least 25× higher than the known LC_{50}, the lethal concentration necessary to kill 50% of a test population (Matten et al., 2008; Siegfried and Hellmich, 2012). Unfortunately, for several Bt crops, the high-dose assumption has not been met, including western corn rootworm in the USA (Gassmann et al., 2012) and pink bollworm in India (Tabashnik et al., 2014). Examples of insufficient Bt refuge area (i.e. a structured area of non-Bt crop) have probably occurred in South Africa for the African (or cotton) bollworm (Helicoverpa armigera), in India for pink bollworm, and in the USA for western corn rootworm, when Bt maize is grown year after year in the same field (Gassmann et al., 2012). Another way to address the lack of a high-dose Bt toxin is the addition of 'pyramided' Bt traits (Burkness et al., 2011; Siegfried and Hellmich, 2012). Indeed, in the USA and for several other countries, many of the Bt maize hybrids and Bt cotton varieties are now sold with two or more Bt traits (Romeis et al., 2008; Tabashnik et al., 2014). The purpose of multi-trait plants is to either provide two or more Bt toxins with unique modes of action against the primary pest species ('pyramided' traits), or to expand Bt efficacy to a broader array of pest species ('stacked' traits) (Siegfried and Hellmich, 2012; Tan et al., 2013; Tabashnik et al., 2013, 2014).

If any of the high-dose or refuge assumptions are compromised, additional aspects of the population biology of the target pest can accelerate the evolution of resistance. For example, limited dispersal of adults emerging from a Bt field can be conducive to non-random mating, in which survivors with genes for resistance are most likely to mate with each other; this scenario is likely to have occurred with western corn rootworm in the USA (Gassmann et al., 2012). Even if a pest is more mobile, and can readily disperse among multiple fields or farm units (e.g. most lepidopteran species), if the initial response to a Bt crop results in larval developmental delays (e.g. H. zea on Cry1Ab maize), it is likely that survivors from Bt crops will emerge later than Bt-susceptible adults emerging from non-Bt fields, thus increasing the odds of Bt-selected survivors mating with other Bt survivors. Tabashnik et al. (2014) provided additional examples where the HDR assumptions are not met for specific pest species.

Understanding the risk of pest resistance is often addressed by population genetic modelling studies that strive to proactively assess the primary factors that influence the evolution of pest resistance (e.g. Gould, 1998; Tabashnik et al., 2013; Onstad, 2014), and laboratory studies to elucidate the inheritance of resistance and mechanisms of resistance, such as unique modes of action among insect midgut binding sites (e.g. Siegfried and Hellmich, 2012; Tabashnik et al., 2013; Onstad, 2014). Models of pest resistance, designed to examine the assumptions of the HDR strategy have been useful in identifying research gaps in pest ecology, as well as in informing IRM policy, given the best information known about Bt crop–pest interactions (Shelton et al., 2002; Tabashnik et al., 2012). Complementary early resistance monitoring programmes are also conducted to track potential changes in the susceptibility of pest species to one or more Bt toxins (Andow et al., 1998; Venette et al., 2000, 2002; Siegfried and Hellmich, 2012), or to estimate the initial frequency of resistance in pest populations (Andow et al., 1998). Results from each level of research are necessary to develop and to inform the development and refinement of IRM plans in the field. At each level of investigation, entomologists, risk assessors and industry personnel play active roles in informing biotechnology regulatory professionals and/ or government policy makers and legislators, as Bt crops are evaluated for com-mercialization.

In addition, extension specialists, entomologists or representatives of the

Ministries of Agriculture, depending on the country, are often responsible for communicating insect resistance management (IRM) guidelines to growers and crop consultants, as to how the technology can best be deployed for long-term value (Ostlie *et al.*, 1997; Romeis *et al.*, 2008; Sappington, 2014). During the dialogue with farmers, the question of how best to integrate the necessary IRM practices within an IPM programme can be a challenge and must therefore be carefully developed and communicated to growers. For example, when many growers first view demonstration plots of Bt crop efficacy, there can be a tendency to rely solely on the new technology and abandon other IPM tactics. In addition, for many grower communities, the concept of requesting a farmer to maintain 5–20% of their crop as a 'non-Bt refuge for insect pests to flourish' can be a difficult challenge. However, for strategies that rely on a non-Bt crop refuge, it is necessary to communicate the long-term gains of the technology versus potential short-term losses in damage and profits (Ostlie *et al.*, 1997; Siegfried and Hellmich, 2012; Onstad, 2014). Moreover, new IRM plans for a given Bt crop are best implemented by farmers if a consistent message about the rationale and implementation details is communicated often, across all production areas on a regional or area-wide basis.

As noted earlier, Bt crop–pest systems that have been particularly successful in delaying or mitigating the risk of pest resistance include: Bt maize and the European corn borer in the USA (Storer *et al.*, 2008; Hutchison *et al.*, 2010; Siegfried and Hellmich, 2012), Bt maize and the Asian corn borer in the Philippines (Yorobe and Quicoy, 2006; Alcantara *et al.*, 2011), Bt cotton and *H. virescens* in the southern US (Adamczyk and Hubbard, 2006), and Bt cotton and pink bollworm in the USA (Naranjo, 2011; Tabashnik *et al.*, 2011). In each of these cases, all the assumptions of the HDR strategy were met; plants produced Bt at a high dose for the targeted pests, non-Bt crop and/or natural refuge areas were maintained as per IRM guidelines (or requirements), the refuge placements were

conducive to random mating by the pest species, and the genetics of resistance for one or more Bt toxins was either recessive or only indicated partial dominance (Tabashnik *et al.*, 2014).

The most robust IRM programmes that will be sustainable for Bt crops, and minimize the risk of pest resistance, are those that are based on all of the available data from laboratory, field and modelling studies, and tend to be conservative in nature. For example, when Bt maize for the European corn borer was commercialized, a 20% non-Bt refuge requirement was eventually approved by the US EPA (Ostlie *et al.*, 1997; Siegfried and Hellmich, 2012), and this was viewed by some as overly conservative. However, this refuge level was adopted as policy by the US EPA because some of the assumptions for adult dispersal (and random mating) were unknown, and initial allele frequencies for resistance were unknown; these, and other uncertainties, cannot fully be accounted for in early IRM simulation modelling (e.g. Gray, 2011, Onstad, 2014). Where research gaps do exist, worst-case scenarios for key assumptions are usually included in IRM models (Onstad, 2014). Also, to err on the side of caution, and until more research gaps can be filled, some of the early Bt products were launched with a conservative view of non-Bt refugia.

During the early years of commercialization (1996–2000), the success of the 20% non-Bt maize refuge was probably aided by the fact that the majority of maize fields continued to be non-Bt, until grower adoption accelerated in the early 2000s. Further, for the European corn borer, both the initial Cry1Ab toxin, and nearly all additional Bt toxins released to date, have provided true high-dose efficacy (e.g. Burkness *et al.*, 2011; Siegfried and Hellmich, 2012; Tabashnik *et al.*, 2014). Alternatively, for western corn rootworm in the USA, most of the Bt traits commercialized so far have not provided a true high dose (Gassmann *et al.*, 2011, 2012). In these situations, it is even more critical to maintain other complementary IPM tactics in the system; for western corn rootworm, for example, maintaining crop rotation is essential. Most of the early examples of corn rootworm resistance consisted of growers who had elected to plant maize after maize for several years (Gassmann *et al.*, 2012).

The most recent trend in IRM for Bt maize is a result of both declining compliance by farmers to implement the previous 20% non-Bt maize structured refugia (e.g. Goldberger *et al.*, 2005) and the availability of stacked or pyramided Bt traits (Burkness *et al.*, 2011; Siegfried and Hellmich, 2012; Sappington, 2014). For example, SmartStax® (Dow AgroSciences and Monsanto) contains three lepidopteran Bt toxins (Cry1F, Cry1A.105 and Cry2Ab2), as well as western corn rootworm Bt toxins (Cry3Bb1 and Cry34/35Ab1) (Storer *et al.*, 2012b). With multiple Bt toxins for a given pest, and assuming unique modes of action (e.g. unique binding sites), most IRM simulation models will show an extended sustainability of the Bt traits compared with that of single toxins (Storer *et al.*, 2012b; Onstad, 2014). Because of the extended life of the pyramided Bt crops, the models will also show that non-Bt refuge size can be reduced. A recent decline in grower compliance for non-Bt block refugia prompted the seed industry to propose a seed blend refuge strategy, known as 'refuge-in-the-bag' (RIB) (Onstad *et al.*, 2011; Storer *et al.*, 2012b; Onstad, 2014), and a send blend with 5% non-Bt seeds was eventually approved (Storer *et al.*, 2012b).

Some researchers have been concerned that a 5% non-Bt blend may be too low given the fact that the surrounding Bt plants produce Bt pollen, and the non-Bt (RIB) ears can easily be cross-pollinated, resulting in a mosaic of Bt and non-Bt kernels in the RIB ear (e.g. Burkness *et al.*, 2011; Sappington, 2014). Consequently, late-season lepidopteran larvae feeding on the RIB ears will be exposed to Bt concentrations that are less than high dose, which compromises the refuge. This scenario, with partial larval mortality, can result in: (i) fewer individuals produced from the 5% 'non-Bt refuge'; and (ii) the individuals that do survive are selected for resistance. These concerns were initially confirmed by Burkness *et al.* (2011) in a study with both the European corn borer and the corn earworm. Subsequent

modelling studies with the corn borer (Kang *et al.*, 2012) indicated that although low-toxin expression in ears of Bt maize can reduce the durability of transgenic corn expressing a single toxin (e.g. Cry1Ab), the durability of pyramided maize hybrids expressing two or more toxins was not significantly reduced over a non-Bt maize block scenario. More recently, however, in a RIB study for the corn earworm using the pyramided SmartStax®, Yang *et al.* (2014) found that Bt cross-pollination to the RIB ears not only resulted in significant larval mortality, but also in low pupal survivorship and only 4% adult survival – an 88% reduction in adult moth numbers compared with the non-Bt block refuge. Given the results today, it is clear that more detailed field studies need to be conducted with all pyramided Bt maize events to fully assess the impact of the 5% non-Bt refuge deployed as an RIB seed mix.

17.5 Insect Resistance Management and Integrated Pest Management

The question of how best to include IRM within IPM programmes has recently led to broader questions about the role of GM crop technology itself as an IPM tactic (e.g. Romeis *et al.*, 2008; Sappington, 2014). Some entomologists have recently engaged in lively debates on the role of Bt crops within an IPM paradigm (Gray, 2011; Sappington, 2014). For example, traditional IPM theory emphasizes the need for pest control action only when justified as pest populations exceed predetermined economic thresholds, or more formally, exceed an economic injury level (EIL). At such levels, pest control action is necessary to prevent or minimize economic loss (Pedigo and Rice, 2006; Gray, 2011; Sappington, 2014). The decision to take pest control action is also typically applied on a field-by-field basis. In contrast, the decision to plant GM crops is usually made several months before the growing season begins, or before pest population levels are known (Pedigo and Rice, 2006; Hutchison *et al.*, 2010; Onstad, 2014). Thus, many scientists view the planting of Bt crops as comparable to the use of prophylactic insecticides – deployed each year, not knowing whether the Bt crop, or the insecticidal Bt traits, will be needed or not. This can result in unnecessary costs and selection pressure on multiple insect pest species.

Conversely, some scientists maintain the view that GM crops reflect an example of host plant resistance (HPR), a widely accepted foundational strategy of IPM theory and practice (Kennedy, 2008; Romeis et al., 2008; Naranjo et al., 2008; Naranjo 2011). As such, HPR can be viewed as a 'preventive' IPM tactic, whereas conventional insecticide use, based on an economic threshold, has been termed 'therapeutic' (Pedigo and Rice, 2006). It should be noted that before the advent of GM crops, and continuing today, a farmer's decision to use a conventional pest-resistant variety also occurs well before pest population levels were known (e.g. Pedigo and Rice, 2006; Kennedy, 2008). Often, what is lost in the debate is the decision-making rationale of the end user, i.e. the growers who must ultimately assess the potential risk of incurring pest crop losses by not planting a Bt crop and weigh the expected 'pay-off' of planting Bt crops before the extent of insect pest pressure is known (Mitchell and Hutchison, 2009; Hutchison et al., 2010; Onstad, 2014).

Briefly, what is unique with the use of Bt crops is that many farmers tend to be risk averse, as they have many potential production risks to consider each year, well beyond losses due to pests (Mitchell and Hutchison, 2009). As such, they often, either directly or indirectly, think of the value of new IPM technologies in terms of a pay-off and the probability of a pay-off, i.e. in how many years out of 5 or 10 years will X method(s) return a profit (net return). They may not expect X technology to have a positive net benefit every year, but they prefer a positive pay-off for a majority of years, e.g. 6 of 10 or 8 of 10 years, depending on their personal risk preferences (Mitchell and Hutchison, 2009). Therefore, given the yield and yield consistency of Bt crops, with the prospects of reduced insecticide use, and

the lower pest densities, it is easy to understand the decisions that farmers make to plant Bt crops. Also, if they view IPM as a risk-reducing activity, and Bt crops as risk reducing, they see the value of planting Bt crops in advance of knowing the extent of pest populations. During the early commercialization of Bt crops, this rationale seemed more logical than the current situation in which population densities of some pests, such as the European corn borer in the Midwest US Corn Belt, continue to remain low (Hutchison et al., 2010; Bell et al., 2012). Regardless of one's definition of IPM, or one's view of Bt crops as a an IPM strategy, Bt crop technology has become one of the most widely adopted agricultural technologies during the past 25 years, and is likely to continue to be widely adopted as long as IRM programmes can be improved and effectively implemented (James, 2013; Onstad, 2014).

17.6 Bt Crops from an Area-wide Perspective

During the past 20 year history of the commercial use of Bt crop research and development, researchers have debated the extent to which the IRM approach is best for selected Bt crops, how IRM is integrated within IPM and to what extent Bt crops themselves are a good fit within traditional IPM paradigms (Romeis et al., 2008; Gray, 2011; Onstad, 2014). Beyond the preventive versus therapeutic paradigms in IPM, there is an increasing body of evidence that viewing the role of Bt crops in IPM requires a landscape, or area-wide, perspective. An area-wide IPM approach to designing IPM programmes has a long history in entomology, and has been adapted for several purposes over the years (Elliott et al., 2008; Koul et al., 2008). Area-wide IPM was first widely conceived and promoted by Edward Knipling, when he served as Head of the US Department of Agriculture (USDA) Entomology Group (Klassen, 2003; Elliott et al., 2008). Knipling (1966) developed an area-wide pest suppression approach with a goal of eradication of the screw-worm fly

(*Cochliomyia hominivorax*), via the release of sterile males. Based on some initial basic research, and following several logistical and budget challenges, the programme was implemented in northern Mexico and southern Texas, and was determined to be a significant success (Knipling, 1966; Klassen, 2003). The essential idea is that, given some knowledge of the population dynamics of the pest species, implementing control tactics over a large area, in a community-wide, cooperative approach, will help to drive down general equilibrium levels (GEL) of the pest species to below the economic injury level (EIL), and in the long-run be more cost-effective than maintaining a conventional field-by-field IPM approach. The other key concept is that most of the IPM tactics used in area-wide programmes typically consist of preventive measures. Table 17.1 includes a summary of factors that characterize the differences between conventional IPM implemented on a field-by-field basis versus an area-wide approach.

The successful deployment of both the sterile male strategy and the theoretical basis for area-wide pest suppression gained considerable momentum among many entomologists in the 1950s–1960s, but was not without controversy (Perkins, 1982; Elliott *et al.*, 2008). One of the criticisms of some early area-wide programmes, including the subsequent cotton boll weevil (*Anthonomus grandis*) eradication programme, was too much reliance on insecticidal control, especially early in the eradication process, or limited insecticide choices which, in turn, generated concerns about resistance, environmental costs and the resurgence of secondary pests due to the loss of biological control agents (e.g. Perkins, 1982; Myers *et al.*, 1998; Koul *et al.*, 2008). Despite these challenges, as new IPM technologies were developed, including 'attract and kill' traps, pheromone-disruption technologies and pest-resistant varieties, the USDA Agricultural Research Service (ARS), and USDA Animal and Plant Health Inspection Service (APHIS) in some states, in collaboration with university entomologists, continued to pursue various area-wide IPM projects through the 2000s. Several high-visibility

projects included those for the Russian wheat aphid (*Diuraphis noxia*), western corn rootworm, codling moth (*Cydia pomonella*), fire ants (*Solenopsis invicta*) and tephritid fruit flies in Hawaii, the gypsy moth (*Lymantria dispar*) and several others (e.g. Koul *et al.*, 2008). As might be expected with such a broad range of targeted pests, all with high damage potential, there was a range of success; see Koul *et al.* (2008) for a thorough review of recent programmes.

Given the landscape-level impacts of Bt maize and Bt cotton, now confirmed for several targeted pest species, the strategy can be viewed with an area-wide focus, even if this was not the initial intent of the technology. For some Bt crop–pest systems, additional complementary IPM tactics have also been used to further enhance the pest control provided by Bt traits, and these additional tactics have probably helped to improve IRM and suppress resistance genes (Tabashnik *et al.*, 2010, 2012). The first report of area-wide suppression with Bt crops was for the pink bollworm in Bt cotton, in Arizona, USA (Carrière *et al.*, 2003); in only 6 years since commercialization, statistically significant reductions were observed in both moth numbers (pheromone traps) and larval infestations. Since then, Wu *et al.* (2008) in China found that several years of Bt cotton use resulted in significantly lower cotton bollworm infestations in nearby non-Bt vegetable crops and in non-Bt maize. An additional study by Lu *et al.* (2012) confirmed that these same reductions in the cotton bollworm resulted in less insecticide use and improved biological control. Also, for Bt cotton, Adamczyk and Hubbard (2006) observed reduced moth catches for the tobacco budworm in the southern USA (Mississippi), which had previously been the major pest of cotton in that region. Blanco (2012) summarized additional data from Louisiana, reporting similar trends in reduced tobacco budworm moth flights.

In both the Eastern US and Midwest US Corn Belts, Storer *et al.* (2008) and Hutchison *et al.* (2010), respectively, documented statistically significant reductions in European corn borer populations. The analysis of larval population reductions for

Table 17.1. Comparison of conventional field-based integrated pest management–insect resistance management (IPM–IRM) versus area-wide IPM–IRM factors that can have an impact on the use of Bt (*Bacillus thuringiensis*) crops.[a]

Conventional strategy	Area-wide strategy
IPM aspects	
Spatial scale: field-by-field focus	Multiple fields: farm, county, state, national or international scope
Audience: individual farmer	Multiple farmers per growing region
Individual farmer decision making	Some individual IPM, but also community based (voluntary or programme based)
Private sector, farmer IPM decisions	Private and industry-based decisions, or private and government (e.g. USDA Agricultural Research Service)-sponsored IPM
Tactics insecticide based, using field-by-field action thresholds, economic threshold (ET), economic injury level (EIL); also cultural and biological controls	Tactics may favour multi-field release of biological control agents, or broader use of host plant resistance (e.g. Bt crops); decision is made before pest density (or ET or EIL) is known
Tactics tend to be therapeutic, with goal to manage local populations after exceeding EIL	Tactics tend to be preventive, with goal to maintain regional pest population dynamics, perhaps to influence geographic range of low densities
Farm size neutral: small and large growers participate	Farm size neutral: small and large growers participate
Landscape aspects of insect movement not explicitly accounted for, or exploited	Landscape aspects such as 'halo effects' accounted for and used to grower advantage
IRM aspects	
Theoretical focus is farm based, with local emphasis on non-Bt refuge compliance and resistance monitoring	Theoretical focus is multi-farm, landscape based, with regional emphasis on refuge compliance and resistance monitoring; can fully take advantage of pest 'source–sink' dynamics
Pest resistance monitoring tends to be field by field, by farm, via farmer, crop consultants (data may not be shared with other farmers)	Pest monitoring and resistance monitoring developed for multiple farms/fields, coordinated by county, state, region, with multiple farmers, collaborators (data shared)
Individual farm-based pest and IRM monitoring may vary	More consistent pest and IRM monitoring likely due to area-wide programme financing
When IPM and IRM are working well, individual farmers benefit	When IPM and IRM are working well, multiple farmers benefit, including those who may not directly participate in the programme
IRM strategies and compliance can vary	IRM strategies and compliance can vary, but there may be more incentive to comply given cooperation among farmers in area-wide integrated pest management (AWIPM)
If IRM strategy loses efficacy, losses may be limited to individual farms (if pest dispersal is limited), but may still depend on compliance of neighbours	If IRM strategy loses efficacy, losses may be experienced by most farmers in the programme (when resistance gene flow and pest dispersal are high)

[a]Concepts based on preventive and therapeutic IPM and area-wide IPM (e.g. Pedigo and Rice, 2006; Elliott *et al.*, 2008; Hutchison *et al.*, 2010; Bell *et al.*, 2012; Onstad, 2014).

three major maize states in the Midwest was complicated by the characteristic episodic dynamics for this pest, in which populations typically followed an outbreak/crash cycle every 7 years. This episodic cycling, reflecting density-dependent mortality in the corn borer, had previously been attributed to the pathogen *Nosema pyrausta* (Onstad and Guse, 1999; Hutchison *et al.*, 2010; White *et al.*, 2014). Notably, since 2009, *Nosema* has continued to cycle with the remaining but ultra-low corn borer populations in the Midwest region (Bell *et al.*, 2012; White *et al.*, 2014). So although it cannot currently maintain itself in host populations in the field via larval–larval transmission, it appears to be maintaining its presence via adult–egg transmission (White *et al.*, 2014). This is important as an additional biological control factor in the event that corn borer populations were to rise again. Also, with little foliar insecticide use in the Midwest Corn Belt there are several generalist predators that are active in the system (see Romeis *et al.*, 2008).

In summary, this brief review suggests that there may be several advantages in developing, or modifying, Bt crop IPM and IRM systems from an area-wide viewpoint. An interesting aspect of assessing the area-wide value of Bt crops is the use of various landscape level models that account for both population dynamics and population genetics for IRM (Gould, 1998). Two such models are noteworthy of mention here, as they can be useful in further area-wide pest management designs. An early model was developed by Kennedy *et al.* (1987) that used *H. zea* feeding on several crops as a case study. Briefly, in this study, one key outcome was to show that crop varieties conferring antibiosis – similar to Bt toxin activity – demonstrated significant population suppression effects. From another perspective, a more general model was developed by Byers and Castle (2005) to compare the long-term pest suppression outcomes when IPM was implemented on a field-by-field basis as each field exceeded the economic threshold (ET), or if IPM was implemented on an area-wide basis. Given an exponential population growth assumption, the latter

strategy of an area-wide approach, in which the treatment decision was based on the average infestation level of all fields, resulted in the most cost-effective outcome and broader pest suppression. Finally, the recent area-wide pest suppression example with the European corn borer (Hutchison *et al.*, 2010) suggests a strong economic incentive for farmers to consider planting a mix of Bt and non-Bt crops on their farms, and to account for the degree to which neighbouring farmers are also planting Bt or non-Bt crops. Farmers (and crop consultants) that are aware of the current effects of area-wide pest suppression can save seed costs by avoiding the Bt technology fees, yet still benefit from the pest suppression effect. A sustainable balance of Bt and non-Bt crops can, therefore, lead to IPM decision making that extends from a single field-to-field basis to the entire farm crop mix, or to a regional crop mix of Bt and non-Bt crops at a landscape level (see also Bell *et al.*, 2012).

While these studies provide a good representation of possible suppression outcomes, there are certainly additional questions that should be addressed, and additional area-wide IPM and IRM strategies that can be evaluated by expanding IRM models for Bt crops to emphasize population dynamics and pest suppression at varying spatial scales. Such models can be modified for Bt crop landscapes to include non-Bt crop refugia, natural refugia in non-Bt crops and various insect-movement scenarios, either within a crop from plant to plant (as per RIP IRM) or among crops when assessing adult migration. Finally, an area-wide approach can account for both short-range dispersal and the long-distance migration associated with synoptic weather conditions. Given the long-distance migration of *H. zea* and other noctuid pests (Westbrook, 2008), it is important to account for the long-distance movement of resistance genes, as has been shown for pyrethroid-resistant alleles in the Midwest field (Hutchison *et al.*, 2007). Future work will need to consider how to achieve pest management and resistance management for the wider range of pest species being targeted within each crop, and the ability to consider this at multiple spatial scales, potentially with coordinated deployment as occurs in area-wide management approaches.

References

Adamczyk, J.J. Jr and Hubbard, D. (2006) Changes in populations of *Heliothis virescens* (Lepidoptera: Noctuidae) and *Helicoverpa zea* (Lepidoptera: Noctuidae) in the Mississippi Delta, from 1986–2005, as indicated by adult male pheromone traps. *Journal of Cotton Science* 10, 155–160.

Alcantara, E., Estrada, A., Alpuerto, V. and Head, G. (2011) Monitoring Cry1Ab susceptibility in Asian corn borer (Lepidoptera: Crambidae) on Bt corn in the Philippines. *Crop Protection* 30, 554–559.

Andow, D.A., Alstad, D.N., Pang, Y.H., Bolin, P.C. and Hutchison, W.D. (1998) Using an F_2 screen to search for resistance alleles to *Bacillus thuringiensis* toxin in European corn borer (Lepidoptera: Crambidae). *Journal of Economic Entomology* 91, 579–584.

Barfoot, P. and Brookes, G. (2014) Key environmental impacts of genetically modified (GM) crop use, 1996–2012. *GM Crops and Food* 5, 149–160.

Bell, J.R., Burkness, E.C., Milne, A.E., Onstad, D.W., Abrahamson, M. *et al.* (2012) Putting the brakes on a cycle: bottom-up effects damp cycle amplitude. *Ecology Letters* 15, 310–318.

Blanco, C. (2012) *Heliothis virescens* and Bt cotton in the United States. *GM Crops and Food* 3, 201–212.

Brent, K.J. (1986) Detection and monitoring of resistant forms: an overview, In: National Research Council (ed.) *Pesticide Resistance: Strategies and Tactics for Management.* National Academies Press, Washington, DC, pp. 298–312.

Burkness, E.C., O'Rourke, P.K. and Hutchison, W.D. (2011) Cross-pollination of nontransgenic corn ears with transgenic Bt corn: efficacy against lepidopteran pests and implications for resistance management. *Journal of Economic Entomology* 104, 1476–1479.

Byers, J.A. and Castle, S.J. (2005) Areawide models comparing synchronous versus asynchronous treatments for control of dispersing insect pests. *Journal of Economic Entomology* 98, 1763–1773.

Carrière, Y., Ellers-Kirk, C., Sisterson, M., Antilla, L., Whitlow, M. *et al.* (2003) Long-term regional suppression of pink bollworm by *Bacillus thuringiensis* cotton. *Proceedings of the National Academy of Sciences of the United States of America* 100, 1519–1523.

Carson, R. (1962) *Silent Spring.* Houghton Mifflin, New York.

Dennehy, T.J., Head, G.P., Moar, W., Greenplate, J., Mohan, K.S. *et al.* (2010) Status of PBW resistance to Bollgard cotton in India. Presentation to: Entomology 2010. ESA 58th Annual Meeting, December 12–15, San Diego, California. Entomological Society of America, Annapolis, Maryland. Available at: http://esa.confex.com/esa/2010/webprogram/Paper 49973.html (accessed 20 August 2014).

Elliott, N.C., Onstad, D.W. and Brewer, M.J. (2008) History and ecological basis for areawide pest management, In: Koul, O., Cuperus, G. and Elliott, N. (eds) *Areawide Pest Management: Theory and Application.* CAB International, Wallingford, UK, pp. 15–33.

Farias, J.R., Andow, D.A., Horikoshi, R.J., Sorgatto, R.J, Fresia, P. *et al.* (2014) Field-evolved resistance to Cry1F maize by *Spodoptera frugiperda* (Lepidoptera: Noctuidae) in Brazil. *Crop Protection* 64, 150–158.

Fleischer, S., Naranjo, S.E. and Hutchison, W.D. (2014) Sustainable management of insect-resistant crops. In: Ricroch, A., Chopra, S. and Fleischer, S.J. (eds) *Plant Biotechnology: Experience and Future Prospects.* Springer, Dordrecht, The Netherlands, pp. 115–127.

Gassmann, A.J. and Hutchison, W.D. (2012) Bt crops and insect pests: past successes, future challenges and opportunities. *GM Crops and Food* 3, 139.

Gassmann, A.J., Petzold-Maxwell, J.L., Keweshan, R.S. and Dunbar, M.W. (2011) Field-evolved resistance to Bt maize by western corn rootworm. *PloS ONE* 6(7): e22629.

Gassmann, A.J., Petzold-Maxwell, J.L., Keweshan, R.S. and Dunbar, M.W. (2012) Western corn rootworm and Bt maize: challenges of pest resistance in the field. *GM Crops and Food* 3, 235–244.

Goldberger, J., Merrill, J. and Hurley, T. (2005) Bt corn farmer compliance with insect resistance management requirements in Minnesota and Wisconsin. *AgBioForum* 8, 151–160.

Gould, F. (1998) Sustainability of transgenic insecticidal cultivars: integrating pest genetics and ecology. *Annual Review of Entomology* 43, 701–726.

Gray, M.E. (2011) Relevance of traditional integrated pest management (IPM) strategies for commercial corn producers in a transgenic agroecosystem: a bygone era? *Journal of Agricultural and Food Chemistry* 59, 5852–5858.

Hutchison, W.D., Burkness, E.C., Jensen, B., Leonard, B.R., Temple, J. *et al.* (2007) Evidence for decreasing *Helicoverpa zea* susceptibility to pyrethroid insecticides in the Midwestern US. *Plant Health Progress.* 19 July 2007. *Symposium Proceedings: Increasing Concerns about* Helicoverpa zea *Susceptibility to Pyrethroids in the Midwestern USA. Annual Meeting of the North Central Branch of the Entomological Society of America, 26 to 29 March 2006, Bloomington, Illinois.* Available at: http://www.plantmanagementnetwork.org/sub/php/symposium/hzea/decrease/ (accessed 30 October 2014).

Hutchison, W.D., Burkness, E.C., Mitchell, P.D., Moon, R.D., Leslie, T.W. *et al.* (2010) Areawide suppression of European corn borer with Bt maize reaps savings to non-Bt growers. *Science* 330, 222–225.

James, C. (2013) *Global Status of Commercialized Biotech/GM Crops: 2013.* ISAAA Brief 46, International Service for the Acquisition of Agri-Biotech Applications, Ithaca, New York.

Kang, J., Onstad, D.W., Hellmich, R.L., Moser, S.E., Hutchison, W.D. *et al.* (2012) European corn borer model for studying the effect of non-transgenic corn ears fertilized by pollen of corn expressing Cry toxin. *Environmental Entomology* 41, 200–211.

Kennedy, G.G. (2008) Integration of insect-resistant genetically modified crops within IPM programs. In: Romeis, J., Shelton, A.M. and Kennedy, G.G. (eds) *Integration of Insect-Resistant Genetically Modified Crops within IPM Programs. Progress in Biological Control, Volume 5.* Springer, Dordrecht, The Netherlands, pp. 1–26.

Kennedy, G.G., Gould, F., dePonti, O.M.B. and Stinner, R.E. (1987) Ecological, agricultural, genetic, and commercial considerations in the deployment of insect-resistant germplasm. *Environmental Entomology* 16, 327–338.

Klassen, W. (2003) Memorial lecture – Edward F. Knipling: titan and driving force in ecologically selective area-wide pest management. *Journal of the American Mosquito Control Association* 19, 94–103.

Knipling, E.F. (1966) Some basic principles of insect population suppression. *Bulletin of the Entomological Society of America* 12, 7–15.

Koul, O., Cuperus, G.W. and Elliott, N. (eds) (2008) *Areawide Pest Management: Theory and Implementation.* CAB International, Wallingford, UK.

Kruger, M., van Rensburg, J.R.J. and Van den Berg, J. (2011) Resistance to Bt maize in *Busseola fusca* (Lepidoptera: Noctuidae) from Vaalharts, South Africa. *Environmental Entomology* 40, 477–483.

Liu, X., Chen, M., Collins, H.L., Onstad, D.W., Roush, R.T. *et al.* (2014) Natural enemies delay insect resistance to Bt crops. *Plos ONE* 9(3): e90366.

Lu, Y., Wu, K., Jiang, Y., Guo, Y. and Desneux, N. (2012) Widespread adoption of Bt cotton and insecticide decreases promote biocontrol services. *Nature* 487, 362–365.

Matten, S.R., Head, G.P. and Quemada, H.D. (2008) How governmental regulation can help or hinder the integration of Bt crops within IPM. In: Romeis, J., Shelton, A.M. and Kennedy, G.G. (eds) *Integration of Insect-Resistant Genetically Modified Crops within IPM Programs. Progress in Biological Control, Volume 5.* Springer, Dordrecht, The Netherlands, pp. 27–39.

Melander, A.L. (1914) Can insects become resistant to sprays? *Journal of Economic Entomology* 7, 167–173.

Mitchell, P.D. and Hutchison, W.D. (2009) Decision making and economic risk in IPM. In: Radcliffe, E.B., Hutchison, W.D. and Cancelado, R. (eds) *Integrated Pest Management: Concepts, Tactics, Strategies and Case Studies.* Cambridge University Press, Cambridge, UK, pp. 33–50.

Myers, J.H., Savoie, A. and Randen, E. (1998) Eradication and pest management. *Annual Review of Entomology* 43, 471–491.

Naranjo, S.E. (2011) Impacts of Bt transgenic cotton on integrated pest management. *Journal of Agricultural and Food Chemistry* 59, 5842–5851.

Naranjo, S.E., Ruberson, J.R., Sharma, H.C., Wilson, L. and Wu, K. (2008) The present and future role of insect-resistant genetically modified cotton in IPM. In: Romeis, J., Shelton, A.M. and Kennedy, G.G. (eds) *Integration of Insect-Resistant Genetically Modified Crops within IPM Programs. Progress in Biological Control, Volume 5.* Springer, Dordrecht, The Netherlands, pp. 159–194.

Onstad, D.W. (2014) *Insect Resistance Management: Biology, Economics and Prediction,* 2nd edn. Elsevier, Amsterdam.

Onstad, D.W. and Guse, C.A. (1999) Economic analysis of transgenic maize and nontransgenic refuges for managing European corn borer (Lepidoptera: Pyralidae). *Journal of Economic Entomology* 92, 1256–1265.

Onstad, D.W., Mitchell, P.D., Hurley, T.M., Lundgren, J.G., Porter, R.P. *et al.* (2011) Seeds of change: corn seed mixtures for resistance management and integrated pest management. *Journal of Economic Entomology* 104, 343–352.

Ostlie, K.R., Hutchinson, W.D. and Hellmich, R.L. (eds) (1997) *Bt Corn and European Corn Borer: Long-term Success through Resistance Management.* North Central Regional Extension

Publication NCR 602, University of Minnesota Extension, East Lansing, Michigan. Updated version available at: http://www.extension.umn.edu/agriculture/corn/pest-management/bt-corn-and-european-corn-borer/ (accessed 29 October 2014).

Park, J.R., McFarlane, I., Phipps, R.H. and Ceddia, G. (2011) The role of transgenic crops in sustainable development. *Plant Biotechnology Journal* 9, 2–21.

Pedigo, L.P. and Rice, M.E. (2006) *Entomology and Pest Management,* 5th edn. Prentice Hall, Upper Saddle River, New Jersey.

Perkins, J.H. (1982) *Insects, Experts, and the Insecticide Crisis: The Quest for New Pest Management Strategies.* Plenum Press, New York.

Peshin, R. and Pimentel, D. (eds) (2014) *Integrated Pest Management: Experiences with Implementation, Global Overview, Vol. 4.* Springer, New York.

Pilcher, C.D., Rice, M.E., Higgins, R.A., Steffey, K.L., Hellmich, R.L. *et al.* (2002) Biotechnology and the European corn borer: measuring historical farmer perceptions and adoption of transgenic Bt corn as a pest management strategy. *Journal of Economic Entomology* 95, 878–892.

Qaim, M. and Kouser, S. (2013) Genetically modified crops and food security. *PLoS ONE* 8(6): e64879.

Qaim, M. and Zilberman, D. (2003) Yield effects of genetically modified crops in developing countries. *Science* 299, 900–902.

Radcliffe, E.B., Hutchison, W.D. and Cancelado, R.E. (2009) *Integrated Pest Management: Concepts, Tactics, Strategies and Case Studies.* Cambridge University Press, Cambridge, UK.

Romeis, J., Shelton, A.M. and Kennedy, G.G. (eds) (2008) *Integration of Insect-Resistant Genetically Modified Crops within IPM Programs, Progress in Biological Control, Volume 5.* Springer, Dordrecht, The Netherlands.

Sappington, T.W. (2014) Emerging issues in integrated pest management implementation and adoption in the North Central USA. In: Peshin, R. and Pimentel, D. (eds) *Integrated Pest Management: Experiences with Implementation, Global Overview, Vol. 4.* Springer, New York, pp. 65–97.

Shelton, A.M., Zhao, J.-Z. and Roush, R.T. (2002) Economic, ecological, food safety, and consequences of the deployment of Bt plants, *Annual Review of Entomology* 47, 845–881.

Shelton, A.M., Fuchs, M. and Shotkoski, F.A. (2008) Transgenic vegetables and fruits for control of insects and insect-vectored pathogens, In: Romeis, J., Shelton, A.M. and Kennedy, G.G. (eds) *Integration of Insect-Resistant Genetically Modified Crops within IPM Programs.* Progress in Biological Control, Volume 5. Springer, Dordrecht, The Netherlands, pp. 249–271.

Shelton, A.M., Olmstead, D.L., Burkness, E.C., Hutchison, W.D., Dively, G. *et al.* (2013) Multi-state trials of Bt sweet corn varieties for control of the corn earworm. *Journal of Economic Entomology* 106, 2151–2159.

Shi, G., Chavas, J.-P. and Lauer, J. (2013) Commercialized transgenic traits, maize productivity and yield risk. *Nature Biotechnology* 31, 111–114.

Siegfried, B.D. and Hellmich, R.L. (2012) Understanding successful resistance management: the European corn borer and Bt corn in the US. *GM Crops and Food* 3, 184–193.

Stern, V.M., Smith, R.F., van den Bosch, R. and Hagen, K.S. (1959) The integrated control concept. *Hilgardia* 29, 81–101.

Storer, N.P., Dively, G.P. and Herman, R.A. (2008) Landscape effects of insect-resistant genetically modified crops. In: Romeis, J., Shelton, A.M. and Kennedy, G.G. (eds) (2008) *Integration of Insect-Resistant Genetically Modified Crops within IPM Programs. Progress in Biological Control, Volume 5.* Springer, Dordrecht, The Netherlands, pp. 273–302.

Storer, N.P., Kubiszak, M.E., King, J.E., Thompson, G.D. and Santos, A.C. (2012a) Status of resistance to Bt maize in *Spodoptera frugiperda*: lessons from Puerto Rico. *Journal of Invertebrate Pathology* 110, 294–300.

Storer, N.P., Thompson, G.D. and Head, G.P. (2012b) Application of pyramided traits against Lepidoptera in insect resistance management for Bt crops. *GM Crops and Food* 3, 154–162.

Tabashnik, B.E., Sisterson, M.S., Ellsworth, P.C., Dennehy, T.J., Antilla, L. *et al.* (2010) Suppressing resistance to Bt cotton with sterile insect releases. *Nature Biotechnology* 28, 1304–1307.

Tabashnik, B.E., Huang, F., Ghimire, M.N., Leonard, B.R., Siegfried, B.D. *et al.* (2011) Efficacy of genetically modified Bt toxins against insects with different mechanisms of resistance. *Nature Biotechnology* 19, 1128–1131.

Tabashnik, B.E., Morin, S., Unithan, G.C., Yelich, A.J., Ellers-Kirk, C. *et al.* (2012) Sustained susceptibility of pink bollworm to Bt cotton in the United States. *GM Crops and Food* 3, 194–200.

Tabashnik, B.E., Brevault, T. and Carrière, Y. (2013) Insect resistance to Bt crops: lessons from the first billion acres. *Nature Biotechnology* 31, 510–521.

Tabashnik, B.E., Mota-Sanchez, D., Whalon, M.E., Hollingworth, R.M. and Carrière Y. (2014)

Defining terms for proactive management of resistance to Bt crops and pesticides. *Journal of Economic Entomology* 107, 496–507.

Tan, S.Y., Cayabyab, B.F., Alcantara, E.P., Huang, F., He, K. *et al.* (2013) Comparative binding of *Bacillus thuringiensis* Cry1Ab and Cry1F toxins in *Ostrinia nubilalis, Ostrinia furnacalis* and *Diatraea saccharalis* (Lepidoptera: Crambidae). *Journal of Invertebrate Pathology* 114, 234–240.

Venette, R.C., Hutchison, W.D. and Andow, D.A. (2000) An in-field screen for early detection and monitoring of insect resistance to *Bacillus thuringiensis* in transgenic crops. *Journal of Economic Entomology* 93, 1055–1064.

Venette, R.C., Hutchison, W.D. and Moon. R.D. (2002) Strategies and statistics of sampling for rare individuals. *Annual Review of Entomology* 47, 143–174.

Westbrook, J.K. (2008) Noctuid migration in Texas within the nocturnal aeroecological boundary layer. *Integrative and Comparative Biology* 48, 99–106.

White, J., Burkness, E.C. and Hutchison, W.D. (2014) Biased sex ratios, mating frequency and *Nosema* prevalence in European corn borer, at low population densities. *Journal of Applied Entomology* 138, 195–200.

Wilson, T.A., Rice, M.E., Tollefson, J.T. and Pilcher, C.D. (2005) Transgenic corn for control of the European corn borer and corn rootworms: a survey of Midwestern farmers' practices and perceptions. *Journal of Economic Entomology* 98, 237–247.

Wu, K.-M., Lu, Y.-H., Feng, H.-Q., Jiang, Y.-Y. and Zhao, J.-Z. (2008) Suppression of cotton bollworm in multiple crops in China in areas with Bt-toxin-containing cotton. *Science* 321, 1676–1678.

Yang, F., Kerns, D.L., Head, G.P., Leonard, B.R., Levy, R. *et al.* (2014) A challenge for the seed mixture refuge strategy in Bt maize: impact of cross-pollination on an ear-feeding pest, corn earworm. *PLoS ONE* 9(11): e112962.

Yorobe, J.M. Jr and Quicoy, C.B. (2006) Economic impact of Bt corn in the Philippines. *The Philippine Agricultural Scientist* 89, 258–267.

Index

Page numbers in **bold** type refer to figures and tables.